Smart Agriculture Practices for Future Food Security

THE EDITORS

Dr. Shahid Ahamad is presently working as Associate Director Research at Directorate of Research,SKUAST Jammu. Earlier he served as Jr. Scientist (Plant Pathology), Sr. Scientist/ Programme Coordinator in KVKs and Dy. Director Research. He is also working as Nodal Officer, PME Cell & Nodal officer, Testing of Products. He granted 5 patents, developed 11 crop varieties, published 250 publications including 15 books, 65 research papers, 35 book chapters. He has been conferred Fellow of Society of Plant Protection Sciences (2005) , New Delhi & Fellow of Indian Phytopathological Society (2015) IARI, New Delhi.

Dr. Suja Nabi Qureshi is working as Scientist/Asst.. Prof KVK , Reasi, SKUAST-Jammu, He has completed his B.Sc.Horticulture from DR BSKKV Dapoli, Maharashtra,M.Sc. Fruit Science and Ph.D. Fruit Science from SKUAST-Kashmir. He has published 02 books and 25 Research Papers.

Dr. Nazam Khan, is working as Associate Professor at Division of Livestock Farm Complex, Faculty of Veterinary Sciences and Animal Husbandry, SKUAST Jammu since 2014. He has one book and over 60 publications and 03 patents in his credit.

Dr. Imran Ahamad Ganai is working as Asstt. Professor at Division of Animal Nutrition, F.V.Sc., R.S. Pura, SKUAST-Jammu. He did B.V.Sc,& A.H. ,M.V.Sc. and Ph.D. from SKUAST Jammu. He has published many research papers in reputed journal and alsp has one patent in his credit.

Dr Vikas Sharma is working as Chief Scientist & Head at Regional Agricultural Research Station, Rajouri of SKUAST-Jammu. He has a vast experience of more than 24 years in the field of Agronomy particularly in the field of natural resource management and climate resilient agriculture in hills. He has published three books and more than 50 research articles in national and international journals of repute.

Smart Agriculture Practices for Future Food Security

– Editors –

Dr. Shahid Ahamad
Associate Director Research,
Directorate of Research, SKUAST Jammu

Dr. Suza Nabi Qurershi
Scientist, Krishi Vigyan Kendra,
Reasi, SKUAST Jammu

Dr. Nazam Khan
Asstt. Professor, Division of Livestock Farm Complex,
F.V.Sc. & AH,R.S.Pura, Jammu

Dr. Imram Ahamad Ganai
Asstt.Professor,Division of Animal Nutrition,
F.V.Sc.,R.S.Pura,SKUAST-Jammu

Dr. Vikas Sharma
Chief Scientist, Regional Agricultural Research Station,
Rajouri, SKUAST Jammu

ASTRAL
Publishing Authors Across the Globe

2025

Daya Publishing House®
A Division of
Astral International Pvt. Ltd.
New Delhi – 110 002

ISBN: 978-93-5919-878-1

Published by : **Daya Publishing House®**
A Division of
Astral International Pvt. Ltd.
– ISO 9001:2015 Certified Company –
4736/23, Ansari Road, Darya Ganj
New Delhi-110 002
Ph. 011-43549197, 8130496929
E-mail: info@astralint.com
Website: www.astralint.com

Printed at : **Replika Press Pvt Ltd**

Preface

A recent survey by the Economist revealed that the world population has increased by 90 per cent in the past 40 years while food productions has increase only by 25 per cent head. Doubling farmers' income is the mission of Government of India to achieve the mission. Indian agriculture awaits transformation through Smart Agriculture Practices for future food security.

In order to increase agricultural production with limited resources, major technological advancements have been implemented throughout much of human history. However, a growing population along with climate change always poses a threat between food supply and demand. The world's population is estimated to reach 9 billion by 2050, which is about a twenty-five percent increase over the current population. However, population growth will be larger, mostly in emerging countries such as Mexico, India, China, and others. Additionally, the urbanization trend is expected to accelerate in developing countries by 2050. Currently, 49% of the world's population lives in urban areas, which is expected to increase to 70% by 2050. Besides, since the living standard is expected to increase in the future, it will further increase food demand, particularly in emerging nations. Due to the continuous increase in the global population, we should be more cautious regarding nutritional values and food quality. To meet future food demands, food production should be doubled by 2050. In particular, grain crops and meat production should be increased from 2.1 billion tons to 3 billion tons, and 200 million tons to 470 million tons, respectively, to meet the world food demand by 2050 .

There has been a long felt need for a book exclusively on the "Smart Agriculture Practices For Future Food Security". An effort has been made to collect information on various Agricultural Technologies viz Plant Pathology, Entomology, Agronomy, Fruit Science, Vegetable Science, Veterinary sciences etc. Chapters in this book have been contributed by noted Agricultural Scientists of SKUAST Jammu namely Dr.Shahid Ahamad, Associate Director Research, Directorate of Research, Dr Nazam Khan, Associate Professor, Division of LFC, F.V.Sc., R.S. Pura, Dr.Imram Ahamad Ganai, Asstt. Professor, Division of Animal Nutrition, F.V.Sc., R.S. Pura, Dr Vikas Sharma, Chief Scientist, Regional Agricultural Research Station, Rajouri, Dr.Anil Bhusan, Professor, Divi.Vegetable Science., Dr. Jagmohan Singh, Prof.& Head, Division of PHM, Dr. B.Brahma, Prof. Division of LPM, Dr. Ranbir Singh, Prof., Division of Plant Pathology, Kamal Sarma, Neeraj Gupta, Vijay Kumar, Dr. Suraj Amrutkar, Vikas Gupta, A.K. Sinha, Dr. Reena, Efath shahnaz, Anjum Ayoub and Muzafar Mir.

I am grateful and indebted to all the learned galaxy of contributors who have spent their considerable time in contributing the book chapters.

A deep sense of respect and gratitude to Dr.B.N.Tripathi, Hon'ble Vice Chancellor, SKUAST-Jammu, Dr. R.K.Samnotra, Ex Director Research and Dr Amrish Vaid, Director Extension.

Last but not the least, I must acknowledge and oblige the M/S Astral International Publishing House, New Delhi who has been always considerate and positive to our contentions and efforts.

Dr. Shahid Ahamad

Contents

Chapter 1

Smart Agricultural Practices for Future Food Security

Ranbir Singh[1] Satendra Kr. Sharma[2] and Sakshi Sharma[3]*

Division of Plant Pathology
Sher-e-Kashmir University of Agriculture Science & Technology, Jammu, 180009

Abstract

This chapter explores the transformative impact of smart agriculture, where technologies like the Internet of Things (IoT), Artificial Intelligence (AI), and data analytics converge to revolutionize farming practices. Commencing with an overview of smart agriculture's role in ensuring global food safety and optimizing resource utilization, the chapter navigates through the significant contributions of Earth observation, particularly satellite remote sensing, in providing extensive data for agricultural research. The narrative then shifts towards agricultural data analytics, employing machine learning to evaluate crop performance, predict economic impacts, and facilitate informed decision-making. Emphasis is placed on the integration of mobile technologies in developing nations as a unique opportunity to connect rural producers with urban consumers, fostering financial inclusion, and enhancing food safety. Digital Soil Mapping (DSM) technology is scrutinized for its role in assessing global soil and climate properties, with a particular focus on its potential in developing countries. The pivotal role of satellite imagery in implementing sustainable agriculture ecosystems is discussed, emphasizing its efficiency and cost-effectiveness compared to traditional field studies. Addressing the global challenge of food security, driven by population growth and urbanization, the chapter underscores the need for innovative agricultural solutions. It introduces the concept of IoT-driven smart agriculture, offering a detailed exploration of its ecosystem, including sensor technology and unmanned aerial vehicles (UAVs). The potential of UAVs in monitoring and action phases of precision agriculture is highlighted, along with challenges and ongoing research to enhance their efficiency. Communication technologies in smart agriculture are explored, taking into account factors such as range, bandwidth, power consumption, and spectrum. The chapter provides insights into widely used communication technologies and their applications in agriculture. The cultivation of innovation in smart agriculture is examined through methods and techniques such as vertical farming, hydroponics, and phenotyping, contributing to efficient resource management and high yields. Precision agriculture takes centre stage, addressing challenges and showcasing the advantages of IoT integration. Greenhouse farming emerges as a transformative practice, blending information technology and IoT applications for precise environmental regulation and enhanced productivity. Urban farming is presented as a practical solution to challenges posed by climate change and water scarcity, emphasizing its adaptability and scalability. Applications of IoT in smart agriculture are detailed, with a specific focus on precision farming, smart irrigation, pest and weed management, precision fertilization, and yield assessment. The chapter concludes by addressing challenges associated with implementing IoT devices in smart agriculture, exploring solutions to enhance UAV battery life and mitigate wireless communication path loss.

In summary, this chapter provides a comprehensive overview of the smart agriculture revolution, showcasing the integration of IoT, AI, and data analytics to foster sustainable crop management practices.

Keywords: *IoT, Smart Agriculture, UAVs, DSM, Artificial Intelligence, Remote sensing*

Introduction

Smart agriculture revolutionizes farming through innovative technologies, utilizing the Internet of Things (IoT), Artificial Intelligence (AI), and agricultural data analysis for enhanced crop quality and quantity (Khanna & Kaur,2019). This approach is pivotal for future food production, ensuring global food safety by optimizing resource use and empowering farmers to increase yields while minimizing inputs (Barkunan *et al.*2019). In developed nations, large-scale implementation of smart farming incorporates AI, IoT, and data analytics, providing valuable insights for yield security and significant crop productivity (Gu *et al.*2017) [4–6]. Earth observation (EO), particularly satellite remote sensing (SRS), plays a crucial role by offering extensive remote sensing (RS) data for agricultural research and applications (Garg *et al.*2018).

Agricultural data analytics, employing machine learning (ML), assesses crop performance in diverse regions and analyses the economic impact of natural disasters on yield production (Tantalaki etal.2019). Integrative and multi-scale AI models are employed to predict crop behaviour under specific field conditions, contributing to informed decision-making in agriculture (Shrivastava, & Marshall-Colon,2019).

In today's world, mobile technologies have reached even the remotest areas in developing nations, providing a unique opportunity to connect rural producers with urban consumers and international investors (Karim *et al.*2013). This widespread access holds the potential to create lasting value, promote financial inclusion, enhance food safety, and empower less privileged farmers to optimize available agricultural resources (Beza *et al.*2018). Preserving soil fertility is crucial for environmental health, with Digital Soil Mapping (DSM) technology providing quick solutions for assessing global soil and climate properties (Lal, R. (2009). While DSM is widely used among soil scientists worldwide, its comprehensive research application in developing countries is still limited (Vågen *et al.*2016). Satellite images play a pivotal role in implementing sustainable agriculture ecosystems and monitoring performance to meet development goals (De Araujo Barbosa et al;.2015). Using these images for mapping agricultural resources globally proves more efficient and cost-effective than traditional field studies (Neigh *et al.*2018). Remote sensing technology, integrating parameters like natural disasters, climate change, carbon biomass, water resources, air quality, ocean management, and food safety, enhances environmental decision-making for sustainable solutions (Kasturirangan,(2007). Scientists can now develop solutions for global benefit by utilizing satellite Earth Observation (EO) data captured through remote sensing technologies (Kansakar,& Hossain,2016). In line with the "2030 Agenda," EO plays a decisive role in monitoring, estimating, and reporting progress towards Sustainable Development Goals (SDGs) (Mas´o 2019). Early indications of crop shortfalls due to adverse weather conditions can inform rapid relief responses, supported by initiatives like the Group on Earth Observations Global Agriculture Monitoring Initiative (GEOGLAM) for global food security (Anderson ,2017).

Food security emerges as a paramount global challenge, given the forecasted world population of 9.7 billion by 2050, constituting 20.6% growth from the current population (United Nations Department of Economic and Social Affairs,April 2020) . The accelerated pace of urbanization further exacerbates this challenge, with a projected 68% of the world population becoming urbanized by 2050, compared to 54% in 2018, leading to a reduction in available arable land (United Nations Department of Economic and Social Affairs. (April 2020).

Moreover, not all Earth's land is arable due to factors such as soil quality, climate, topography, and high variability within homogeneous land. Compounding this, the rate of arable land decline exceeds the rate of recovery due to pollution, soil erosion, and land degradation (Faiçal *et al.*2017). These challenges underscore the imperative for the adoption and advancement of agriculture as a solution to address these pressing global issues.

1. IoT Driven Smart Agriculture: Unleashing Potential through Internet-Based Things

The smart agriculture practice revolves around an IoT ecosystem, incorporating various technologies and equipment seamlessly integrated into systems, as illustrated, outlining the IoT system architecture. The process commences with data acquisition from sensors, and subsequently, the data is transmitted to the cloud architecture. In the cloud, decisions are made to execute operations in the field, offering valuable insights for end-user applications in smart agriculture. A distinguishing feature of this system is that all components operate independently, eliminating the need for human-to-human or human-to-machine interaction. In the context of smart agriculture, several common components contribute to making this entire process seamless and integrated, ensuring efficient and automated operations.

2. IoT Sensor Technology and Its Components

Smart agriculture relies heavily on sensor technology as it plays a pivotal role in gathering and measuring various environmental factors that could impact crop yield. The accuracy of sensor data acquisition is fundamental to the success of precision agriculture, enabling crop- and soil-specific management (Muhammad,2021). Equipment and vehicles, including tractors, harvesters, unmanned aerial vehicles, and sensor devices, are equipped with remote sensing facilities such as Geographic Information System (GIS) and Global Positioning System (GPS) for precise and autonomous site-specific operations. A diverse range of IoT sensors is available for monitoring applications, categorized into two types. The first type consists of intelligent multipurpose imagery sensors that can be embedded in unmanned aerial vehicles (UAVs), rails, and fixed-position components, involving remote sensing (Senthilnath,2017). When combined with deep learning, these sensors reach their full potential, capable of tasks like soil and vegetation/crop mapping, crop phenology, crop height estimation, yield estimation, fertilizer impact, biomass measurement, water stress detection, drought conditions, pest detection and management, weed detection, and greenhouse monitoring (Feng *et al.*2019). The second type of sensor is more specific to use cases and can be deployed at various locations in the field. Common sensors include airflow, soil moisture, electrochemical, capacitive humidity, position, mechanical, optical, and temperature sensors. The working principles of these sensors and their use cases. Additionally, several factors make IoT sensors suitable for smart agriculture, such as

- Computational efficiency,
- Cost-effectiveness,
- Coverage,
- Durability,
- Memory capacity,
- Portability,
- Power efficiency, and reliability.

3. Unmanned Aerial Vehicles (UAVs) Technology-based smart Agriculture

Apart from the IoT ecosystem, UAVs are an emerging and self-existing technology that is a combination of various other technology stacks such as robotics, onboard computing, artificial intelligence (AI) (Verma *et al.*2020), information and communication technology (ICT), IoT, and battery. The reason behind the popularity of UAVs is that it is filling the gap of limitations of remote sensing imaging through satellite because of weather and cloud penetration and on- ground limitations of robots because of uneven plains, obstacles, and speed. UAVs provide imaging with high resolution using hyperspectral, multispectral, and red-green-blue (RGB) cameras (Sierra Wireless,2020,May).It entails more accurate details of the field at a much cheaper cost. UAVs prove effective in both monitoring and action phases of applications within precision agriculture. In the monitoring phase, common applications include soil and crop mapping, sampling (The Economist,2019, April), yield forecasting (FAO,2018), weed detection (Bogue,2019) and pest and disease detection (Angrist, & Lavy,2009), along with soil and crop stress assessment Andrade-Pacheco *et al.* 2019). Moving to the action phase, UAV applications encompass sowing seeds (Wu & Moraghan,2018), spraying herbicides , pesticides , and fertilizers .

In the realm of smart agriculture practices, UAVs manifest two primary types, as illustrated, fixed-wing UAVs and rotary-wing UAVs. Fixed-wing UAVs, resembling airplanes, exhibit similar flight characteristics. They are particularly advantageous for extensive areas due to their capacity for long-range, high-speed, high-altitude operations, and crash tolerance. On the other hand, rotary-wing UAVs, including helicopters and multirotor types like quadcopters (Lu *et al.*2019), hexacopters (Bencherif,& Novikov,2019), and octocopters (Mekala *et al.* 2018), are adept at hovering, similar to helicopters. Rotary-wing UAVs offer advantages such as easy setup and operation, low-altitude flight, precise location operation ability, no wind planning requirements, and full autonomy in daily agriculture operations.

Despite these advantages, UAVs come with certain limitations. Technical constraints include low battery efficiency and flight time, payload capacity, communication distance, and efficiency. Fixed-wing UAVs can communicate up to 100 kilometers, with an average flight time of approximately 5 hours (Singh *et al.*2018). To address battery efficiency challenges, ongoing research focuses on developing more efficient hybrid batteries and implementing advanced battery management and optimization techniques (Tao *et al.*2019).

4. Communication Technologies in Smart Agriculture

Smart agriculture relies on ICT as its backbone, enabling the transmission of data to databases or clouds for further computation and analysis. Communication technologies in smart agriculture vary based on factors such as communication range, data rate bandwidth, power consumption, licensed or unlicensed spectrum, frequency band, and subscription prices. Each communication technology excels in different application scenarios, depending on the specific requirements.

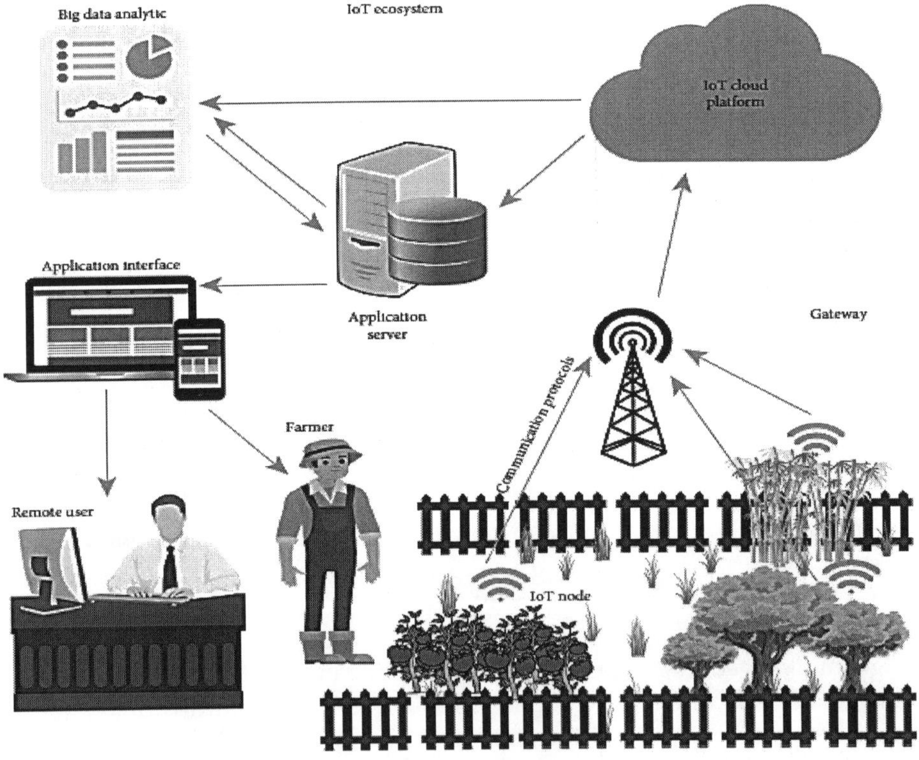

Figure 1: IoT ecosystem for agriculture.

Wireless Communications and Mobile Computing

For instance, Zigbee is well-suited for greenhouse agriculture monitoring, while Narrowband Internet of Things (NB-IoT) and Long-Range (LoRa) are more suitable for field precision agriculture (Yildirim *et al.*2016). Choosing the right communication technology involves considering multiple factors, with unlicensed spectrum technologies offering better bandwidth but presenting challenges like radio frequency interference, insecure communication, infrastructure setup costs, and low-range connectivity. Examples of unlicensed spectrum technologies include Radio Frequency Identification (RFID), Bluetooth, Wi-Fi, and Zigbee. On the other hand, licensed spectrum technologies are reliable, provide access to large areas, ensure security, and have lower infrastructure costs but require a subscription for data transmission and offer lower data rate bandwidth compared to unlicensed ones. Researchers indicate that Zigbee, Wi-Fi, and cellular technologies are popular among researchers for agriculture applications, with Zigbee at 45%, Wi-Fi at 25%, and cellular or multihopping technologies at 20% in agriculture-related experiments (Biswas *et al.*2019). Additionally, emerging Low-Power Wide-Area (LPWA) technologies like NB-IoT and Long-Term Evolution Machine (LTE-M) may garner more attention, as they are slated to be part of the 5G standard and will be the only LPWA-supported 5G technology according to the 3rd Generation Partnership Project (3GPP), the standard group specifying 5G and other wireless networking standards (Chowdhury *et al.*2018).

5. Cultivating Innovation: Methods and Techniques for Smart Agriculture

Over centuries, addressing food requirements has driven the adoption of various advanced agriculture techniques. With the advent of IoT, innovative methods such as vertical farming, hydroponics, and phenotyping have become integral, leveraging IoT to enhance performance. These techniques prove cost-effective and contribute to efficient resource management encompassing input, labor, and operational resources while ensuring high yields. In the realm of smart agriculture, technology applications extend to geospatial and temporal sampling and mapping, disease and pest monitoring, as well as smart irrigation and fertilization. Essential technologies and equipment include sensors, UAVs, IoT-based machinery, and communication systems, demonstrating a harmonious integration of modern technology in agricultural practice.

6. Precision Agriculture: Nurturing Smart Farming for Optimal Yields

Precision agriculture, while having historical roots, faced viability challenges for small, medium, and even large farmers in developing countries like India. The inevitability of challenges such as climate change, gaps in food demand and supply, urbanization, and diminishing arable land prompted the emergence of IoT, ushering in a new era for precision agriculture. This involves a convergence of existing technologies, including WSN, RFID Gateways, cloud computing, communication protocols, middleware components, and end-user interfaces (Singh *et al.*2018).

The core objective of precision agriculture is the efficient utilization of natural resources while safeguarding the environment. Implementation involves four key steps: characterizing the extent and scale of variability in soil and crop attributes, interpreting the significance and causes of variability, managing variability spatially and temporally, and monitoring outcomes resulting from variability management practices (Mohan *et al.*2018). The effectiveness of these steps is significantly enhanced through the integration of IoT.

The adoption and implementation of technology in precision agriculture outline the key advantages of IoT in precision agriculture (PA). A starting point for precision agriculture adoption is yield monitoring, utilizing gathered data to develop spatial and temporal feature databases for effective land management, interpretation, and yield mapping.

7. Greenhouse Innovations: Transforming Agriculture with Smart Farming Practices

Greenhouse farming shares similarities with precision farming but stands out due to its unique characteristics and objectives. It thrives within closed or isolated spaces where environmental variables are precisely regulated and managed by sophisticated smart systems. While not a recent concept, the infusion of information technology and IoT applications has elevated greenhouse farming, particularly in areas such as temperature control, humidity management, and shed monitoring.

The critical distinction lies in the precise and continuous monitoring and control, a feat only achievable through the seamless integration of IoT and smart systems. Despite occupying a smaller space compared to traditional open-field agriculture, greenhouse farming boasts significantly higher productivity. The Netherlands serves as a prime example, emerging as the second-largest exporter of agricultural goods through the adept use of greenhouse farming and hydroponics for a diverse range of crops (Karihaloo & Bhavani,2017).

Beyond conventional farming landscapes, greenhouse farming presents a transformative opportunity to harness unconventional spaces. This includes repurposing desert areas for sustainable agriculture, marking a paradigm shift in agricultural sustainability (Mude *et al.*2018). The strategic alignment of greenhouse farming with advanced technologies marks a pivotal moment in the evolution of smart agriculture, promising efficient resource utilization and enhanced productivity in a controlled environment.

8. Smart Solutions for Urban Farming: Cultivating Sustainability through Technology

Urban farming represents a groundbreaking concept, encompassing various methods such as rooftop farming, indoor farming, vertical farming, hydroponics, aquaponics, and aeroponics. As the global population increasingly gravitates towards urban areas, these regions become significant consumers of food products. However, urban agriculture emerges as a strategic response to challenges posed by climate change, water scarcity, and the complexities of sustaining food security.

In urban landscapes, the significant distance between food producers and consumers results in transportation and supply chain expenses, impacting food quality and contributing to environmental pollution from transport vehicles. Urban farming emerges as a viable solution by enabling individuals to cultivate fresh and affordable food near their living spaces. This form of agriculture relies heavily on precise environmental controls and systems, allowing for continuous cultivation throughout the year, irrespective of seasonal and weather variations.

The potential of urban farming is exemplified by initiatives such as Paris's largest rooftop farm, expected to feature 30 different types of plants, and the ambitious plan to utilize over 100 hectares of rooftops for farming. Similarly, in London, abandoned underground spaces, reaching 100 feet below the ground, are repurposed for vertical farming. These examples highlight the adaptability and scalability of urban farming in addressing the challenges of sustainable food production (Ragul et al,2018). As urban areas continue to evolve, the integration of innovative agricultural practices becomes crucial for fostering a resilient and sustainable food system. Urban farming stands at the forefront, offering a practical solution to the growing demand for fresh, locally produced food while mitigating the environmental impact associated with conventional food supply chains.

9. Applications of IoT in Smart Agriculture

In the era of smart agriculture, the transformation towards data-driven processes is evident, facilitated by continuous monitoring through on-site IoT devices. The applications of smart agriculture made possible exclusively through the integration of IoT, encompass geospatial and temporal mapping and sampling (Raut, & Gavhane, 2019), intelligent drip and sprinkler irrigation, monitoring and control of pests and pathogens, precise assessment of yields, targeted fertilization, and the maintenance of environmental conditions. A detailed exploration of these applications is presented below:

10. Precision Farming: Harnessing Geospatial and Temporal Mapping for Smart Agriculture Practices

Precision agriculture is a transformative approach that plays a pivotal role in crop field assessment and mapping. This technique harnesses the power of geospatial and temporal sampling, offering applications such as efficient weed management systems, accurate water stress assessments, and precise vegetation indexes. Additionally, it facilitates spatial variability assessment through Geographic Information Systems (GIS) (Kim, & Kim, 2016).

The implementation of geospatial techniques involves various methods, including remote sensing, aerial surveys using planes, and remote imagery obtained through Unmanned Aerial Vehicles (UAVs). While these methods were initially expensive and less efficient, advancements in satellite remote sensing and cloud distortion correction have made UAVs a cost-effective and highly efficient tool. Farmers in developing countries can now embrace UAVs as the initial step in adopting precision farming practices, revolutionizing agriculture with enhanced accuracy and resource optimization.

11. Smart Irrigation Strategies for Efficient Resource Management

In the face of escalating water scarcity, particularly in regions like India transitioning to water-stressed conditions, the adoption of IoT and smart systems introduces weather-adaptive smart irrigation. This innovative approach seeks to curtail the consumption of the precious resource of water (Venkatesan *et al.*2019). The fundamental concept behind smart irrigation is its capacity to irrigate only when necessary, tailoring its actions based on the specific crop and soil stress levels. This not only ensures efficient water utilization but also contributes to sustainable agricultural practices.

Utilizing UAVs adds another layer of efficiency to address the variability factor in water stress. Specifically, UAVs can be employed for precise sprinkler irrigation, delivering targeted and timely water distribution on the spot (Istepanian, & Al-Anzi,2016). This integration of unmanned aerial vehicles enhances the precision and effectiveness of irrigation practices.

12. Pest and Weed Management System

Pests, weeds, and pathogens pose significant threats to crops, potentially causing up to a 30% reduction in productivity, with weeds alone accounting for this substantial impact (Pantelopoulos,& Bourbakis, 2010) . Moreover, the use of pesticides and herbicides raises concerns, as it not only diminishes profits but also degrades the quality of agricultural products, presenting a considerable challenge for consumers.

In response to these challenges, IoT and smart systems come to the forefront with advanced pest and weed management solutions. These technologies enable the early detection of diseases, pests, and weeds in crops, providing timely information to farmers. Additionally, they facilitate the precise targeting of pests and pathogens through the judicious application of pesticides and herbicides. The use of smart vehicles further enhances the efficiency of these pest management strategies (Georgiou *et al.*2016). The integrated approach ensures not only higher agricultural productivity but also addresses environmental and health concerns associated with indiscriminate pesticide use.

13. Precision Fertilization

IoT's role in agriculture extends to precision fertilization, a pivotal application with dual benefits of cost savings and environmental conservation. Imbalanced fertilization can lead to multifaceted damage, where excessive fertilizer may result in nutrient runoff, soil salinity, plant rot, decreased productivity, extra costs, and contribute to climate change through evaporation. Conversely, insufficient nutrients can hinder plant growth and productivity.

The precise proportions of different fertilizer elements such as nitrogen (N), potassium (K), phosphorus (P), and water are crucial, contingent upon factors like plant type, soil composition, and weather conditions. Failure to tailor these proportions appropriately can impede crop productivity. Variability in these factors demands precision monitoring and mapping of both land and crops.

Smart IoT-based agricultural systems offer optimal estimations of nutrient requirements, reducing labor and input costs. By providing a tailored approach to fertilization, these systems contribute to sustainable farming practices, ensuring resource efficiency and minimizing environmental impact.

14. Yield Assessment

The initial step involves precise and continuous monitoring of both biotic and abiotic factors, made possible through the integration of IoT, Wireless Sensor Networks (WSNs), and Unmanned Aerial Vehicle (UAV) imagery. These technologies generate vast amounts of unstructured data, forming the basis for informed decision-making.

The acquired data plays a crucial role in early disease prediction, crop forecasting, and harvest planning. Leveraging these applications empowers farmers to reduce labor and operational costs, conduct error-free assessments of diseases and pests, estimate revenue and profit, and strategically plan harvesting periods for increased profitability over the long term.

15. Challenges and Solutions of Implementing IoT Devices in Smart Agriculture

One of the primary challenges associated with using IoT devices, particularly in the context of agriculture, revolves around the limited battery life. This issue is particularly pronounced when employing Unmanned Aerial Vehicles (UAVs) in agricultural practices, impacting the overall flight time. Researchers have extensively explored potential solutions to address this limitation, recognizing it as an open area for innovation (Gaikwad *et al.*2017)].

To extend the flight time of UAVs, some proposed solutions focus on ultra-light weight Wireless Power Transfer (WPT) systems (Anand, & Singh, (2019). These flexible systems are designed to accommodate changes in air-gap geometry and can charge UAVs midair, significantly increasing their flight time to approximately 7 minutes. This approach enables wireless charging with a power output of 10 W.

Another approach involves developing wireless charging systems for UAVs utilizing Capacitive Power Transfer (CPT) technology (Zou *et al.*2016). This system allows for charging UAVs over wide areas, employing a circuit, transformer, and inductors on the emitting side semiconductor elements for a DC-DC converter, and charge-controlling IC on the receiving side. This prototype system operates at around 12W, achieving more than 50% efficiency. Efficiency and high-power transfer capability led to the proposal of a wireless charging system for UAVs in agriculture fields using magnetic resonant coupling (Mondal *et al.* 2018). This approach utilized specific coil configurations in both transmitter and receiver circuits, demonstrating maximum transfer power and efficiency in experimental settings. In addition to battery life, wireless communication path loss in the agricultural environment poses another significant hurdle, requiring accurate modeling. Improved models, such as the hybrid exponential and polynomial models with particle swarm optimization, have been proposed and tested (Mishra, & Singh,(2018). These models aim to enhance the coefficient of determination (R2) of the regression line, reducing mean absolute error (MAE) and improving wireless data communication reliability.

Addressing path loss challenges in Wireless Underground Sensor Networks (WUSNs) for precision agriculture, a system based on predicting the Complex Dielectric Constant (CDC) has been developed (Choudhary *et al.*2019). Known as WUSN-PLM, this system outperforms existing path loss models, demonstrating improved precision and balanced accuracy on real cheap sensors in various communication scenarios.

References

Anand, D., & Singh, D. (2019). Advanced precision agriculture through IoT and drone- based systems. In Internet of Things: Concepts, Technologies, and Applications (pp. 163–183). Springer, Cham.

Anderson, K., Ryan, B., Sonntag, W., & Kavvada, A. (2017). Earth observation in service of the 2030 agenda for sustainable development. Geo-spatial Information Science, 20(2), 77–96.

Andrade-Pacheco, R., *et al.* (2019). Computer vision algorithms for plant phenotyping in 3D reconstructed cotton crops. Sensors, 19(17), 3769.

Angrist, M., & Lavy, V. (2009). The effects of high-stakes high school achievement awards: evidence from a randomised trial. NBER Working Paper No. 14945. https://www.nber.org/papers/w14945 at low altitude UAV aerial spraying in rice cultivation. Computers and

Barkunan, S. R., Bhanumathi, V., & Sethuram, J. (2019). Smart sensor for automatic drip irrigation system for paddy cultivation. Computers & Electrical Engineering, 73, 180–193.

Bencherif, M. A., & Novikov, A. (2019). High-level architecture for cognitive IoT devices. Journal of Ambient Intelligence and Humanized Computing, 10(9), 3367–3382.

Beza, E., Reidsma, P., Poortvliet, P. M., Belay, M. M., Bijen, B. S., & Kooistra,

Biswas, M., Singh, D., Jeon, G., Choi, J. W., & Park, J. H. (2019). Efficient data analysis for cloud-based smart homes: A deep learning approach. Information Sciences, 488, 112–124.

Bogue, R. (2019, November). The top 10 emerging technologies of 2019. World Economic Forum. https://www.weforum.org/agenda/2019/11/ the-top-10- emerging-technologies-of-2019/

Choudhary, S., Prakash, R., & Deo, A. (2019). IoT based smart agriculture monitoring system using cloud. In 2019 3rd International Conference on Electronics, Communication and Aerospace Technology (ICECA) (pp. 424–429). Coimbatore, India.

Chowdhury, A. N., Aboelaze, M., & El-Sayed, H. (2018). An intelligent system for monitoring and control of precision irrigation based on IoT technology. In 2018 IEEE 4th Middle East Conference on Biomedical Engineering (MECBME) (pp. 97–101). Tunis, Tunisia.

De Araujo Barbosa, C. C., Atkinson, P. M., & Dearing, J. A. (2015). Remote sensing of ecosystem services: A systematic review. Ecological Indicators, 52, 430–443. Electronics in Agriculture, 167, 105045.

Faiçal, B. S., Freitas, H., Gomes, P. H., *et al.* (2017). An adaptive approach for UAV- based pesticide spraying in dynamic environments. Computers and Electronics in Agriculture, 138, 210–223.

Feng, X., Yan, F., & Liu, X. (2019). Study of wireless communication technologies on the Internet of Things for precision agriculture. Wireless Personal Communications, 108(3), 1785–1802.

Food and Agriculture Organization of the United Nations. (2018, November). Digital agriculture: A FAO perspective. http://www.fao.org/3/ca2383en/ca2383en.pdf

G. J., Hain, C. R. (2018). Smallholder crop area mapped with wall-to-wall World View sub-meter panchromatic image texture: A test case for Tigray, Ethiopia. Remote Sensing of Environment, 212, 8–20.

Gaikwad, T., Kale, K., & Swami, S. (2017). An advanced agricultural and wildlife monitoring system using wireless sensor network. In 2017 International Conference on I-SMAC (IoT in Social, Mobile, Analytics and Cloud) (I-SMAC) (pp. 32–36). Palladam, India.

Garg, B., Aggarwal, S., & Sokhal, J. (2018). Crop yield forecasting using fuzzy logic and regression model. Computers & Electrical Engineering, 67, 383–403.

Georgiou, P., Toumazou, C., & Wong, A. C. (2016). Continuous patient monitoring: state-of-the-art and future directions. Journal of Medical Engineering & Technology, 40(4), 196–211.

Gu, J., Yin, G., Huang, P., Guo, J., & Chen, L. (2017). An improved back propagation neural network prediction model for subsurface drip irrigation system.Computers & Electrical Engineering, 60, 58–65.

Istepanian, R. S. H., & Al-Anzi, T. (2016). m-Health 2.0: new perspectives on mobile health, machine learning and big data analytics. Methods of Information in Medicine, 55(6), 459–462.

Kansakar, P., & Hossain, F. (2016). A review of applications of satellite earth observation data for global societal benefit and stewardship of planet earth. Space Policy, 36, 46–54.

Karihaloo, B. L., & Bhavani, R. V. (2017). Development of IoT-based integrated system for agriculture field monitoring. In 2017 IEEE Calcutta Conference (CALCON) (pp. 68–73). Kolkata, India.

Karim, L., Anpalagan, A., Nasser, N., & Almhana, J. (2013). Sensor-based M2M agriculture monitoring systems for developing countries: State and challenges. Networks and Protocols, 5(3), 68–86.

Kasturirangan, K. (2007). Space technology for humanity: A profile for the coming 50 years. Space Policy, 23(3), 159–166.

Khanna, A., & Kaur, S. (2019). Evolution of Internet of Things (IoT) and its significant impact in the field of precision agriculture. Computers and Electronics in Agriculture, 157, 218–231.

Kim, K., & Kim, H. (2016). Secure authentication protocol for IoT-based healthcare applications. In 2016 IEEE/RSJ International Conference on Intelligent Robots and Systems (IROS) (pp. 5373–5378). Daejeon, South Korea.

L. (2018). Exploring farmers' intentions to adopt mobile Short Message Service (SMS) for citizen science in agriculture. Computers and Electronics in Agriculture, 151, 295–310.

Lal, R. (2009). Soil degradation as a reason for inadequate human nutrition. Food Security, 1(1), 45–57.

Lu, Y., Zhang, J., Du, K., & Duan, X. (2019). A distributed method for managing precision agriculture tasks in cloud environment. Concurrency and Computation: Practice and Experience, 31(21), e5064.

Mas´o, J., Serral, I., Domingo-Marimon, C., & Zabala, A. (2019). Earth observations for sustainable development goals monitoring based on essential variables and driver-pressure-state-impact-response indicators. International Journal of Digital Earth, 1–19.

Mekala, M. S., Viswanathan, P., Mekala, M. S., & Perumal, V. (2018). A survey: smart agriculture IoT with cloud computing. In 2017 International Conference on Microelectronic Devices, Circuits and Systems (ICMDCS) (pp. 1–7). Vellore, India.

Mishra, A., & Singh, D. (2018). An intelligent IoT framework for smart agriculture. In 2018 2nd International Conference on I-SMAC (IoT in Social, Mobile, Analytics and Cloud) (I-SMAC) (pp. 46–52). Palladam, India.

Mohan, S., Patil, R., Deshpande, A., & Hooli, M. (2018). IoT based smart agriculture and field monitoring system using MQTT protocol. In 2018 International Conference on Data Management, Analytics and Innovation (ICDMAI) (pp. 396–401). Pune, India.

Mondal, B., Jain, D., & Puthal, D. (2018). Design and development of a smart IoT-based greenhouse farm for agricultural practices. In 2018 IEEE Calcutta Conference (CALCON) (pp. 1–6). Kolkata, India.

Mude, A. N., Amatya, A., & Bueti, C. (2018). A comprehensive study of IoT sensors used in agriculture. In 2018 2nd International Conference on Inventive Systems and Control (ICISC) (pp. 1022–1026). Coimbatore, India.

Muhammad, M. N., Wayayok, A., Mohamed Shariff, A. R., Abdullah, A. F., & Husin, E. M. (2019). Droplet deposition density of organic liquid fertilizer

Neigh, C. S., Carroll, M. L., Wooten, M. R., McCarty, J. L., Powell, B. F., Husak,

Pantelopoulos, A., & Bourbakis, N. G. (2010). A survey on wearable sensor- based systems for health monitoring and prognosis. IEEE Transactions on Systems, Man, and Cybernetics, Part C (Applications and Reviews), 40(1), 1–12.

Ragul, S. B., Viswanath, A., & Lakshmipriya, S. (2018). IoT-based smart agriculture monitoring and controlling system using arduino and cloud. In 2018 2nd International Conference on I-SMAC (IoT in Social, Mobile, Analytics and Cloud) (I-SMAC) (pp. 1–5). Palladam, India.

Raut, S. K., & Gavhane, P. (2019). IoT-based smart agriculture monitoring system using raspberry pi. In 2019 3rd International Conference on Electronics, Communication and Aerospace Technology (ICECA) (pp. 424–429). Coimbatore, India.

Senthilnath, J., Kandukuri, M., Dokania, A., & Ramesh, K. N. (2017). Application of imaging platform for vegetation analysis based on spectral- spatial methods.Computers and Electronics in Agriculture, 140, 8–24.

Shrivastava, S., & Marshall-Colon, A. (2019). Big Data in Agriculture and Their Analyses. Encyclopedia of Food Security and Sustainability. https:// doi. org/10.1016/b978-0-08-100596-5.22191-4

Sierra Wireless. (2020, May). The Path to 5G with LTE-M and NB-IoT. https:// www.sierrawireless.com/resources/iot-blog/iot-blog/2020/05/ the-path-to-5g-with-lte-m-and-nb-iot/

Singh, D., Reddy, G. K., & Tripathi, S. (2018). A cloud computing based approach for IoT enabled WSN for real-time environmental monitoring. Journal of King Saud University - Computer and Information Sciences. https://doi. org/10.1016/j.jksuci.2018.07.019

Singh, D., Tripathi, S., & Kumar, N. (2018). An internet of things approach for motion-based object tracking in smart homes. Computers, Materials & Continua, 54(1),19–40.

T. (2016). Mapping of soil properties and land degradation risk in Africa using MODIS reflectance. Geoderma, 263, 216–225.

Tantalaki, N., Souravlas, S., & RoumelloTis, M. (2019). Data-driven decision making in precision agriculture: The rise of big data in agricultural systems. Journal of Agricultural and Food Information, 1–37.

Tao, F., Qi, Q., & Wang, L. (2019). Digital twin-driven product design, manufacturing and service with big data. The International Journal of Advanced Manufacturing Technology, 100(5–8), 1385–1396.

The Economist. (2019, April). The new arms race in agricultural technology. https://www.economist.com/special-report/2019/04/27/the-new-arms-race-in-agricultural-technology

United Nations Department of Economic and Social Affairs. (April 2020). The world population projected to live in urban areas by 2050, says UN. https:// www.un.org/development/desa/en/news/population/2018- revision-of-world-urbanization-prospects.html

United Nations Department of Economic and Social Affairs. (April 2020). Key facts-world urbanization prospects-population division-United Nations. https://population.un.org/wup/Publications/Files/WUP2018-KeyFacts. pdf

Vågen, T. G., Winowiecki, L. A., Tondoh, J. E., Desta, L. T., & Gumbricht,

Venkatesan, S., Prabakaran, B., & Jeyapriya, R. (2019). An IoT-based health care system for patients monitoring and medication reminder. Materials Today: Proceedings, 18, 3284–3289.

Verma, S., Bhatia, A., Chug, A., & Singh, A. P. (2020). Recent advancements in multimedia big data computing for IoT applications in precision agriculture: opportunities, issues, and challenges. In Multimedia Big Data Computing for IoT Applications (pp. 391–416). Springer, Singapore.

Wu, X., & Moraghan, J. T. (2018). The internet of things in agriculture for sustainable rural development. Sustainability, 10(8), 2766.

Yildirim, A. B., Abidi, M. H., & Radecka, K. (2016). Cognitive IoT-Cloud integration for smart healthcare: case study for epileptic seizure detection. In 2016 IEEE International Conference on Communications (ICC) (pp. 1–6). Kuala Lumpur, Malaysia.

Zou, Y., Ji, S., & Han, Y. (2016). A survey on wireless sensor networks-based agricultural monitoring and decision support systems. Computers and Electronics in Agriculture, 123, 276–291.

Chapter 2

Transforming Agriculture: Leveraging Decision Support Systems for Sustainable Growth

Veena Sharma

Agromet Section, SKUAST-J, Main Campus, Chatha
veena_sharma29@rediffmail.com

Abstract

Agriculture stands as a cornerstone in providing our essential needs such as food, clothing, and various materials crucial for our existence. Technological progress has significantly altered the operational landscape of industries. Historically, traditional agriculture heavily depended on manual laboUr. However, with the ongoing advancements in computer and communication technologies, the agricultural sector has undergone a substantial transformation. To support the ever-growing world population with limited farm land, researchers investigate the implementation of Decision Support System.The Decision Support System (DSS) aids farmers in making timely decisions to maximize productivity while minimizing inputs and associated costs. Integrated with simulation models, soil, weather, and crop databases, as well as statistical tools and other information technologies, the DSS offers a relevant, user-friendly system for managing crop cultivation. It provides site-specific agronomic practices and recommendations to farmers without adverse environmental impacts. Parameters such as weather conditions, soil characteristics, water levels, and incidences of pests and diseases are taken into account within the DSS, facilitating effective crop management.

Introduction

The agricultural sector is subject to various influences, including climate change, water availability, soil fertility, and economic variables (Udias *et al.*, 2018). Seasons, among these factors, significantly affect agricultural crop production. In recent decades, farmers have encountered numerous challenges stemming from unpredictable climate changes and issues with both the quantity and quality of crop production. These challenges often arise due to the lack of soil testing facilities or farmers' unawareness of optimal crop management decisions. Given the crucial role of soil fertility in crop growth, periodic soil testing is essential for determining fertilizer usage and irrigation needs.

Furthermore, the limited availability of farmland and its gradual reduction pose severe challenges that require immediate solutions to sustain the ever-growing global population. Addressing these challenges is crucial, and the integration of modern computing techniques, such as Decision Support Systems (DSS) with the Internet of Things (IoT), holds promise in finding solutions.

Technological advancements, including Big Data, Artificial Intelligence (AI), and IoT, aim to enhance agricultural efficiency, sustainability, and environmental impact (Lampridi *et al.*, 2019). Machine learning, IoT, and AI have revolutionized various aspects of the agriculture industry, including yield prediction (Haque *et al.*, 2020), disease detection (Darwish *et al.*, 2020), crop recognition (Zhang *et al.*, 2020), and water management (Goldstein *et al.*, 2018). Low-cost sensor-based IoT technologies have been developed to optimize production efficiency, improve quality, and minimize environmental impacts by efficiently managing resources like water and energy (Ferrández *et al.*, 2016).

The emergence of smart farming practices incorporates big data for predictive and operational purposes (Wolfert *et al.*, 2017). Additionally, cloud-based precision agriculture applications are beneficial for optimizing irrigation management (López-Riquelme *et al.*, 2017).

Decision Support System

A decision support system (DSS) is an application based on computer technology that collects, arranges, and evaluates data to facilitate informed decision-making for management, operations, and planning purposes. The DSS can operate entirely through automated processes or with human involvement, and in certain instances, it may utilize a combination of both approaches. This system conducts analyses on information and may even generate decisions on behalf of the user. The realm of DSS application spans across agricultural production and marketing, with a focus on fostering sustainable development through the application of relevant concepts, techniques, and principles.

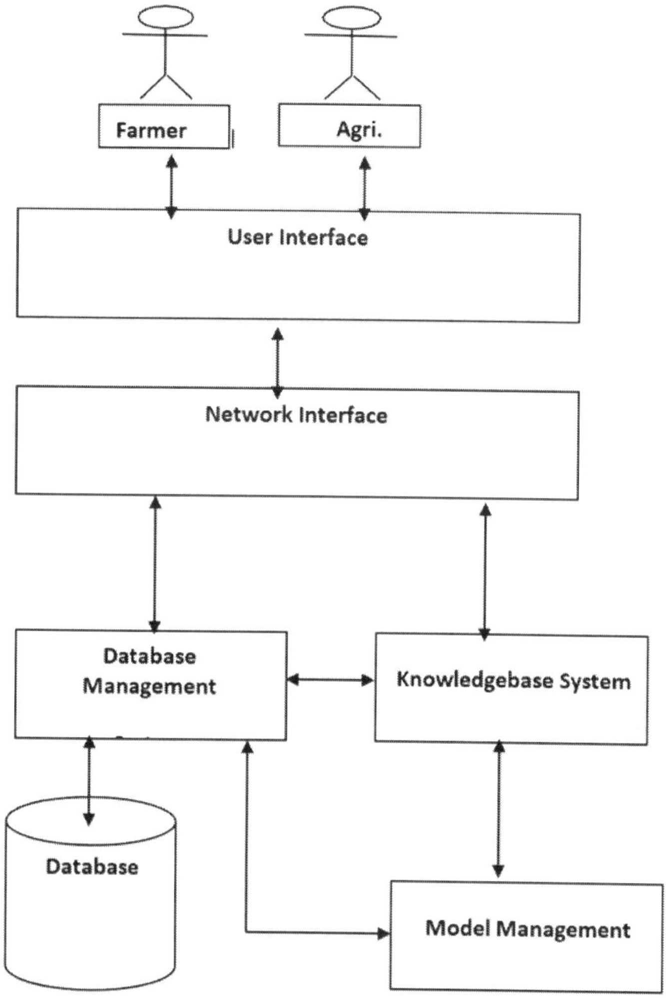

Figure: Conceptual framework for decision support system in crop planning

Three basic components of a DSS architecture are:

- Database (or knowledge base data collection)
- Model (actual system)
- User interface.

The users themselves are also important components of the structure. Further DSS components may be classified as:

- Inputs: Factors, numbers, and things to analyze
- User learning and expertise: Inputs requiring hand-operated analysis
- by the user
- Outputs: Converted data from which DSS "decisions" are generated.
- Decisions/Results: Results generated by the DSS based on user tests.

A Decision Support System (DSS) has the capacity to provide farmers with a range of suggestions to enhance their decision-making and achieve improved outcomes using available information. Various existing DSS applications in agriculture employ mathematical optimization techniques and focus on specific analyses, such as water distribution [Bonfante *et al.*, 2019], land allocation [Basso *et al.*, 2013], weather and climate forecasting [Han *et al.*, 2017], and collaborative learning processes within the agricultural sector [Thorburn *et al.*, 2011]. These systems intake diverse forms of information and utilize models and data processing to generate tailored recommendations. DSSs can also be tailored to support farmers and other relevant stakeholders [Nelson *et al.*, 2002], including governmental bodies and agricultural advisors, based on the availability of data. However, effective stakeholder engagement is crucial for the success of DSS implementation. This involvement can be achieved by employing simplified models to facilitate informed and timely decision-making [Soltani *et al.*, 2013].

Various information systems, expert systems, and decision support systems have been devised, developed, and put into action to aid farmers. The table below provides a summary of the names and purposes of these agricultural systems.

Table 1: DSS and ES for Agriculture Management

DSS and ES for Agriculture Management		
S. No.	**Name**	**Purpose**
1	DSSAT	DSS for agricultural technology transfer-Decides type of seed to grow, crop yield prediction, how much to irrigate, frequency of fertilizer application.
2	CROP-9-DSS	Identification of pests and disease control ,fertilizer management, water management of 9 crops of Kerala
3	CROPLOT	An Expert System that determined suitability of a plot for a given crop
4	CALEX	A DSS for cotton irrigation management.
5	ESIM	Expert system for irrigation management..
6	CropSyst	A model for water –Nitrogen interaction in wheat crop. Agricultural water management system regime

DSS and ES for Agriculture Management		
S. No.	**Name**	**Purpose**
7	CROPGRO	DSS for increasing soybean yield in water limiting environment based on climatic data
8	AQUAMAN	Web based DSS for irrigation scheduling of peanut
9	TUNGGUL	Development model DSS for rain water management in semi-arid area.
10	DAIRY MGT	A DSS for management of the Dairy business
11	APSIM	APSIM is the System developed to simulate biophysical process in farming systems for economic and ecological outcomes of management practice in the face of climatic risk.
12	Android Based ADSS	Android based DSS to select the crop variety according to whether conduction and availability of water.
13	Intelligent fertilization DSS Based on Knowledge Model and Web GIS Decision for Fertilization	The study constructed the basic frame of decision support system for fertilization.
14	CROPWAT	system for irrigation management.
15	IPM	This DSS is mainly used by the farmers of West Bengal to get climate date to decide the crop .this system has provided comprehensive awareness of the paste management.
16	CROPMAN	A DSS used by farmers in Punjab to get site specific climate data. This system has helped farmers to enhance the yield by changing the transport schedule from May to June.
17	MKRISHITM	A DSS applied by the farmers in Maharashtra t acquire the climate data for particular crop. The system helped the farmers to schedule the scouting of paste and nutrients.
18	MKRISHITM	A DSS applied by the farmers in Maharashtra t acquire the climate data for particular crop. The system helped the farmers to schedule the scouting of paste and nutrients.
19	Rayat Mitra	The web portal created by Department of Agriculture to provide Agricultural Extension services to farmers and to transfer the latest technical knowledge to the farming community, introduction of high yielding varieties, laying demonstrations, imparting training to farmers to improve skills & knowledge to boost up the agricultural Production and Productive.

DSS and ES for Agriculture Management		
S. No.	**Name**	**Purpose**
20	Farmer web portal	This web portal to make available relevant information and services to the farming community and private sector through the use of information and communication technologies, to supplement the existing delivery channels provided for by the department. Farmers' Portal is an Endeavour in this direction to create one stop shop for meeting all informational needs relating to Agriculture, Animal Husbandry and Fisheries sectors production, sale/ storage of an Indian farmer.
21	Bhoomi	This portal is the project of on-line delivery and management of land records in Karnataka. It provides transparency in land records management with better citizen services and takes discretion away from civil servants at operating levels.
22	eSagu	A agricultural DSS that helped the farmers of Tamil Nadu to farm specific decisions in order to enhance the yield of by assessing the crop specific climate data from the system.

Decision support systems (DSS) can be very useful in agriculture to assist farmers and other stakeholders in making better decisions. Below are several illustrations demonstrating the application of DSS in agriculture:

DSS in Crop Productivity Improvement

DSS finds extensive application in various agricultural management tasks, with crop productivity being a prominent focus area where significant results have been achieved through its utilization. One such DSS, known as "Crop Environment Resource Synthesis (CERES) - Wheat," is a component of DSSAT. This system has been effectively employed to simulate wheat crop growth and development under varying climatic, water, and nitrogen conditions in semi-arid and subtropical regions (Arora *et al.*, 2007). Analysis from the model indicates that grain yield and water productivity are influenced by soil water holding capacity. Subsequently, this model was expanded with the Cropping System Model (CSM) into "DSSAT-CSMCERES-Wheat 4.0," facilitating the estimation of crop yield, evapotranspiration, crop water productivity (CWP), and irrigation water productivity (IWP) (Timsina *et al.*, 2008).

Another simulation-based DSS model, the Cropping System Simulation Model (CropSyst), integrates with Geographic Information System (GIS) and adopts a similar approach to simulate the growth and development of various herbaceous crops using periodic biomass and Leaf Area Index (LAI) data. This model aids in formulating water-saving strategies and policies for rice crops by incorporating crop management practices such as transplanting dates, seed types, and irrigation methods (Jalota *et al.*, 2009). Both the CropSyst and CERES models prove effective for on-farm management activities like irrigation and soil nitrogen management (Singh *et al.*, 2000).

Additionally, the CROPGRO - Soybean model, part of DSSAT, has successfully simulated the effects of climatic changes on soybean growth and yield (Lal *et al.*, 1999). It has been utilized to estimate potential yields in water-limited and non-limited areas, as well as to assess yield gaps in major soybean regions of India (Bhatia *et al.*, 200).

DSS in Crop Water Requirements Management

DSS is also applied to careful management of water resources, crop water requirements. Crop water requirements depend on the factors such as evaporation, evapotranspiration, meteorological factors such as solar radiation, air temperature, humidity, wind speed etc. (Allen *et al.*, 1998). CROPWAT DSS helped to decide irrigation scheduling for different crop patterns and to calculate crop water requirements (Stancalie *et al.*, 2010). Along with climatic data it helped in assessment of reference evapotranspiration under temperate conditions (Rajput and

DSS Based on Climatic Data

Climatic factors such as temperature, precipitation, and sunlight duration are pivotal in determining crop productivity. The DSSAT Cropping System Model (DSSAT-CSM) is utilized to investigate the impact of climatic conditions on productivity within rice-wheat systems. Comparing observed and simulated yields of rice and wheat indicates that integrating biotechnological advancements and precision farming practices can enhance crop productivity in conjunction with climatic variables (Subash and Mohan, 2012).

A simulation-based cropping model named "Crop Production and Management (CROPMAN)" is employed to analyze the effects of transplanting dates and weather parameters on yield, evapotranspiration, and water productivity over multiple years and various crops. Findings suggest that shifting the transplanting dates of rice from mid-May to later in June can increase yields in Punjab (Chahal *et al.*, 2007). The model also examines grain yields of chickpea crops in semi-textured soils within rice-chickpea cropping patterns, indicating increased yields compared to rice-wheat patterns. Moreover, it identifies higher irrigation water requirements in low rainfall environments with coarse-textured soils compared to medium-textured soils and areas with higher rainfall (Jalota *et al.*, 2006).

Evapotranspiration, a key component of the hydrological cycle, significantly influences crop yield, groundwater recharge, and land use planning. The Decision Support System for Evapotranspiration (DSS-ET) is developed using climatic data from various ecosystems in India to estimate evapotranspiration. Serving as both a research and teaching tool, the DSS aids in parameter estimation and addresses missing data (Bandyopadhyay *et al.*, 2012).Pest and

Disease Management

At a regional scale, Geographic Information Systems (GIS) coupled with necessary environmental data and epidemic simulation models offer geographic delineation of disease and insect pest risk zones. These zones aid in strategic decision-making regarding deployment strategies for crop varieties and estimating the duration of host plant resistance. Historical climate data from specific locations prove valuable in assessing site suitability for particular diseases. Efforts are underway to integrate disease prediction systems with online weather and weather-interpolation platforms.

Traditional yield loss studies have traditionally established the relationships between nitrogen application rates, disease severity, seasonal variations, and grain yield. While these studies provide a qualitative understanding of host-pathogen-environment interactions and inform disease management strategies, their recommendations are limited in validity. This is because they are heavily influenced by factors such as disease onset, rate of spread, farm management practices, environmental conditions, and their complex interactions. Physiologically based simulation models offer a valuable tool for comprehending the mechanisms of damage and analyzing their impacts on rice crop growth and yield.

Increased Efficiency of Multi-environment Evaluation

Development and release of a variety is a complex process that may extend over a period of 10-15 years. Once the breeding lines have become homozygous, they are bulked and then tested in observational, replicated and multilocation yield trials.These multi-site trials are expensive and need several crop seasons/ locations to understand geno type by environment (GxE) interactions.

Since the systems approach integrates different components of agroeco systems, it can play a great role in increasing the efficiency of plant breeding process, in particular multi- environment testing of geno types. Crop models together with GIS can facilitate biological characterization of the physical environment (geography, soil, climate, etc.), and thus define key environment aldomains for which improved varieties are to be developed. Alternatively, the same methodology can be used to determine the adaptation domains of geno types. A modeling approach can also provide timates of yield probability in target environments based on the understanding of G x E interactions. Such studies can help in reducing the number of sites/season required for field evaluation and thus increase the efficiency of the whole process of variety development.

A very simple role for crop modeling would be the provision of an index of site quality for use in traditional GxE analysis. This could be achieved by simulating performance of number of check varieties over a wider range of weather conditions at each site. Regression of observed test variety lines on the simulated index would give an idea of stability. Regression of the same check variety yield son the simulated index would give an idea of the effect of specific weather conditions. Definition of breeding environments could also be achieved by crop models by modeling a set of probe geno types known to discriminate between breeding environments. A target region could be divided into breeding domains in which different sets of germplasm could be tested.

By modeling a range of probable geno types and selected environments known to discriminate between the geno types, it is possible that the crop parameters determining the specific interaction could be identified. Hypothetical varieties could then be modeled combining the crop parameters conferring most advantage as an indication of suitable traits and breeding targets.

Patel, 2006). It also gave good results when applied to simulate different crop water requirements as per the need of crop under different planting dates and probable canal water supplies (Raja, 2010). Along with the GIS mapping model it was used to spatially analyze and study crop water requirements of rice. It also helped to study the spatial variation of climatic water balance, probabilistic monthly monsoon rainfall and mapping of cold periods in the region (Verma and Kar, 2005).

DSS in Irrigation Scheduling

Irrigation scheduling stands as a crucial activity in agriculture. A Decision Support System (DSS) tool named Simulation of Water and Salt (SWASALT) was calibrated and validated for irrigated areas within semi-arid regions. This tool proved instrumental in averting on-farm waterlogging and soil salinization issues (Singh and Singh, 1997). Furthermore, the SWASALT model was employed to compute Water Management Response Indicators (WMRI), aiding in the optimization of on-farm irrigation schedules by minimizing percolation losses into groundwater across different soil types (Singh and Singh, 1996).

An integrated GIS-based model, incorporating rainfall, soil, water use, canal flow, soil water balance, and groundwater flow models, has been utilized as an effective DSS tool. This model has contributed to increased crop production by optimizing cropping patterns in accordance with groundwater availability (Chowdary *et al.*, 2003).

DSS in Advisory System Advisory

Decision Support Systems (DSS) have emerged as integral tools in Indian agriculture, facilitating significant advancements in farm productivity and management. One such example is e-Sagu, a farm-specific DSS developed by IIIT Hyderabad and Media Lab Asia. This system has revolutionized farm productivity by delivering tailored agro-expert recommendations directly to farmers, spanning all stages of crop cultivation from sowing to harvesting. By reducing cultivation costs and enhancing both productivity and quality of agricultural commodities, e-Sagu has proven to be a game-changer in Indian agriculture (Reddy and Ramaraju, 2009).

Similarly, "mKRISHITM," a DSS tool developed by Tata Consultancy Services, has been deployed in Borgaon village, Maharashtra, to provide grape farmers with guidance on proper nutrient and pest management via mobile phones (Pande *et al.*, 2021). Integrated Pest Management (IPM) is another crucial aspect of agriculture, and the development of the "Cell Phone" IPM DSS has contributed to sustainable plant protection by continuously educating farmers and subsequently improving crop productivity (Patel *et al.*, 2012).

Moreover, the DSS tool developed by the Indian Meteorological Department (IMD) serves to formulate and disseminate biweekly weather forecast-based agro-advisories, aiding farmers in making informed decisions regarding agricultural activities. These examples underscore the vital role that DSS plays in enhancing agricultural practices and outcomes in India.

Conclusion

Being a key part of sustainable agriculture, DSS will become more and more robust. The increased connectivity of technologies and the multiplication of smart devices on the field stimulates the accumulation and storage of data. By integrating and analysing the captured data, DSS are meant to become more accurate as well as reliable. Last but not least, actors from the sector have the ambition to make DSS more ergonomic and user-friendly. Companies already offer mobile interfaces on smartphones and tablets, that are easy to handle and can be used in real time, directly on the field. Such developments allow more farmers to use DSS, since, at the end of the day, they are the final recipients and users of new technologies which will revolutionize agriculture.

References

Allen, R.G., Pereira, L.S., Raes, D., Smith, M. 1998. "Crop evapotranspiration - Guidelines for computing crop water requirements - FAO Irrigation and drainage paper 56", FAO - Food and Agriculture Organization of the United Nations, Rome.

Arora, V.K., Singh, H., and Singh B. 2007. "Analyzing wheat productivity responses to climatic, irrigation and fertilizer-nitrogen regimes in a semi-arid sub-tropical environment using the CERES-Wheat model", Agricultural Water Management, Volume 94, 1–3, Pp 22-30.

Bandyopadhyay, A., Bhadra, A., Swarnakar, R.K., Raghuwanshi, N.S., Singh,

R. 2012. "Estimation of reference evapotranspiration using a user-friendly decision support system: DSS_ET", Agricultural and Forest Meteorology, 154–155,19-29.

Basso, B., Cammarano, D., Fiorentino, C., Ritchie, J.T. 2013. Wheat Yield Response to Spatially Variable Nitrogen Fertilizer in Mediterranean Environment. Eur. J. Agron. 51, 65–70.

Bhatia, V.S. P. Singh, S.P. Wani, G.S. Chauhan, A.V.R. Kesava Rao, A.K. Mishra, K. Srinivas, "Analysis of potential yields and yield gaps of rain-fed soybean in India using CROPGRO-Soybean model", Agricultural and Forest Meteorology, 148(8–9).

Bonfante, A., Monaco, E., Manna, P., De Mascellis, R., Basile, A., Buonanno, M., Cantilena, G., Esposito, A., Tedeschi, A., De Michele, C., *et al.* 2019. LCIS DSS—An Irrigation Supporting System for Water Use Efficiency Improvement in Precision Agriculture: A Maize Case Study. Agric. Syst. 176, 102646.

Chahal, G.B.S., Sood, A., Jalota, S.K., Choudhury, B.U., Sharma, P.K. 2007. "Yield, evapotranspiration and water productivity of rice (Oryza sativa L.)–Wheat (Triticum aestivum L.) system in Punjab (India) as influenced by transplanting date of rice and weather parameters", Agricultural Water Management, 88 (1–3), 14-22.

Chowdary, V.M., Rao, N.H., Sarma, P.B.S. 2003. "GIS-based decision support system for groundwater assessment in large irrigation project areas", Agricultural Water Management, 62(3), 229-252.

Darwish, A., Ezzat, D., Hassanien, A.E. 2020. An Optimized Model Based on Convolutional Neural Networks and Orthogonal Learning Particle Swarm Optimization Algorithm for Plant Diseases Diagnosis. Swarm Evol. Comput. 52, 100616.

Ferrández-Pastor, F.J., García-Chamizo, J.M., Nieto-Hidalgo, M., Mora-Pascual, J., Mora-Martínez, J. 2016. Developing Ubiquitous Sensor Network Platform Using Internet of Things: Application in Precision Agriculture. Sensors.16, 1141.

Goldstein, A., Fink, L., Meitin, A., Bohadana, S., Lutenberg, O., Ravid, G. 2018. Applying Machine Learning on Sensor Data for Irrigation Recommendations: Revealing the Agronomist's Tacit Knowledge. Precis. Agric. 19, 421–444.

Han, E., Ines, A.V.M., Baethgen, W.E. 2017. Climate-Agriculture-Modeling and Decision Tool (CAMDT): A Software Framework for Climate Risk Management in Agriculture. Environ. Model. Softw. 95, 102–114.

Haque, F.F., Abdelgawad, A., Yanambaka, V.P., Yelamarthi, K. 2020. Crop Yield Prediction Using Deep Neural Network. In Proceedings of the 2020 IEEE 6th World Forum on Internet of Things (WF-IoT), New Orleans, LA, USA, 2–16 June 2020, pp. 1–4.

Jalota, S.K., Singh, K.B., Chahal, G.B.S., Gupta, R.K., Chakraborty, S., Sood, A., Ray, S.S. and Panigrahy, S. 2009. "Integrated effect of transplanting date, cultivar and irrigation on yield, water saving and water productivity of rice (Oryza sativa L.) in Indian Punjab: Field and simulation study", Agricultural Water Management, 96 (7).

Jalota, S.K., Sood, A., Harman, W.L. 2006. "Assessing the response of chickpea (Cicer aeritinum L.) yield to irrigation water on two soils in Punjab (India): A simulation analysis using the CROPMAN model", Agricultural Water Management, 79 (3), 312-320.

Kar, G., Verma, H.N. 2005. "Climatic water balance, probable rainfall, rice crop water requirements and cold periods in AER 12.0 in India", Agricultural Water Management, 72, 15-32.

Lal, M., Singh, K.K., Srinivasan, G., Rathore, L.S., Naidu, D and Tripathi, C.N. 1999. "Growth and yield responses of soybean in Madhya Pradesh, India to climate variability and change" , Agricultural and Forest Meteorology, 93, (1) 25.

Lampridi, M.G., Sørensen, C.G., Bochtis, D. 2019. Agricultural Sustainability: A Review of Concepts and Methods. Sustainability 11, 5120.

López-Riquelme, J.A., Pavón-Pulido, N., Navarro-Hellín, H., Soto-Valles, F., Torres-Sánchez, R. A. 2017. Software Architecture Based on FIWARE Cloud for Precision Agriculture. Agric. Water Manag. 183, 123–135.

Nelson, R.A., Holzworth, D.P., Hammer, G.L., Hayman, P.T. 2002. Infusing the Use of Seasonal Climate Forecasting into Crop Management Practice in North East Australia Using Discussion Support Software. Agric. Syst. 74, 393–414.

Pande, A., *et al.* 2021. "m-KRISHI Market Facing Innovation", TATA Consultancy Services TCS Innovation Lab, Mumbai.

Patel, L., Mondal, C., Maitra, N. 2012. "Cell Phone – A Decision Support for sustainable plant protection of the district South 24 Parganas, West Bengal", Global Journal of BioScience&BioTechnology, Volume .1 (1), 59-64.

Raja, W. 2010. "Validation of CROPWAT 8.0 for Estimation of Reference Evapotranspiration using Limited Climatic Data under Temperate Conditions of Kashmir", Research Journal of Agricultural Sciences. 1(4) : 338-340.

Rajput, T.B.S. and Patel, N. 2006. "Determination of the Optimal Date for Sowing of Wheat in Canal Irrigated Areas using FAO CROPWAT Model", Water Technology Centre, IARI, New Delhi, India.

Reddy, P. and Ramaraju, G. 2009. "e-Sagu: The Next Generation IT-based Query-less, Cost-Effective and Personalized Agro-Advisory System", Media Lab Asia Project, IIIT, Hyderabad.

Singh, A. K., Tripathy, R. and Chopra, U. K. 2008. "Evaluation of CERES-Wheat and CropSyst models for water–nitrogen interactions in wheat crop", Agricultural Water Management, 95(7).

Singh, R., Singh, J. 1996. "Irrigation planning in cotton through simulation modeling", Irrigation Science, 17(1).

Singh, R., Singh, J. 1997. "Irrigation planning in wheat (Triticum aestivum) under deep water table conditions through simulation modeling", Agricultural Water Management, 33 (1).

Soltani, A., Stoorvogel, J.J., Veldkamp, A. 2013. Model Suitability to Assess Regional Potato Yield Patterns in Northern Ecuador. Eur. J. Agron. 48, 101–108.

Stancalie, G., Marica, A., Toulios, L. 2010. "Using earth observation data and CROPWAT model to estimate the actual crop evapotranspiration", Physics and Chemistry of the Earth, Parts A/B/C, 35(1–2).

Subash, N., Ram Mohan, H.S. 2012. "Evaluation of the impact of climatic trends and variability in rice–wheat system productivity using Cropping System Model DSSAT over the Indo-Gangetic Plains of India", Agricultural and Forest Meteorology, 164.

Thorburn, P.J., Jakku, E., Webster, A.J., Everingham, Y.L. 2011. Agricultural Decision Support Systems Facilitating Co-Learning: A Case Study on Environmental Impacts of Sugarcane Production. Int. J. Agric. Sustain. 9, 322–333.

Timsina, J, Godwin,D., Humphreys, E., Singh,Y., Singh, B., Kukal,S. and Smith,

D. 2008. "Evaluation of options for increasing yield and water productivity of wheat in Punjab, India using the DSSAT-CSM-CERES-Wheatmodel", Agricultural Water Management, Volume 95, Issue 9, September.

Udias, A., Pastori, M., Dondeynaz, C., Carmona Moreno, C., Ali, A., Cattaneo, L., Cano, J. A. 2018. Decision Support Tool to Enhance Agricultural Growth in the Mékrou River Basin (West Africa). Comput. Electron. Agric. 154, 467–481.

Wolfert, S., Ge, L., Verdouw, C., Bogaardt, M.J. 2017. Big Data in Smart Farming—A Review. Agric. Syst. 153, 69–80.

Zhang, W., Liu, H., Wu, W., Zhan, L., Wei, J. 2020. Mapping Rice Paddy Based on Machine Learning with Sentinel-2 Multi-Temporal Data: Model Comparison and Transferability. Remote Sens. 12, 1620.

Chapter 3

Smart Livestock Farming in India for Future Food Security

Kamal Sarma[1] *and Jonali Devi*[2]

[1]*Professor, Division of Veterinary Anatomy,* [2]*Professor & Head, Division of Veterinary Physiology & Biochemistry, Faculty of Veterinary Sciences & Animal Husbandry, S.K. University of Agricultural Sciences & Technology, Jammu*

Global population explosion, demand for agriculture and livestock products, changing food habits and agro-climatic conditions are stressing on the need for alternate strategies and approaches for food security and sustainable agricultural development. By the year 2050, there is an expectation that the world's population will grow to nearly 9 billion; thus increasing our need for food by more than 100 percent. Presently, 1 in 8 people, or 842 million, struggle with hunger every day. Further, roughly 1 billion people in the world are food insecure, means they lack access to a sufficient quantity of affordable, nutritious food. In one hand agriculture has evolved in order to meet these intensely growing food demands, in the other hand the farmers will need to increase food production by 70-100 percent to meet global nutrition needs. It is known that there is relatively little available land on which to cultivate food. The Food and Agriculture Organization of the United Nations (FAO) projections indicate that 80 percent of the additional food required to meet demand of growing population in 2050 will need to come from land already under cultivation. As a consequence, our farmers and food producers have to produce those higher yields using the same (or less) acreage than they use today while relying on fewer natural resources.

Livestock plays a major role in Indian economy. Approximately 20.5 million people depend upon livestock for their livelihood in India. Livestock contributed 16% to the income of small farm households against an average of 14% for all rural households. Again, livestock provides livelihood to two-third of rural community and it provides employment to about 8.8 % of the population in India. The vast Livestock sector of India contributes 4.11% of GDP and 25.6% of total Agriculture GDP.

The livestock sector of India is one of the largest in the world with 11.6 percent of the world's total livestock population (BAHS-Basic Animal Husbandry Statistics, 2014). In recent years, the livestock sector has emerged as ansignificant segment of an expanding agricultural sector in the Indian economy (Tisdell and Gali, 2000). It is wel known that the livestock farming is a very important socio-economic activity in Indian agriculture, as milk is the second largest agricultural commodity contributing to GNP (Gross National Products), next to rice (Sarkar and Ghosh, 2010).

India is	World's highest livestock owner at about 535.78 million
	First in the total buffalo population in the world - 109.85 million buffaloes
	Second in the population of goats - 148.88 million goats
	Second largest poultry market in the world Second largest producer of fish and also second largest aquaculture nation in the world
	Third in the population of sheep (74.26 million)
	Fifth in in the population of ducks and chicken (851.81 million)
	Tenth in camel population in the world - 2.5 lakhs

Significance of Livestock for Providing Food and Non-food Items to the People

1. Food: The livestock provides various food items such as milk, meat and eggs for human consumption. India is number one milk producer in the world and produced about 176.34 million tons of milk in a year (2017-18). Similarly, it produced about 95.22 billion of eggs, 7.70 million tonnes of meat in a year. The value of output of livestock sector was estimated to be Rs. 9,17,910 crores at current prices during 2016-17 which was about 31.25% of the value of output from agricultural and allied sector. The value of output from livestock was about 31.11% of the value of the output from total agriculture and allied sectors. During the financial year 2017-18, the total fish production of India was estimated at 12.61 million Metric tonnes.

2. Fibre and skins: The livestock equally contributes to the production of wool, hair, hides, and pelts. Leather is one of the most important products which has a very high export potential. India produced about 41.5 million Kg of wool per annum during 2017-18.

3. Draft: Bullocks are still acting as the back bone of Indian agriculture. Despite of increasing advancements in the use of mechanical power in Indian agricultural farming systems, the Indian farmer especially in rural areas still depend upon bullocks for various agricultural operations. The bullocks are saving a lot on fuel which is a necessary input for using mechanical power like tractors, combine harvesters and other appliances that use fuel.Other animals like camels, horses, donkeys, ponies, mules etc are being extensively used to transport goods in different parts of the country in addition to bullocks. Donkeys and mules are also especially used by the Indian armed forces in high altitudes. In situations like hilly terrains mules and ponies also serve as the only alternative to transport goods.

4. Dung and other animal waste materials: Dung and other animal wastes act as very good farm yard manure and the value of it is the tune of several crores of rupees. Additionally, the cow dung is also used as fuel (bio gas, dung cakes), and for construction as poor man's cement.

5. Income: Livestock is a source of additional and emergency income for many families in India especially the resource poor who maintain only few heads of animals. Milch cows and buffaloes provide regular income to the livestock farmers through sale of milk. Again, livestock like sheep and goat serve as sources of income during emergencies to meet exigencies like marriages, treatment of sick persons, children education, repair of houses etc in rural India. Therefore, livestock are considered as 'moving banks' because of their potentiality to be disposedof during emergencies. They serve as capital and in cases of landless agricultural labourers and often it is the only capital resource they possess. Livestock serve as a property and in case of emergencies they serve as guarantee to avail loans from the local sources such as money lenders in the villages.

6. Employment: A huge number of people in India depend upon agriculture for their livelihoods being less literate and unskilled. But agriculture being seasonal in nature could provide employment for a maximum of around six months in a year. For this, the landless and less land holding people depend upon livestock for utilizing their labour during lean agricultural season, and this is how the livestock rearing offers them a source of income. For many unemployed educated youths also, livestock farming is providing a source of employment.

7. Cultural: Livestock has got cultural significance also by offering security to the owners and also by adding to their self-esteem especially when they are owning prized animals such as pedigreed bulls, dogs and high yielding cows/ buffaloes etc. Among many tribal communities in India, there is a ritual of offering or gifting livestock during their marriages.

8. Social security: The animals also offer social security to the owners in terms of their status in the society. The families which own animals are better placed in the society than those who do not. Rearing of animals is a part of the

Indian culture from the time immemorial. Animals are used for various socio religious functionse.g.,cows are used for house warming ceremonies; rams, bucks and chicken for sacrifice during festive seasons; bulls and cows are worshipped during various religious functions, etc.

9. Sports / recreation: The animals like rams, bulls, cocks, etc.are also used by the people for competition and sports. Although there is a ban on these animal competitions, cock fights, ram fights and bull fights (jallikattu) are quite common during festive seasons.

10.Companion animals: Dogs are well known for their faithfulness and are being used as companions since time immemorial. When the nuclear families are increasing in number and the old parents are forced to lead solitary life, the pets like dogs, cats provide the needed company to the latter thus making them lead a comfortable life.

Contribution of Livestock Sector in Indian Economy

The livestock sector of India is one of the largest in the world. The total milk production in our country grew at 6.50 percent and it reached 187.75 million tonnes in the year 2018-19, which was estimated as more than double the growth of world milk production. This record production of milk improved the per capita availability of milk to 394 gm per day per person, which was considerably higher than recommended rate of availability by Indian Council of Medical Research (ICMR) i.e., 280 gm per day per person. Uttar Pradesh, Rajasthan, Madhya Pradesh, Andhra Pradesh and Gujarat are among the top five milk producing states, constituting 53.10 percent of the total milk production of the whole country. Nearly 35 percent of the milk production of India is contributed by indigenous buffalos followed by 26 percent by crossbred cattle. On the other hand, the total meat production reached 8.11 million tonnes in the year 2018-19 with a growth rate of 6 percent. Out of the total meat production in India, contribution of poultry meat is 50 percent (4.06 million tonnes) followed by buffalo meat (19.50 percent). The poultry meat production showed a growth rate of 7.80 percent. Uttar Pradesh, Maharashtra, West Bengal, Andhra Pradesh and Telangana are among the top five meat producing states in the country, constituting 56.90 percent of the total meat production. In 2018-19, egg production reached 103.32 billion with an annual growth rate of 8.50 percent. Presently, the per capita availability of eggs is approximately 79 eggs per person per year, which is much less than the recommended level i.e., 180 eggs per person per year by ICMR. The states like Andhra Pradesh, Tamil Nadu, Telangana, West Bengal and Haryana altogether contributed 65 percent of the total egg production of India and are recognised as the top five eggs producing states. The total egg production from commercial poultry farms is 84.91 billion (82.20 percent) and from backyard poultry is 18.41 billion (17.80 percent). The total wool production in the year 2018-19 was 40.42 million kgs. Telangana, Karnataka and Gujarat are the top five wool producing states of the country and altogether contributed 78.50 percent of the total wool production.

Major Challenges of Livestock Sector

1. Low Productivity

Although India possesses highest livestock population and it ranks number one in milk production in the globe, the productivity, particularly of ruminants has been extremely low, turning this precious asset of the poor people into a liability. It is worthy to mention here that, over 60 percent of the rural households rear large ruminants, mostly for milk and partly for bullock power, although the average milk yield is significantly low.

2. High economic losses due to animal diseases

The diseases in livestock cause major economic loss to the farmers. With the improvement in the quality of livestock through cross-breeding program, the susceptibility of these livestock to various diseases including exotic diseases has been increased. The insufficient coverage of vaccination is continuously resulting into economic losses due to various animal diseases. Estimations of losses due to different diseases are not easy because it is very difficult to report all diseases at all places. The direct losses estimation is based on reported diseases and it indicated that average annual economic losses due to diseases like Haemorrhagic Septicaemia (HS), Foot and Mouth Disease (FMD), Brucellosis, Peste des Petits Ruminants (PPR), Classical Swine Fever were in around Rs. 5255 crores in the year 2014, Rs. 20000 crores in 2016, Rs. 20400 crores in 2015, Rs. 2417 crores in2016 and Rs. 429 crores in the year 2016, respectively. It indicated that farmers in India faces a direct loss of almost Rs. 50,000 crores every year due to the five fully preventable (with vaccination) diseases although the government of India along with state governments spend equally good amount of funds on vaccination against these five diseases that costs approximatelyevery year. But adequate results are not achieved due to substandard vaccines, inefficient vaccination, ill-education of livestock farmers etc.

3. Inadequate infrastructure and human resources for support services

Indian livestock sector is facing the problem of poor infrastructure and human resources. As on 31 March 2017, the number of veterinary institutions stood at 65242. As per the recommendation of the National Commission on Agriculture (NCA)-1976, One Veterinary Institution is to be provided for every 5,000 cattle units (one cattle unit =1 cow or 1 buffalo or 10 sheep or 10 goats or 5 pigs or 100 poultry) to ensure proper and adequate veterinary health care. On the same line, the Veterinary Council of India (VCI) has recommended that for every 5000 livestock population there is need of one veterinarian for effective delivery of veterinary services to the farmers. It is reported that there are a total of 67651 veterinarians in India, whereas estimated requirement by VCI is around 1.1-1.2 lakh (Damodaran, H. 2015) in the whole country. This highly inadequate human resources resulted in poor and insufficient veterinary services to the farmers. For the purpose of improvement of cattle breed and in turn milk production, Artificial Insemination (AI) technology was adopted in India in 1970 onwards. However, in the last 50 years, due to several such constraints, the average conception rate through AI is not going more than 30-40 percent at field level and the share of milk from crossbred cows to total milk production is not exceeding beyond 26 percent.

4. Shortage of feed and fodder

India has only 2.29 percent of land area of the world, but is maintaining nearly 17 percent of world human population and 10.70 percent of livestock (more than 535.82 million)which posing a huge pressure on land, water and other resources. India is having only 5 percent of its cultivable land under fodder production. Again, area under permanent pastures and grazing lands comprises a mere 3.30 percent of the total area, and it has also been declining steadily. Among different resources of livestock fodder, crop residues are major one and these feeding resources are major ingredient meeting more than 50 percent of the livestock sector demand in the country. At prevailing livestock productivity and production, livestock sector is facing severe feed and fodder shortage in the country. In a report titled "Revisiting National Forage Demand and Availability Scenario", released during August 2019 by the ICAR-Indian Grassland and Fodder Research Institute (IGFRI) has reported that there is a deficit of 23.40 percent in the availability of dry fodder, 11.24 percent in green fodder and 28.90 percent for concentrates in India (Roy *et al.*, 2019). There are already proven high yielding varieties of fodder and technologies such as silage making, hay making and urea- molasses treatment for crop residue is in practice. However, adoption of such technologies is very poor in many of the states, especially in rural areas. According to the Indian Ministry of New and Renewable Energy (MNRE) report, India generates on an average 500 million tons of crop residue per year and a majority of this crop residue is in fact used as fodder and fuel for other domestic and industrial purposes. But there is still a surplus of 140 million tons, out of which 92 million tons is burned each year, which can be effectively used for animal fodder (Bhuvaneshwari, *et al.*, 2019).

5. Inadequate public institution support

In India, the livestock sector did not receive the policy and adequate financial attention as it deserved. The sector is granted with only about 12 per cent of the total public expenditure on agriculture and allied sectors, which is disproportionately lesser than its contribution to agricultural GDP. The sector has been also neglected by financial institutions. The share of livestock in the total agricultural credit has not ever exceeded four percent in the total credit. The government mechanisms to protect animals against various risk are not strong enough. At present, only 6 percent of the animal heads (excluding poultry) are provided insurance cover. Livestock extension machinery has remained grossly neglected in the past. However, in May 2019, Government of India has constituted separate ministry i.e., Ministry of Fisheries, Animal Husbandry and Dairying at central level to address various challenges of livestock sector, which is definitely going to boost the livestock sector.

6. Inadequate processing and value addition

Processing and value addition of livestock product is being regarded as the potential tool for sustainability of livestock production. In India, the milk processing industry has expanded at a Compound Annual Growth Rate (CAGR) of 14.80 percent as of 2018 (Laura Wood, 2019). Moreover, the meat industry is one of the most important part of food processing industry. The processing rate of kara beef (buffalo meat)and poultry meat are approximately 21% and 6 %, respectively. The major reason for inadequate processing and value addition in meat product chiefly lack of necessary infrastructure. As on 2019, in the country, there are only 1377 slaughter houses, 68 abattoirs with proper meat processing facilities approved by Agricultural and Processed Food Products Export Development Authority (APEDA), 32 APEDA registered meat processing plants, 11 APEDA registered stand-alone abattoirs, 9 carcass utilization centres, and 25000 small scale meat retail shops apart from a few private companies. There is a forecast in the report of FAO(2011)entitled "Mapping Supply and Demand for Animal-Source Foods to 2030" that the demand for various livestock products will increase by 80-100 percent by the year 2030 and out of the total increase, nearly 60 percent of demand will be due to change in consumption pattern and frequency of intake of the people and remaining 40 percent demand increase will be due to population growth. It clearly indicates the importance of value addition and processing of livestock products (Robinson, T.P. &Pozzi, F. 2011).

7. Issues in marketing of livestock and livestock products

Paving the way to the markets is critical to speed up the commercialization of livestock production. Lack of access to markets may act as a disincentive to farmers for adoptinglatest technologies and quality inputs. The change in the present market is not uniform. The changes are either species specific or products. But there are phenomenal changes from informal to formal market system seen currently in dairy and poultry sectors. However, nearly 60 percent of milk in the country is sold by the unorganized sector. On the other hand, sheep, goat and beef remains in the informal sector without much investment from private firms. If it is to be expressed concisely, the Indian livestock and livestock product market are mostly underdeveloped, irregular, uncertain, and lack transparency and often dominated by informal market intermediaries who exploit the producers.

8. Lack of attention of small ruminates

The small ruminants (sheep and goats) are generally reared by small, marginal and landless farmers, who cannot afford to own large ruminants. However, most of these small ruminants which are dependent on free grazing without any investment on supplementary feeding and health care, do not make significant contribution to the income due to low pruductivity. As the demand for meat is expected to grow high during the next two decades, the present system of unsustainable husbandry practices, which highlights the status of these species getting deprived of technological and managerial support services, needs to be addressed.

9. Inadequate attention towards extension services

The delivery system of livestock services has chiefly three components viz. providing technical services to the animals,giving technical inputs and educating the livestock farmers. Providing various services to animals includes vaccination, deworming, breeding and disease management for which technical inputs such as vaccines, medicines, semen, AI guns, syringes and needles, etc. needs to be supplied. The livestock farmers have no option but to depend on vets or para-vets for all the above-mentionedservices and, to some extent, supply of technical inputs. Unfortunately, the third component, ie. educating livestock farmers on various aspects of livestock management, (feeding, vaccination, disease management, breeding etc.) is totally neglected. Often, supply of inputs and providing services are considered as an extension service ignoring the fact that "education of farmers is the core of livestock extension service" (Rao, 2013). The main focus of any extension services requires building the capabilities or capacities of the farmers so that they can take care of their animals and crop in addition tooffering knowledge on transfer of technology and strengthening of various infrastructure and support services. Earlier, the State Departments of Animal Husbandry (SDAH) were part of State Department of Agriculture but with the changing times, this department got bifurcated and now every state of India has independent department for Animal Husbandry. However, since beginning, the orientation of the SDAH remained confined towards giving healthcare and breeding services rather than focus on production. Again, in most of the states, the designation of veterinarians is Veterinary Officer (VO), Veterinary Assistant Surgeon (VAS) rather than Livestock Development Officer (LDO) which normally psychologically restricts their role on livestock development and building farmer's capacities. The veterinarians are supposed to educate the farmers on scientific management practices, sustainability of livestock farming, ways to meet paucity of animal feed and fodder, marketing of livestock, processing of livestock products, environmental issues, etc. Thus, the development role of veterinarians on livestock production aspect is not adequately explored in India. In spite of above challenges, the Indian livestock sector is performing well, however in view of fulfilling the increasing demand of livestock products, these needs to be addressed.

Smart Livestock Farming

To mitigate all the above-mentioned challenges and drawbacks of the livestock sector, it is the need of hour to shift towards the smart livestock farming. The objective of Smart livestock farming is to achieve more productive, efficient and sustainable farm operations with the effective use of available digital technologies. The largest potential of it lies in individual animal monitoring and analysis, which is called the PrecisionLivestock Farming (PLF).

The smart livestock farming consists of the following components-

1. Artificial Intelligence (AI)

The capacity of a computer or robot to perform tasks commonly associated with intelligent beings. It is composed of learning, reasoning and self-correction. It also generates insights from data more quickly and accurately than humanly actions and is able to act automatically on that insight.

2. Cyber-Physical Systems (CPS) and Internet of Things (IoT)

A CPS is a mechanism controlled and monitored by computer-based algorithms. The network of connected CPS, people and devices creates the IoT where different devices communicate and work in coordination with each other and with humans in real-time, thereby enabling decentralized and automatized decision-making adjustments or actions to reduce waste at their end.

3. Platforms

A platform is a business model which creates value by facilitating exchange between two or more interdependent groups, which are usually the consumers and producers. Examples of certain plate forms are - Facebook and Uber. Such platforms also provide opportunities for the livestock meat sector, e.g., e-Sourcing platforms for transports along the supply chain, more efficient distribution of manure from local animal to local crop farmers, etc.

4. Information and Communication Technologies (ICT)

These are the technologies which provide access to information through telecommunication, including the internet, wireless networks, mobile phones and other communication media.

5. Big Data Analytics

It includes the process of collecting, organizing, and analyzing data to discover patterns, trends, correlations, consumer preferences and other information.

6. Cloud Computing

It is the practice of using a network of remote servers hosted on the internet to store, manage and process data, rather than using local servers or personal computers.

7. Crowdsourcing

It is the practice of obtaining input for a task or project by enlisting services of a large number of people, paid or unpaid, typically via the internet. In the livestock sector, crowdsourcing is especially interesting for farmoperations like emission reduction, efficient feed composition and animal health, as these topics are of a growing public interest.

8. Transparent supply chains

Digitalization offers great potential in creating real-time transparency along the supply chains and across company borders, owing to the increased amount and accuracy of available data. Across different sectors and industries, corporations can create platforms where the different stakeholders can easily communicate and exchange data, information and knowledge to improve the supply chain, thereby creating a competitive advantage.

9. Flexible production systems

Real-time transparency along the supply chain also lays the foundation for establishing flexible and adaptable production systems. Better data availability, together with predictive analytics, enables real-time optimization of multi-functional production assets in small loops, thereby allowing flexible production planning and control. This provides enormous potential to the meat sector, since demand can be volatile at particular times.

Role of Start-ups for Sustainable Livestock Farming in India

India is emerging as a worldwide destination for Innovation, Incubation and Start-ups providing great opportunities to the young generation of the country. Government of India has come up with exciting programmes to support start-up ecosystem and to nurture innovations and has launched "Start-up India action Plan" for bringing in policies and frameworks that can support and augment the start-up eco-system. India ranks third in Start-up ecosystem after United States and UK and is presently emerging as a hub for innovation and has a huge population of youth with untapped ideas. Innovations can ignite social change and solve problems.In the country, Bangalore ranks first in innovation and start-up destination. India already has over 450 aggrotech start-ups, and the number is still increasing day by day, with a growth rate of approximately 25% yearly. Each aggrotech start-up uses its own way to help farmers to increase their productivity and revenue, some focus on supply chain process optimization, some focus on online market trading, while many of them focus on precision farming. At present there are 103 Livestock Tech start-ups in India. List of the 10 most exciting onesis summarised below-

Name of the Start-up	Location	Year of start
Licious	Bangalore	2015
Fresh to Home	Bangalore	2015
Stellapps	Bangalore	2011
Animall	Bangalore	2019
ZappFresh	Delhi	2015
Tropical Animal Genetics	Gurugram	2014
Egg Oz	Bihar Sharif	2017
Krimanshi	Jodhpur	2015
Moo Farm	Gurugram	2014
Identis	Hyderabad	2010

Some star-ups in India are engaged in monitoring aspects. The top 5 start-ups engaged in Livestock monitoring in India are-

Anitrack – Cattle Monitoring

Serket – Pig Monitoring

Ewetrack – Sheep Monitoring

Eruvaka Technologies – Fish Monitoring

Faromatics – Poultry Monitoring

New technologies can enhance or improve the productivity capacity, welfare, or management of animals and livestock. The concept of the 'connected cow' is a result of more and more dairy herds being fitted with specific sensors to monitor their health and in turn increase productivity.Some of the Examples of Farming Innovations are:

1. Ground-breaking protein discovery platform:

Nuritas from Dublin uses AI and machine learning for invention of high potential bioactive peptides and proteins. Such discoveries lead to patented peptipes with medicinal and nutritious properties. This method may be used on waste streams of traditional foodstuffs. Waste can thus be turned into medicine and food again, by isolating the valuable high potential peptides.

2. 3D-printing vegan milk

The US start-up Perfect Day has developed an animal free milk of the same name that is vegan while still tasting like the fresh dairy product. The animal-free milk has been produced with laboratory-bred yeast, natural ingredients and fermentation processes. The team 3D-prints bovine DNA sequences onto the yeast, triggering the production of casein proteins that are also found in natural milk. The proteins are mixed with corn sugar and other ingredients in fermentation tanks and it is left for times to develop.

3. Small device determines vitamin deficiencies

"Vitameter", is the Canadian start-up,which is a simple blood test that allows users to check themselves regularly for vitamin deficiencies. For this, a drop of blood is taken and then tested it in the device. The "Vitameter" then determines the concentration of all standard and important nutrients in the blood. If needed, users can then manage deficiencies with supplements. "Vitameter" is particularly useful for people with food intolerances who are unable to feed themselves optimally, as it saves them from doctors' visits.

4. Robot tends and guides cattle herds

Scientists at the Australian Centre for Field Robotics at the University of Sydney have come up with a robot named "SwagBot". It is being designed to help cattle breeders. According to the results of a pilot test on a farm, the robot was able to herd cattle independently, clear weeds and pull heavy loads. The advanced version of the robot will also be developed to make it able to check the health of cattle by using sensors to analyse body temperature and movements.

5. Cow sends sms alerts to farmer

Start-up Food Blockchain XYZ has formulated a plan to create Food Supply Chain 2.0 by using blockchain. The start-up expects to tackle six major challenges like -lack of insight, central party reliance, misleading labels, the inability for small producers to compete, supply chain opacity and a lack of sustainability. Food Blockchain XYZ will introduce series of interconnected sensors for providing comprehensive food quality assurance, without the intervention of humans.

6. Software for optimum feeding on farms

The Dresden-based company fodjan has developed software called "fodjan smart feeding" which enables farmers to work out the right food rations for their livestock. Lot of feeding goals are taken into consideration, including cost savings and animal health. For planning purposes, the software does it possible to set up feeding calendars for animals and even check out the respective stocks of food. Besides all the feed-related features, "fodjan smart feeding" helps farmers to manage their farms more efficiently in a better way too.

NGOs in Livestock Services

Many NGOs are also engaged in India for boosting the livestock farming with an effort for ensuring national food security. Few are as listed below-

- Bhartiya Agro Industries Foundation (BAIF)
- J.K. Trust Gram Vikas Yojana
- Professional Assistance for Development Action (PRADAN)

Prospects for Sustainable Livestock Farming for Future Food Security: Strategic Option for Challenges

Union Budget 2022-23 is expected to boost the dairying and livestock sector (https://www.downtoearth.org.in/tag/livestock) with a host of measures to make it sustainable amid the ongoing COVID-19 pandemic. Dairy is the single-largest agricommodity in India. It contributes approximately five per cent to the national economy and directlyemploys 80 million dairy farmers. A boost in economic activities, increasing per capita consumption of milk and milk products, changing dietary preferences and rising urbanisation in the country, has driven the dairy industry to grow by 9-11 per cent during the year 2021-22.

1. Global best practice roll-out

Best practices are regarded as the methods or techniques that have consistently shown results superior to those achieved by other means. Large yield and productivity gaps exist across the livestock sector, especially between the industrialized advanced and developing countries. Digitalization can make a roll-out in to an easier, faster, and less expensive by leveraging transparent supply chains, cloud computing, improved networks and global knowledge platforms.

2. Precision feeding

Precision feeding can be a highly effective tool in enabling a reduction of feed intake per animal at the same time maximizing individual growth rates in livestock. It enables the intake of the right amount of feed, in the right nutrient composition, at the right time and for each animal individually.

3. Smart waste management

About 20% of all meat goes to waste every year. Although most waste is generated at the consumer end, the root causes lie across the whole supply chain, e.g., poor packaging or bad meat quality. Smart waste management offers high efficiency to first identify such root causes and second, to improve existing processes and thereby decrease waste amounts. The intelligent collection and use of data is the key for understanding what is produced and what is thrown away. Through this, players along the supply chain can make focused adjustments to reduce waste at their end as far as possible.

4. Animal health and welfare monitoring

Various health issues and diseases can decrease production efficiency of livestock by up to 33%. The trend towards more and more intensification of livestock farming systems in one hand increases productivity, but on the other hand can also have adverse effects on animal health and welfare, and also increases the risk of rapid and far-reaching disease outbreaks within stocks. Through ICT and the IoT, comparatively more performance-related data can be collected from the animals than the traditional methods by the use of cameras and image recognition softwares, wearables, as well as weight or sound monitoring, etc. Additionally, data from livestock facilities can also help to improve animal health, for example through climate, air quality and ventilation monitoring and by using drones.

References

Anonymous, 2017 Anonymous (2017 November 17). Status and Opportunities for Meat Processing Industries in India. SME Venture, Industry Dynamics, Food processing. Retrieved from https://www.smeventure.com/status-opportunities-meat-processing/

BAHS-Basic Animal Husbandry Statistics, 2014. BAHS. Government of India (GOI). 2014. Basic Animal Husbandry Statistics. Ministry of Agriculture. Department of Animal Husbandry, Dairying and Fisheries.

Damodaran, H. 2015 Damodaran H. (2015 December 15). 30 crore cattle and rising, but where are the country's vets? The Indian Express, India Others. Retrieved from https://indianexpress.com/article/india/india-others/30-crore-cattle-and-risingbut-where-are-the-countrys-vets/.

Laura Wood, 2019 Laura Wood (2019 March 19). The Dairy & Milk Processing Market in India, 2018- 2019 & 2023. Business Wire. Retrieved from https://www.businesswire.com/news/home/20190322005336/en/Dairy-MilkProcessing-Market-India-2018-2019-2023

Rao, SVN. 2013 Rao, SVN. (2013 August 26). Why Collaboration Matters for Livestock Development?, AESA Blog 06, Livestock Extension. Retrieved from https://www.aesanetwork.org/wp-content/uploads/2013/08/AESA-BLOG-6.pdf

Robinson, T.P. &Pozzi, F. 2011 Robinson, T.P. &Pozzi, F. (2011). Mapping supply and demand for animal-source foods to 2030. Animal Production and Health Working Paper. No. 2. Rome. pp.10, 48. Retrieved from http://www.fao.org/3/i2425e/i2425e00.pdf

Roy, A. K. et al,. 2019 Roy, A. K., Agrawal, R. K., Bhardwaj, N. R., Mishra, A. K. & Mahanta, S. K. (2019 August). Revisiting National Forage Demand and Availability Scenario. Retrieved fromhttps://www.researchgate.net/publication/336230799_Revisiting_National_For age_Demand_and_Availability_Scenario

Bhuvaneshwari, S, Hettiarachchi, H. and Meegoda, J. N. 2019. Crop Residue Burning in India: Policy Challenges and Potential Solutions. International Journal of Environment Research and Public Health 6(5), 832; https://doi.org/10.3390/ijerph16050832

Tisdell and Gali, 2000Tisdell, C. and Gali, J. 2000. Trends and developments in India's livestock industry. Econ. Ecol. and Env. Working paper 43 (Aug).

Chapter 4

Smart Agricultural Practices for Management of Vegetable Diseases

Sakshi Sharma, Ranbir Singh, Ayushi Sharma and L.V. Ravishankar

The importance of vegetables in providing balanced diet and nutritional security has been realized world over. Vegetables are now recognized as health food globally and play important role in overcoming micronutrient deficiencies and providing opportunities of higher farm income. The worldwide production of vegetables has tremendously gone up during the last two decades and the value of global trade in vegetables now exceeds that of cereals. Hence, more emphasis is being given in the developing countries like India to promote cultivation of vegetables.

New pathological threats have been seen in the last few years on vegetable crops under changing climatic conditions. It has been observed that the vegetable crops have now become more prone to viral, bacterial and fungal diseases. In some of the vegetable crops like tomato, spotted wilt virus, bacterial wilt and fungal wilts have been seen to be of common occurrence. The situation, however, is entirely different under greenhouse conditions. Under protected conditions, the severity and incidence of powdery mildew and root rots were found more alarming. Among viral diseases of tomato, tospo virus and leaf curl virus have occupied a significant position. Sclerotium, Rhizoctonia and Pythium root rots of tomato and capsicum are of significant importance in greenhouses or protected structures like playhouses. In bacterial diseases, Ralstonia solanacearum has become more devastating than ever before. Vegetable Crops suffer badly from various fungal, bacterial and viral diseases. Identification of diseases under field conditions with their characteristic symptoms is crucial for developing the appropriate management strategy. Only through the integration of various practices such as cultural, biological, and chemical right from the field preparation to crop harvest can be effective in disease management. No pathogen is completely eradicated from a natural ecosystem, but it's population can be reduced below economy injury level. In the recent past, the term management has replaced control, which is the appropriate term for safe coexistence in an ecosystem. Therefore, the prime importance to manage the vegetable diseases are to adopt the traditional as well as modern techniques. Among the modern techniques smart agricultural practices are very important which can be used for accurate diagnosis and management of vegetable diseases.

Smart Agricultural Practices: Smart agriculture, with the aid of artificial intelligence (AI), is playing a pivotal role to ensure agriculture sustainability (Shaikh *et al.*2021). Smart agriculture, an innovative approach to farming, leverages advanced technologies to enhance efficiency and sustainability in the agricultural sector. Incorporating tools such as IoT devices, sensors, and data analytics, smart agriculture enables farmers to make informed decisions based on real-time data. This technology facilitates precise monitoring of crop conditions, soil health, and weather patterns, optimizing resource use and minimizing environmental impact. Automation, drones, and robotics contribute to tasks such as planting, irrigation, and harvesting, streamlining operations and increasing productivity. Smart agriculture holds the promise of revolutionizing traditional farming practices, making them more precise, data-driven, and adaptive to the challenges of a rapidly changing world.

Vegetable diseases are responsible for enormous yield losses and for threatening global food production (Ning *et al.* 2017); hence, proper detection and reliable diagnostic methods for identifying the etiological agents of disease are essential to conserving time and money by preventing or limiting crop damages (Singh *et al.* 2018). Classically, diseases were recognized based on traditional methods; these methods, often subjective, were strictly dependent on the observer and though time-consuming overall, were prone to inaccuracy. Additionally, human scouting is expensive and, in many cases, impractical due to human error and/or the occurrence of cryptic when not mild symptoms, making diagnosis at early stages impossible (Qin *et al.* 2021). Therefore, a technologically driven agricultural revolution is important to permanently solve the problems mentioned earlier at a reasonable cost with little environmental impact. With the continuous adoption of recent advanced technologies such as Internet of Things devices, intelligent algorithms, sophisticated sensors, and modern machines, agriculture has changed. It is currently changing from being accomplished by human workers to using smart agricultural machines and robots. Smart agricultural machines and robots have been developed which detect plant diseases early on and at the same time monitor their long-distance movement (Cui *et al.* 2018). Many researchers have used high-resolution imagery collected from satellites, airplanes, on-the-ground machines, and drones to identify agricultural diseases. Satellites and airplanes can cover vast areas in a short amount of time. However, satellites and airplanes have poor spatial and temporal image resolutions compared to drones and are highly susceptible to weather conditions that can affect overflight (Qin *et al.* 2021, Cui *et al.* 2018, and Martinelli *et al.* 2015).Therefore, aerial remote sensing (RS) using drones (Unmanned Aerial Vehicles (UAV) or Unmanned Aerial Systems (UAS)) with intelligent visual systems may be an efficient and inexpensive way for farmers to detect crop and plant diseases in a variety of agricultural fields, from the most intimate greenhouse to the largest farm (Mahlein, 2016).

Digital (red, blue, and green or RBG), multispectral, hyper-spectral, fluorescent, and thermal infrared-based imaging sensors paired with effective algorithms mounted on drones can efficiently detect, differentiate, and quantify the severity of the symptoms induced by various pathogens under field conditions (Bauriegel and Herppich, 2014; Sun *et al.* 2019; Khot *et al.*2016, Gua *et al.* 2021; Sarkar *et al.* 2016; Castrignano *et al.* 2020; Shahi and Xu, 2023; Franceschini *et al.* 2017; Yamamoto *et al.* 2023; Abdulridha *et al.* 2020). Appropriate and reliable evaluation of crops' phytosanitary status, intended as the observation of occurrence and outbreak of plant diseases, is very important, as timely estimation of disease incidence, symptom severity, and the resulting impacts on economically important crops is decisive for managing agronomical interventions such as pesticide application time. The methods for disease detection have been categorized into direct and indirect methods. Direct methods, known as "old generation" methods, include traditional (symptomology, microscopy, and incubation method), molecular diagnostic methods (e.g., polymerase chain reaction (PCR), rapid fragment length polymorphisms (RFLP), real-time PCR, loop-mediated isothermal amplification (LAMP), recombinase polymerase amplification (RPA), and point-of-care diagnostic methods), and serological methods. However, due to their slowness and low capacity, these methods are not well-suited for implementation in the field, delaying early detection and response to disease outbreaks. To effectively prevent and control future outbreaks, a quick and high-throughput approach for the early detection of plant diseases must be developed. Traditional methods usually follow the evaluation of characteristic disease symptoms and visible signs of the pathogens. The evaluation of disease symptoms is performed by trained experts and can be affected by temporal variations. Moreover, traditional methods strictly depend on individual experience, and these methods become accurate and reliable only if the guidelines and standards for assessment are properly followed. Microscopic identification depends on the observation of pathogen inoculum (e.g., mycelia, spores, and fruiting bodies). For microscopic methods, specific dichotomous keys and identification manuals are available; however, due to the need to cultivate the pathogens on using artificial selective media before proceeding to identification, this method is too time-consuming. Molecular and serological methods are commonly utilized in quarantines departments and research institutes for detecting and identifying phytopathogens, and can be applied directly in the greenhouse or the field. For example, to assess the presence of the potato viruses Phytophthora infestans, Ralstonia salanacarum, Ervinia amylovora, Papillus mosaic virus, and Tomato Mosaic Virus, a lateral flow-through version of ELISA is often used. The major drawbacks of molecular and serological methods are that they are time-consuming and require trained operators; in addition, it should be mentioned that the amount of pathogen inoculum does not always positively correlate with the severity of the disease. Furthermore, these methods are particularly unreliable at the asymptomatic stages of plant pathogens (Torres *et al.* 2013), even though they are very sensitive, accurate, and effective; unfortunately, they are unsuitable for monitoring cryptic pathogens that have entered the plants before showing visible symptoms. On top of that, the sampling method from the field to the laboratory is laborious and should be properly sampled. Additionally, few diseases can be detected in only a few plants.

Indirect methods, known as "New Generation", essentially exploit biomarker-based techniques such as metabolite profiling from plant–pathogen interactions as well as stress-based detection techniques such as imaging and spectroscopy using drones. Recently, various indirect methods have been launched, in particular drones, which can estimate disease more accurately compared to molecular, serological, and microbiological diagnostic techniques.

In summary, methods of crop disease monitoring are being improved through different RS technologies. Integrating drone-mounted spectral sensor data with spectroscopy, fluorescence, and thermal imaging data, along with other non-RS based methods to provide more accurate and fruitful plant disease detection and diagnosis, remains a work progress.

There is need to develop more effective disease surveillance and experts system for farmers advisory system in the management of vegetable diseases. There is thus a need to concentrate on diseases of economic importance after undertaking accurate disease diagnosis for multiple diseases control based on the principles of evasion, exclusion, eradication, protection and crop improvement through conventional and molecular breeding programmes.

Besides smart agricultural practices some other methods which should be adopted by the farmers to protect their vegetable crops for various diseases are:

Resistant Varieties: Vegetable crops are exposed to a wide range of potential parasites, therefore, are more prone to various biotic and abiotic stresses. Among biotic stresses, frequent occurrences of fungal, bacterial and viral diseases are major cause of reduced productivity and quality of vegetables. Cultivation of resistant or tolerant cultivars is one of the best options to minimize the losses due to disease occurrence.

Crop Rotation

A cost-effective and efficient way to control some diseases is to rotate one crop with another that are resistant to the pathogen. However, knowledge of the pathogen's life cycle is necessary for the effective application of crop rotation in disease control. Crop rotation is often used to manage diseases by growing non-host plants until the pathogen in the soil either dies or its population is brought down to a level where the crop harm is minimal.

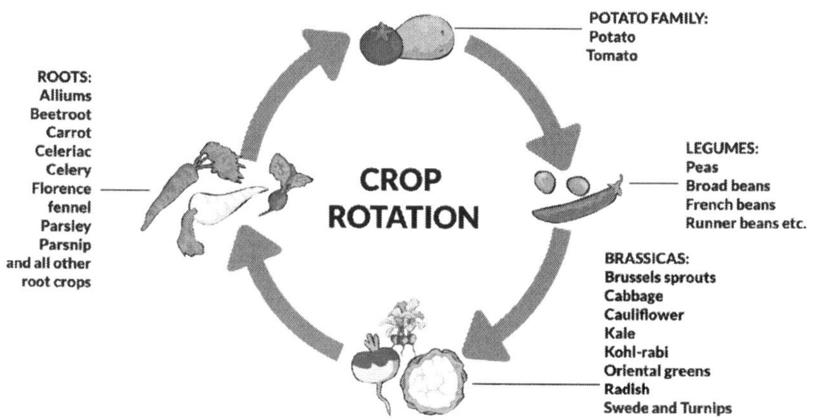

Biological Control

The management of plant disease almost based on chemical pesticides. The huge uses of pesticides is ecologically harmful, continuously use of pesticides lead to resistance development in pathogens against pesticides. Now necessary to provide pesticide free technologies for disease control. It has been observed that biological control has become as effective alternative method for the control of plant diseases to avoid the huge application agrochemicals in agriculture. Biocontrol agents have a many advantages such as growth inhibiting potential against pathogens, rhizosphere competence and ability to produce high quality of antibiotics. (Singh D).

Bio control organisms work by competing with the pathogen for space and nutrients, by parasitism or predation, by inducing the plant's natural defense system, and/or by the production of antimicrobial substances. Some of the examples of bio-control agents are listed as under

- Streptomyces lydicus reduce root and seed rot severity in peas and results in significantly higher final emergence.
- Compete Plus (Six species of Bacillus, Streptomyces griseoviridis, Trichoderma harzianum plus organic nutrients) significantly reduces incidence of potato tubers with both Black Scurf (Rhizoctonia solani) and common scab (Streptomyces scabies).

- Plant growth promoting rhizobacteria significantly reduces incidence of root rot (Pythium, Rhizoctonia).
- Compost Tea significantly reduces potato tubers with both Black Scurf and Common Scab and reduced scab severity.
- Bi-nucleate Rhizoctonia are effective against diseases caused by Rhizoctonia and significantly reduces Black Scurf and stem canker on potatoes.
- Coniothyrium minitans is applied to the soil before cropping and can significantly reduce lettuce drop caused by Sclerotinia species.
- Bacillus subtilis as a seed or soil treatment that significantly reduces Black Scurf and stem canker on potato, resulted in significantly less post emergence disease and significantly higher biomass in spinach challenged by soil-borne diseases.
- Muscador albus is a novel biocontrol organism that acts as a bio fumigant by producing gaseous compounds. It has shown good efficacy against storage insect pests of apples and potatoes.
- Trichoderma virens significantly reduces Black Scurf incidence and Common Scab severity on potato, results in significantly less post emergence disease on spinach from soil-borne disease.(Anonymous, 2013)
- Application of Bacillus subtilis (BS2)as seed treatment (10g/kg) and soil application (2.5 kg mixed in 50 kg FYM/ha) was effective in reducing the Fusarium (Fusarium oxysporum f.sp. lycopersicae) wilt (>40%) and enhanced the yield (>30%) in tomato. This also enhanced the lycopene content and shelf-life. In addition it was also effective against collar rot (Sclerotium rolfsii) of cowpea reducing incidences by> 60% and enhanced the yield by >50%.
- In tomato, an application of talc based formulation of Trichoderma harzianum (BATF 43-1) as seed treatment (10g/kg) and soil application (2.5 kg mixed in 50kg of FYM/ha) was effective in controlling damping off, Pythium aphanidermatum (>60%) and collar rot, Sclerotium rolfsii (>45%) diseases and enhancing the yield of tomato (>25%).

Conclusion

Smart agriculture represents a transformative approach to farming that leverages cutting-edge technologies to enhance efficiency, sustainability, and productivity in the agricultural sector. The integration of Internet of Things (IoT), artificial intelligence (AI), data analytics, and other advanced technologies enables farmers to make data-driven decisions, optimize resource use, and mitigate environmental impact. Smart agriculture brings forth numerous benefits, including precision farming, real-time monitoring, and automation of various processes. These advancements empower farmers to better manage crops, monitor soil health, and optimize irrigation, leading to increased yields and reduced resource wastage. Additionally, the use of predictive analytics helps farmers anticipate challenges such as pests and diseases, allowing for proactive and targeted interventions. Moreover, the connectivity provided by smart agriculture fosters collaboration and knowledge-sharing among farmers, agricultural researchers, and industry stakeholders. This collaborative ecosystem contributes to continuous innovation and the development of sustainable agricultural practices. While the adoption of smart agriculture holds great promise, it is essential to address challenges such as initial investment costs, technological literacy among farmers, and data privacy concerns. Policymakers, industry leaders, and researchers play a crucial role in supporting the widespread implementation of smart agriculture by fostering education, incentivizing adoption, and ensuring ethical and responsible use of technology. In essence, smart agriculture represents a pivotal shift towards a more resilient, efficient, and sustainable future for global agriculture. By harnessing the power of technology, farmers can navigate the complexities of modern agriculture with greater precision and effectiveness, ultimately contributing to food security, environmental conservation, and economic development.

Future Prospects

The future prospects for smart agriculture are exceptionally bright, driven by ongoing advancements in technologies like AI, IoT, and precision farming. With the integration of 5G connectivity, block chain for supply chain transparency, and the evolution of precision agriculture, farmers can expect more efficient and sustainable practices. Climate-smart farming, data analytics, and predictive modeling will play pivotal roles in adapting to changing environmental conditions. The rise of urban and vertical farming, government support, and the Farm-as-a-Service model further contribute to a transformative landscape. As global collaboration and knowledge sharing intensify, smart agriculture is poised to revolutionize the sector, ensuring increased productivity, resource optimization, and resilient food production systems for the future.

References

Abdulridha, J.; Ampatzidis, Y.; Kakarla, S.C.; Roberts, P. Detection of target spot and bacterial spot diseases in tomato using UAV-based and benchtop-based hyperspectral imaging techniques. Precis. Agric. 2020, 21, 955–978.

Agrios, GN. 2005. Plant Pathology. Fifth edition. Academic Press, Massachusetts, 922p.

Anonymous, 2013. https://ag.umass.edu/vegetable/fact-sheets/biological-control-of-plant-diseases.

Bauriegel, E.; Herppich, W.B. Hyperspectral and chlorophyll fluorescence imaging for early detection of plant diseases, with special reference to Fusarium spec. infections on wheat. Agriculture 2014, 4, 32–57.

Castrignanò, A.; Belmonte, A.; Antelmi, I.; Quarto, R.; Quarto, F.; Shaddad, S.; Sion, V.; Muolo, M.R.; Ranieri, N.A.; Gadaleta, G.; *et al.* Semi-Automatic Method for Early Detection of Xylella fastidiosa in Olive Trees Using UAV Multispectral Imagery and Geostatistical-Discriminant Analysis. Remote Sens. 2020, 13, 14.

Chen, J.-W.; Lau, Y.Y.; Krishnan, T.; Chan, K.-G.; Chang, C.-Y. Recent advances in molecular diagnosis of Pseudomonas aeruginosa infection by State-of-the-Art, Genotyping Techniques. Front. Microbiol. 2018, 9, 1104.

Christiansen, M.P.; Laursen, M.S.; Jørgensen, R.N.; Skovsen, S.; Gislum, R. Designing and testing a UAV mapping system for agricultural field surveying. Sensors 2017, 17, 2703.

Coakley, Stella Melugin, Scherm, Harald and Chakraborty, Sukumar. 1999. Climate change and plant disease management. Annu. Rev. Phytopathology. 37:399–426.

Cui, S.; Ling, P.; Zhu, H.; Keener, H.M. Plant Pest Detection Using an Artificial Nose System: A Review. Sensors 2018, 18, 378.

Franceschini, M.H.D.; Bartholomeus, H.; van Apeldoorn, D.; Suomalainen, J.; Kooistra, L. Assessing changes in potato canopy caused by late blight in organic production systems through Uav-Based Pushbroom imaging spectrometer. Int. Arch. Photogramm Remote Sens. Spat. Inf. Sci. 2017, XLII-2/W6, 109–112.

Ge, Y.; Thomasson, J.A.; Sui, R. Remote sensing of soil properties in precision agriculture: A review. Front. Earth Sci. 2011, 5, 229–238.

Ghini, Raquel, Hamada, Emília and Bettiol, Wagner. 2008. Climate change and plant diseases. Sci. Agric. 65: 98-107.

Guo, A.; Huang, W.; Dong, Y.; Ye, H.; Ma, H.; Liu, B.; Wu, W.; Ren, Y.; Ruan, C.; Geng, Y. Wheat Yellow Rust Detection Using UAV-Based Hyperspectral Technology. Remote Sens. 2021, 13, 123.

Herrmann, I.; Bdolach, E.; Montekyo, Y.; Rachmilevitch, S.; Townsend, P.A.; Karnieli, A. Assessment of maize yield and phenology by drone-mounted superspectral camera. Precis. Agric. 2020, 21, 51–76.

Khot, L.R.; Sankaran, S.; Carter, A.H.; Johnson, D.A.; Cummings, T.F. UAS imaging-based decision tools for arid winter wheat and irrigated potato production management. Int. J. Remote Sens. 2016, 37, 125–137.

Kuska, M.; Wahabzada, M.; Leucker, M.; Dehne, H.-W.; Kersting, K.; Oerke, E.-C.; Steiner, U.; Mahlein, A.-K. Hyperspectral phenotyping on the microscopic scale: Towards automated characterization of plant-pathogen interactions. Plant Methods 2015, 11, 28.

Mahlein, A. Present and future trends in plant disease detection. Plant Dis. 2016, 100, 1–11.

Mahlein, A.-K. Plant Disease Detection by Imaging Sensors—Parallels and Specific Demands for Precision Agriculture and Plant Phenotyping. Plant Dis. 2016, 100, 241–251.

Martinelli, F.; Scalenghe, R.; Davino, S.; Panno, S.; Scuderi, G.; Ruisi, P.; Villa, P.; Stroppiana, D.; Boschetti, M.; Goulart, L.R.; *et al.* Advanced methods of plant disease detection. A review. Agron. Sustain. Dev. 2015, 35, 1–25.

Ning, Y.; Liu, W.; Wang, G.L. Balancing Immunity and Yield in Crop Plants. Trends Plant Sci. 2017, 22, 1069–1079.

Pandey, K.K., Nagendran, V., Tripathi A.N. , Manjunath M., Rai A.B. and Singh B. , Division of Crop Protection, ICAR - Indian Institute of Vegetable Research.

Qin, J.; Wang, B.; Wu, Y.; Lu, Q.; Zhu, H. Identifying Pine Wood Nematode Disease Using UAV Images and Deep Learning Algorithms. Remote Sens. 2021, 13, 162.

Sarkar, S.K.; Das, J.; Ehsani, R.; Kumar, V. Towards autonomous phytopathology: Outcomes and challenges of citrus greening disease detection through close-range remote sensing. In Proceedings of the 2016 IEEE International Conference on Robotics and Automation (ICRA), Stockholm, Sweden, 16–21 May 2016; pp. 5143–5148.

Schmale, I.I.I.D.G.; Ross, S.D. Highways in the sky: Scales of atmospheric transport of plant pathogens. Annu. Rev. Phytopathol. 2015, 53, 591–611.

Shahi, T.B.; Xu, C.-Y.; Neupane, A.; Guo, W. Recent Advances in Crop Disease Detection Using UAV and Deep Learning Techniques. Remote Sens. 2023, 15, 2450.

Shaikh, F. K., Memon, M. A., Mahoto, N. A., Zeadally, S., and Nebhen, J. (2021). Artificial intelligence best practices in smart agriculture. IEEE Micro, 42(1), 17-24.

Shi, Y.; Thomasson, J.A.; Murray, S.C.; Pugh, N.A.; Rooney, W.L.; Shafian, S.; Rajan, N.; Rouze, G.; Morgan, C.L.S.; Neely, H.L.; *et al.* Unmanned Aerial Vehicles for High-Throughput Phenotyping and Agronomic Research. PLoS ONE 2016, 11, e0159781.

Singh, A.K.; Ganapathysubramanian, B.; Sarkar, S.; Singh, A. Deep Learning for Plant Stress Phenotyping: Trends and Future Perspectives. Trends Plant Sci. 2018, 23, 883–898.

Singh, D., Saha, P., Chongloi, K. L., Gupta, A. K., Ansari, M. A., & Prakash, N. (2022). Management of Plant Diseases through Application of Biocontrol Agents in Climate Smart Agriculture-Review. Innovative Approaches for Sustainable Development: Theories and Practices in Agriculture, 231-246.

Singh, R.S. 1984. Introduction to the principles of Plant Pathology. Oxford and IBH Publishing Co. Pvt. Ltd., Calcutta, 534

Sun, Q.; Sun, L.; Shu, M.; Gu, X.; Yang, G.; Zhou, L. Monitoring Maize Lodging Grades via Unmanned Aerial Vehicle Multispectral Image. Plant Phenomics 2019, 2019, 5704154.

Tallapragada, P.; Ross, S.D.; Schmale, D.G., III. Lagrangian coherent structures are associated with fluctuations in airborne microbial populations. Chaos Interdiscip. J. Nonlinear Sci. 2011, 21, 033122.

Taylor J. A., 2023. Precision agriculture. Elsevier. 4: 710-725. https://doi.org/10.1016/B978-0-12-822974-3.00261-5

Torres-Sánchez, J.; López-Granados, F.; De Castro, A.I.; Peña-Barragán, J.M. Configuration and specifications of an unmanned aerial vehicle (UAV) for early site specific weed management. PLoS ONE 2013, 8, e58210.

Yamamoto, S.; Nomoto, S.; Hashimoto, N.; Maki, M.; Hongo, C.; Shiraiwa, T. Monitoring spatial and time-series variations in red crown rot damage of soybean in farmer fields based on UAV remote sensing. Plant Prod. Sci. 2023, 26, 36–47.

Zhang, D.; Zhou, X.; Zhang, J.; Lan, Y.; Xu, C.; Liang, D. Detection of rice sheath blight using an unmanned aerial system with high-resolution color and multispectral imaging. PLoS ONE 2018, 13, e0187470.

Chapter 5

Smart Agriculture Practices-Urgent Need for Sustaining Food Security, Safety and Quality

*Jagmohan Singh and Suhani Sajad**

Division of Food Science and Technology
Sher-e-Kashmir University of Agricultural Sciences and Technology of Jammu
Chatha, India
Correspondingemailid:suhanisajad94@gmail.com

1. Introduction

Climate change is emerging as a major threat to agriculture, food security and livelihood of millions of people in many places of the world(kumaretal.,2019). According to several studies, changes in temperature, rainfall patterns, and the frequency and intensity of extreme climatic events like floods and drought might have a considerable impact on agriculture production. According to estimates of historical and future climate changes effects on cereal crop yields in various regions,yield losses for different crop scan rangefromupto35percentfor rice, 20 percent for wheat, 50 percent for sorghum, 13 percent for barley, and 60 percent for maize, depending on the location, scenarios, and projected areas (Porter *et al.*, 2014). Changes in crop cultivation suitability and associated agriculture biodiversity, decrease in input use efficiency, and prevalence of pests and diseases are some of the major causes of climate change impacts on agriculture. Agriculture, nutrition, and climate change are all interrelated. Temperature and precipitation, as well as the frequency and seriousness of extreme weather events, are all impacted by climate change and fluctuation. Agriculture will be impacted by temperature rises, heat waves, and droughts, with the biggest implications being reduced crop yields and animal productivity, as well as decreases in fisheries and agro forestry in areas already susceptible to food insecurity (FAO, IFAD, UNICEF, WFP, and WHO, 2017). (Lemi *et al.*, 2019).The effects of climate change on agriculture will, in turn, have significant implications for food security, and thus human diets and nutrition. As climate change affects the ability to move food from production to markets, access to diverse, high-quality diets may become more limited. Reduced access to sufficient nutrient-dense foods will subsequently lead to impaired nutritional status and diminished resiliency, particularly in low-income communities (United Nations System Standing Committeeon Nutrition Secretariat, 2010). Even in the "business-as-usual" models, nutrition and health outcomes are likely to worsen (Fanzo *et al.*, 2018).

2. Impact of Climate on Food Security, Safety and Quality

Climate change is a major concern for India because the majority of India's population depends on climate-sensitive sectors such as agriculture,forestry, and fishing for livelihood(Dev *et al.*, 2010). Climate change has adversely affected India's agriculture sector and is an emerging threat to food security. Agriculture yields in India are very sensitive to climate changes such as unnatural high or low precipitation, erratic temperatures and humidity. There is evidence that the frequency and severity of extreme weather events (floods, droughts, and heat waves) are increasing, which harms India's food security and productivity. Studies have shown that the output of wheat decreases by 4 to 5 million metric tonnes for every 1 degree Celsius increase in temperature. The effects of climate change on food crops, beverage crops(tea,coffee),medicinal plants, dairy products, and fish arewell documented (Sarkar *et al.*, 2019). The current challenges facing Indian food security involve problems impacting the food-producing sector and production, transportation, storage, and access to food.

Ensuring food safety and quality under changing climate conditions is one of the major challenges of this era. There are about 925 million food-insecure people in the world about 16 percent of the population in developing countries. The global population will increase from 7 billion currently to over 9 billion people by 2050, creating a demand for a more diverse diet that requires additional resources to produce. To meet the demand for food, fuel, fiber, and fodder, competition for land, water, and energy will increase. According to several forecasts, the world's food needs will need to double by 2050 (Smith *et al.*, 2013), in addition to maintaining and, where possible, enhancing the resilience of natural ecosystems. Climate change impacts on agriculture will have consequences for food security globally. Food security includes four components: availability, stability, access, and utilization of food. Food security is affected by a variety of supply and demand-side pressures, including economic conditions, globalization of markets, safety and quality of food,land-use change, demographic change, and disease and poverty. Climate change is expected to have numerous effects on food security within the complex global food chain. The distribution of food- and water-borne diseases, food trade, and food distribution will all be impacted by projected rising temperatures, changing weather patterns, and an increase in the frequency of extreme weather events in addition to affecting agricultural production.

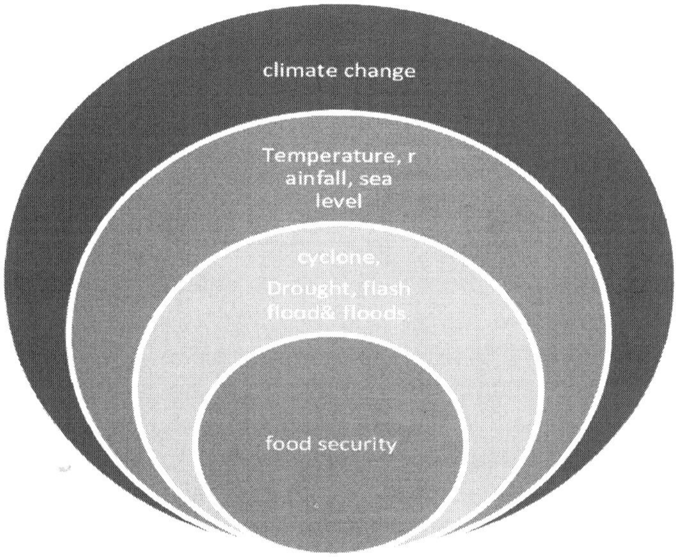

3. Smart Agriculture

Smart agriculture is the concept of doing farming innovatively, using the latest technologies to increase the quantity and quality of agricultural products. It is the future of the food production sector as well as novel processes to confirm food safety of the world. It enables farmers to reduce their resources as well as increase their yields. Thesaved resources include fertilizers, manpower, seeds and water. Smart farming is being conducted on a large scale using the IoT, Artificial Intelligence (AI)and agricultural data analysis in developed nations. Agricultural data analytics offers farmers useful and relevant insight for smart agriculture that leads to increased crop productivity and yield security. The development and evolution of earth observation (EO) technology, particularly satellite remote sensing (SRS), has provided remote sensing (RS) data on a very large scale for agricultural research and other applications (Khanna *et al.*, 2019).

3.1 Climates Mart Agriculture

Agriculture is the most important economic sector of many developing countries. Agricultural production systems are expected to produce food for a global population that will amount to 9.1 billion people in 2050 and over 10 billion by end of the century (Branca *et al.*, 2011). In order to secure and maintain food security agricultural systems need to be transformed to increase the productive capacity and stability of smallholder agricultural production. Agriculture production systems require adaptation to these changes to ensure the food and livelihood security of farming communities. Climate-smart agriculture (CSA) is an integrative approach to address these interlinked challenges of food security and climate change, Climate smart agriculture (CSA) as explained by Food and Agricultural Organizations (FAO) during the Hague Conference on Agriculture, Food Security and Climate Change in 2010, takes part in the achievement of sustainable progress. By jointly addressing food safety and climate challenges, it combines three forms of sustainable progress (financial, community-based, and ecological), because of this, it was built upon the following three pillars: i) Sustainably increasing agricultural productivity and incomes; ii) adapting and building resilience to climate change; iii) lowering and/or eliminating greenhouse gases emissions, where possible; Hence, CSA is a method to enhance scientific rule and investment setting to achieve sustainable agricultural advancement to ensure the supply of food under climate change (Capone *et al.*, 2014).Climate smart agriculture is a way to get sustainable developments as well as green economy goals. It intends towards food availability and takes part to conserve natural assets. It is closely associated with the perception of improved growth, as FAO develops it for crop yields.

Thus CSA is a technique to improve the scientific rule and investment set to attain sustainable agricultural progress to ensure the food availability under climate change. Climate smart agriculture is a way to get sustainable developments as well as green economy goals. It intends towards food availability and takes part to conserve natural assets. FAO's development of it for crop yields, is directly related to the perception of better growth. Consideration of the impacts of world climate variations on sustainable green growth may help to predict and follow better farming techniques to improve agricultural yields. Adaptation is a process that involves the management of risks generated by climate change, including variability. Adaptation to the climate change is an important issue for researchers, program managers and policy setters who are conscious that climate variations are happening and are dangerous to challenge community and environmental sustainability. Sustainability is a crucial guiding principle for international environmental policy and decision making in the twenty first century (Braimoh etal., 2010). The worst issues of climate change could be controlled by successful adaptation, which would probably be less than the cost of impacts that would occur without adjustments. CSAs aim to boost the farmer's ability to adapt to climate change along with enhancing farmers' income by bringing about strong linkage/collaboration between farmers, researchers and local partners such as KVKs, SAUs, and ICAR institutes. The four main components of a climate smart village are climate smart technologies, climate information services, village development plans and local knowledge and institutions.

4. Nano biosensors

A Nano biosensor is a modified biosensor, a compact analytical device incorporating a biologically sensitized element onto a physico chemical transducer with a miniature structure when compared to conventional biosensors. This can be used for sensing the variety of fertilizers, herbicide, pesticide, insecticide, pathogens, etc. and support sustainable agriculture for enhancing crop productivity (Singh *et al.*, 2017). Nano material applications have therefore proven to be beneficial for our civilization. Due to the ongoing usage of ever-increasing levels of herbicides, pesticides, and heavy metals, total land and water resources for agriculture are decreasing daily and there has been claimed to be a significant loss in agricultural output. Recently nanomaterial-based technology emerged as an evolving field to revolutionize agricultural systems (sustainable) which enhance the quality of agricultural products. Nano biosensors are non-invasive,sensitive, and specifically designed sensors manufactured using novel nanobiotechnology approaches (Kalyani *et al.*, 2021). Nanobiosensors create real-time response signals that are easily collected and analyzed (Omara etal., 2019,Srivastava *et al.*, 2018). Different nanomaterials, including nanotubes, nanowires, nanoparticles, nanocrystals, and nanocomposites, are used in the nanobiosensors. In the agricultural industry, these small, compact and portable devices can allow the farmers to monitor and control the soil conditions on-site. Nanobiosensors have been used to analyze the fertility, pH, moisture content, mineral concentrations, and detection of pests and deficiencies of minerals in the soil before the onset of diseases. However, the use of nanobiosensors in the food industry is also growing rapidly. For example,nanobiosensors have been utilized inthe food industry to assure food safety through out production and packaging and identify allergens and/orinfections (Sridharetal.,2021; Neethirajan *et al.*, 2018). In case of the food industry,various other challenges exist such as food security, safety, availability and utilization (Calicioglu *et al.*, 2019, Dasgupta *et al.*, 2017). Thus, nanobiosensors exhibit superior attributes such as easy operations, high sensitivity along with rapid detection for pesticide residues.

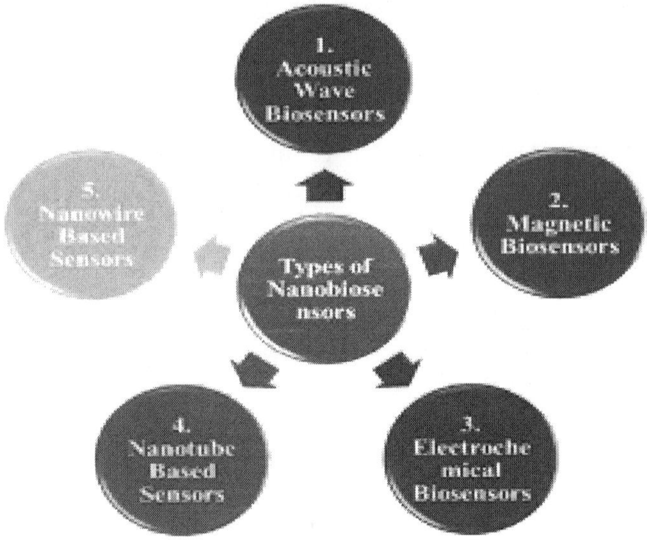

Metallic nanoparticles offer a wide range of uses in the field of food technology due to their unique shape- and size-dependent characteristics. The optical, chemical, mechanical, antimicrobial, and electronic properties of metallic nanoparticles make them unique for combination with enzymes, drugs, ligands, antibodies, colorimetric and fluorimetric agents, and other types of biomolecules. As a result, these nanoparticles can be used for a variety of tasks, including advanced diagnostic assays, radiation, thermal ablation, gene and medication delivery, optical imaging, efficient antibacterial activity, and the decontamination of hazardous chemicals (Srivastava *et al.*,2018). Moreover, metallicna noparticles have a huge impact on food duction, consumption, and the ability to quickly respond to pathogens, packaging issues, pesticides, and toxic residues, by detecting microbial weakening of the quality of food as well as contaminants. Of all the metallic nanoparticles currently in use, the colloidal gold nanoparticle is found to be the most commonly applied in a variety of functions, because of its flexible size range of about10–250 nm; it also possesses shape-dependent optical and mechanical propertiesas well as being generally complex in nature.

5. Internet of Things(IoT)

Internet of things (IoT) driven system can be used to ensure food security by providing transparency, traceability and accountability, resulting in reduced food wastages and ensuring food quality from the stage of harvesting to consumption (Nukala *et al.,* 2015). The use of IoT in the food supply chain will improve food safety and quality through monitoring and surveillance of food conditions along the food supply chain, and sharing the data obtained with consumers and supervisors. In order to detect adulterants in milk through a variety of sensors, including gas sensors, temperature sensors, and RFID readers,

(Bouzembrak *et al.*, 2019) developed an IoT system for food fraud.(Ahmed *et al.*, 2018) developed an IoT system to monitor and assess the quality of meat and seafood items across the supply chain and control food quality. Several sensors have been used to detect food spoilage, such as temperature and humidity sensors. Different ways are explored in which IOT driven by various disruptive technologies, such as Traceability, Artificial Intelligence (AI), RFID, etc., can be used as enablers for sustainable food security.

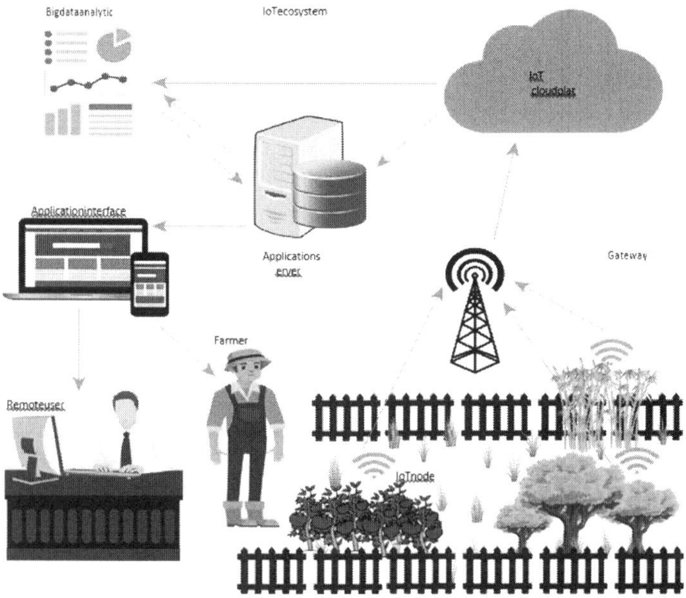

5.1 Traceability via Blockchain Technology

Blockchain technology can be used in various fields and diverse areas where IoT applications are involved because it is a distributed data structure that is duplicated and shared across network users (Ferrag *et al.*, 2018). Blockchain technology and machine learning techniques can be integrated for safesmart agriculture. Therefore, blockchain technology promises a future secure and transparent system for the exchange of supply chain and logistics information across supply networks as shown in the figure. AgriBlock IoT (Caro *et al.*, 2018) is a decentralized blockchain-based traceability solution for managing the agri-food supply chain. Using two separate blockchain implementations—Ethereum and Hyperledger Sawtooth—it is capable of combining existing IoT devices into the food supply chain for transparent, fault- tolerance, immutable, and auditable records. (Casado-Vara *et al.*, 2018) proposed a blockchain approach to improving the current agricultural, supply chain. BRUSCHETTA is a blockchain- based system for tracking the entire process of production of Extra Virgin Olive Oil by allowing customers to access a copy of the full immutable product history.

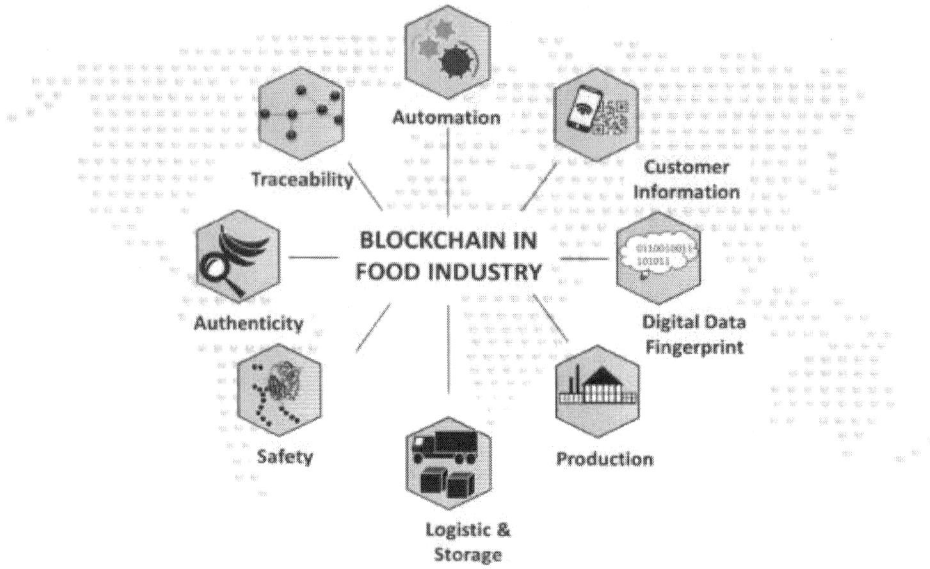

5.2 Artificial Intelligence (AI)

AI in the food industry is becoming increasingly important due to its ability to help save food, improve the hygiene of production sites, and clean up more processing equipment quickly; therefore, there are many cases of use of artificial intelligence and machine learning in the food industry. Automated systems can, in a few seconds, collect hundreds of data on a single food product and quickly make an assessment of it (Barth *et al.*, 2017). A system, for example, can collect and process data from hundreds of individual ingredients, as they move quickly on a conveyor belt; these systems can significantly reduce labor costs and waste. Application of AI technologies within the food industry;

- AI for sorting food—one of the most advanced AI applications in the food industry is TOMRA Sorting Food, which uses optical sensor-based solutions with machine learning capabilities. TOMRA uses cameras and sensors to enable customers to order meals based on their tastes, saving time and money during production while also raising the standard of the final product (Silver *et al.*, 2016).
- Food industry supply optimization—AI helps in supply chain management, and technology can support companies to test and monitor food safety products at everystage of the supply chain (Chacón *et al.*, 2019).
- AI to ensure hygiene standards—A successful case is represented by the KanKan company (Las Vegas, USA) which has worked on creating intelligent solutions to improve hygiene conditions in China. The system, which can also be used in restaurants, uses cameras to monitor

workers and uses facial recognition and object recognition software, to determine whether workers wear hats and masks as required by food safety laws. If it detects a violation, the software extracts the screen images for review (Garbie *et al.*,2010).

Thus, it is possible to say that technology has the potential to create a healthier and more thriving food industry for workers and consumers. Moreover, new technology (such as Blockchain) enables agri-food supply chains to be more connected, enhancing food security and traceability (Tripoli *et al.*, 2018).

5.3 RFID Tags for monitoring the quality of harvested fruits and vegetables

Harvest quality and safety are a great concern for sellers and consumers. Due to their limited shelf life, the freshness of the hand-picked fruits and vegetables needs to be thoroughly monitored. The freshness of the fruits/vegetables is highly dependent upon the relative humidity, environment temperature, and specific gas concentration. Recently, RFID tags are being explored to monitor the freshness of vegetables. The tags are fabricated with different types of sensors to measure different parameters. RFID sensors are now used to measure the temperature and relative humidity of packaged vegetables. The tags operate on the HF (13.56 MHz) frequency spectrum and employ an add-on microchip. By using temperature and relative humidity sensors, the proposed RFID sensors could determine the texture and quality of the products by measuring the transpiration process and the water loss process for the fruits and vegetables (Le *et al.*, 2016). The figure presents a typical example of a fruit & vegetable monitoring operation with RFID sensors.

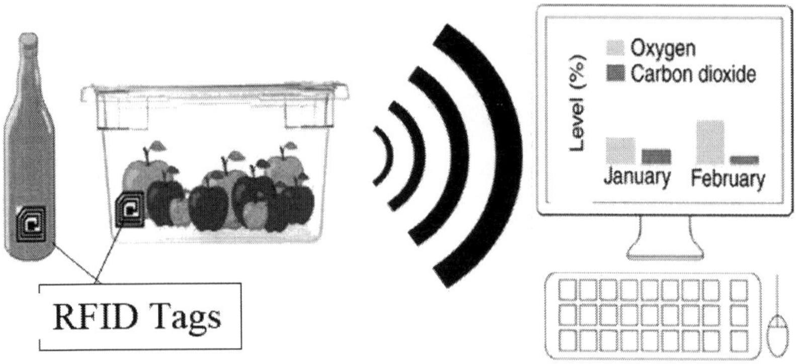

To assess the freshness of the vegetables, a semi-passive, ultra-low power(8.9mW) RFID add-on microchip sensor was created. To understand the freshness and logistics of the fruit (apples), the sensor includes sensors for temperature, relative humidity, gas concentration, and light intensity (Zampolli *et al.*, 2008). An RFID sensor was combined with a temperature sensor and a shelflife monitoring model to determine the vine-ripened tomatoes' expiration date and keep track of the food inventory.

Grains like wheat, rice, peanuts, mustard, and lentils tend to absorb moisture under normal atmospheric temperature when they are stored in the warehouse or storage inventories of farms. The accumulation of moisture content causes a fungus called Aflatoxin, which can deteriorate the quality of the food grains and also alters the weight and texture of the grains, affecting the economic value for the farmers, sellers, and consumers (Cantwell *et al.*, 2017). Thus, it is paramount to monitor the moisture content of the food grains. A low-cost RFID sensor integrated with a chemical sensor (CuMOF sensor) which can track the gunny bags and can monitor the moisture content of the grains. Silo bags are storage bags made up of plastic materials, which have a high level of gas tightness (O2and CO2). This increases the respiration of the biotic components such as insects and fungi, increases the carbon dioxide level and spoils the grains inside the silo bags. Hence, an RFID sensor integrated with several sensors and a web platform was developed to monitor the changes in the parameters (Rayhana *et al.*, 2021).

Conclusion

This chapter has evaluated the importance of Smart agriculture in promoting sustainable agricultural development and ensuring food security, food safety, food quality and mitigating the negative impacts of climatic changes on agricultural productivity in India. The climate adversities associated with changing climate are projected to cause huge losses in agricultural production. In these situations, ensuring the food and nutritional security of the ever-increasing population is a difficult task. Due to several challenges in the future for the agriculture and food sector and various factors such as climate change, population growth, technological progress,and the state of natural resources (water, etc.), it is urgent to use digital technologies in different stages of the agriculture supply chain such as the use of sensors,artificial intelligence, RFID tags for improved monitoring of food and water, for agriculture food product traceability. Feeding 10 billion people in 2050 sustainably will require changes in our food chains. This will require consumers and food businesses to adapt their quality requirements and specifications. If the global food systems could change in this direction, the global food security, safety and quality will improve and be more resilient. Using modern IT technology offers the best promise of more efficient source reduction,reprocessing, and recycling of food. Food quality and authenticity evaluation using miniature spectral cameras has become popular in the industry and efforts are under way to bring this capability to consumers through their smartphones. The industry is also exploring the benefits of blockchain technology for traceability in case of pathogen out breaks and to ensure food safety.

Eventually, the economic, environmental, and social key performance indices that nano biosensors, the internet of things, artificial intelligence, blockchain management, and RFID technologies may have an impact on higher productivity, lower production costs, and higher quality (e.g. improved public health, consumer demand-driven, quality of life improvement). The pace of innovations is astounding and tasks that seemed impossible a few years ago have now been implemented with great success. Embracing technology innovations and putting them to an advantage is important for the success of the modern agriculture and food industry.

References

Ahmed,I.,Lin,H.,Zou,L.,Li,Z.,Brody,A. L., Qazi, I. M.,and Sun, L. (2018). An overview of smart packaging technologies for monitoring safety and qualityof meat and meat products. Packaging Technology and Science, 31(7), 449-471.

Barth, H., Ulvenblad, P. O., and Ulvenblad, P. (2017). Towards a conceptual framework of sustainable business model innovation in the agri-food sector: A systematic literature review. Sustainability, 9(9), 1620.

Bouzembrak,Y.,Klüche,M.,Gavai,A.,andMarvin,H.J.(2019).Internet of Things in food safety: Literature review and a bibliometric analysis. Trends in Food Science &Technology, 94, 54-64.

Braimoh,A.K.,&Osaki,M.(2010).Land-use change and environmental sustainability. Sustainability Science, 5(1), 5-7.

Branca, G., McCarthy, N., Lipper, L., and Jolejole, M. C. (2011). Climate-smart agriculture: a synthesis of empirical evidence of food security and mitigation benefits from improved cropland management. Mitigation of climate change in agriculture series, 3, 1-42.

Calicioglu, O., Flammini, A., Bracco, S., Bellù, L., and Sims, R. (2019). The future challenges of food and agriculture: An integrated analysis of trends and solutions. Sustainability, 11(1), 222.

Cantwell, M. I., Albornoz, K., and Hong, G. (2017, October). Ammonia accumulation is a useful indicator of the postharvest freshness and quality of spinach and kale. In VI International Conference Postharvest Unlimited 1256 (pp. 303-310).

Capone, R., Bilali, H. E., Debs, P., Cardone, G., andDriouech, N. (2014). Food system sustainability and food security: connecting the dots. Journal of FoodSecurity, 2(1), 13-22.

Caro, M. P., Ali, M. S., Vecchio, M., and Giaffreda, R. (2018, May).Blockchain-based traceability in Agri-Food supply chain management: A practical implementation. In2018IoTVerticalandTopicalSummitonAgriculture-Tuscany(IOTTuscany) (pp. 1-4). IEEE.

Casado-Vara,R.,Prieto,J., De la Prieta,F., and Corchado, J. M. (2018). How blockchain improves the supply chain: Case study alimentary supply chain. Procediacomputer science, 134, 393-398.

Chacón Ramírez, E., Albarrán, J. C., and Cruz Salazar, L. A. (2019, October). The control of water distribution systems as a holonic system. In International Workshop on Service Orientation in Holonic and Multi-Agent Manufacturing(pp. 352- 365).Springer, Cham.

Dasgupta, S., Auth, T., and Gompper, G. (2017).Nano-and microparticles at fluid and biological interfaces. Journal of Physics: Condensed Matter, 29(37), 373003.

Dev, S. M., and Sharma, A. N. (2010). Food security in India: Performance, challenges and policies.

Fanzo, J., Davis, C., McLaren, R., andChoufani, J. (2018). The effect of climate change across food systems: Implications for nutrition outcomes. Global food security, 18, 12-19.

Ferrag, M. A., Derdour, M., Mukherjee, M., Derhab,A., Maglaras, L., &Janicke, H. (2018).Blockchain technologies for the internet of things: Research issues and challenges. IEEE Internet of Things Journal, 6(2), 2188-2204.

Garbie, I. H. (2010).Enhancing the performance of industrial firms through implementation of lean techniques. In IIE Annual Conference. Proceedings (p. 1). Institute of Industrial and Systems Engineers (IISE).

Kalyani, N.,Goel,S.,and Jaiswal,S.(2021).On-site sensing of pesticides using point-of-care biosensors: a review. Environmental Chemistry Letters,19(1), 345- 354.

Khanna, A., and Kaur, S. (2019). Evolution of Internet of Things (IoT) and its significant impact in the field of Precision Agriculture. Computers and electronics in agriculture, 157, 218-231.

Kumar,S., Thombare, P., and Kale, P. (2019). Climate smart agriculture: Challenges, implications,innovations for achieving food and nutrition security. AGRICULTURE & FOOD: e-Newsletter, 1(9), 267-271.

Le, G. T., Tran, T. V., Lee, H. S., and Chung, W. Y. (2016). Long-range battery less RF sensor for monitoring the freshness of packaged vegetables. Sensors and Actuators A: Physical, 237, 20-28.

Lemi, T., &Hailu, F. (2019). Effects of climate change variability on agricultural productivity. Int. J. Environ. Sci. Nat. Resour, 17, 14-20.

Neethirajan, S., Ragavan, V., Weng, X., & Chand, R. (2018). Biosensors for sustainable food engineering: challenges and perspectives. Biosensors, 8(1), 23.

Porter, J. R., Xie, L., Challinor, A. J., Cochrane, K., Howden, S. M., Iqbal, M. M., ... &Travasso, M. I. (2014). Food security and food production systems.

Rayhana,R.,Xiao,G.,&Liu,Z.(2021).RFID sensing technologies for smart agriculture. IEEE Instrumentation & Measurement Magazine, 24(3), 50-60.

Sarkar, A., Dasgupta, A., & Sensarma, S. R. (2019). Climate Change and Food Security in India: Adaptation Strategies and Major Challenges. In Sustainable Solutions for Food Security (pp. 497-520).Springer, Cham.

Silver, J., Harris, P., Fern, G., Kosier, E., Kay, M., & Mitchell, J. (2016).Plastic Packaging Recycling Using Intelligent Separation Technologies for Materials. IDW.

Singh, R. P. (2017). Application of nanomaterials toward development of nanobiosensors and their utility in agriculture. In Nanotechnology(pp. 293-303).Springer, Singapore.

Smith, P. (2013). Delivering food security without increasing pressure on land. Global FoodSecurity, 2(1), 18-23.

Sridhar, A., Ponnuchamy, M., Kumar, P. S., &Kapoor, A. (2021). Food preservation techniques and nanotechnology for increased shelf life of fruits, vegetables, beverages and spices: a review. Environmental Chemistry Letters,19(2), 1715- 1735.

Srivastava,A. K.,Dev, A.,&Karmakar,S. (2018). Nanosensors and nano biosensors in food and agriculture. Environmental Chemistry Letters, 16(1), 161-182.

Srivastava, A. K., Dev, A., & Karmakar, S. (2018). Nanosensors and nanobiosensors in food and agriculture. Environmental Chemistry Letters, 16(1), 161-182.

Tripoli,M., & Schmidhuber, J. (2018).Emerging Opportunities for the Application of Blockchain in the Agri-food Industry.

Zampolli, S., Elmi, I., Cozzani, E., Cardinali, G. C., Scorzoni, A., Cicioni, M., & Becker, T. (2008).Ultra-low-power components foran RFID Tag with physical and chemical sensors. Microsystem technologies, 14(4), 581-588.

Chapter 6

Food Fortification: Nutritional Aspect, Benefits and Limitations

Neeraj Gupta, Seerat Gupta and Fozia Hameed

Division of Food Science and Technology, Chatha, SKUAST-Jammu, J&K UT, 180009

Abstract

Millions of people in low and middle income nations suffer from deficiencies in one or more micronutrients like iron, zinc, folic acid, vitamin D and vitamin A, which jeopardize their physical and mental health. Micronutrient deficiencies are generally caused by a lack of nutrient-dense food consumption as well as nutrient losses from poor diet, illnesses, and menstrual blood loss.Incorporation of vitamins and minerals in commonly consumable foods during processing to boost their nutritional value is referred as food fortification. In India, fortified foods like rice, wheat flour, milk, oil, and salt are commonly consumed.

Vehicles for fortification include the LSFF, biofortification, and point-of-use fortification.Governments often initiate, require, and control LSFF, a nutrition-specific intervention, with the goal of addressing or preventing micronutrient deficiencies in groups that are at higher risk.A promising, practical, and affordable method of providing micronutrients to people that might not have easy access to diverse meal and other micronutrient intervention is biofortification. To reduce children's micronutrient deficits, point-of-use fortification with powdered vitamins and minerals is a very crucial strategy. High benefit:cost ratios are guaranteed by low unit prices for food fortification, which promotes the fortification of foods with iron, iodine, zinc etc. A considerable proportion of infant and toddler fatalities are prevented by fortification with iron, vitamin A, and zinc, making it a particularly alluring preventive healthcare strategy. It can be used to treat mild, moderate, and severe malnutrition as well as to prevent it.

Introduction

Vitamins and minerals are essential nutrients which play a crucial role in better health of living organisms. Vitamins and minerals come under category of micronutrients which are necessary for several important body functions, but our body can't make them on their own. Inadequate intake of nutrients and nutrient losses due to poor diets, infections and blood loss during menstruation are responsible for micronutrient deficiencies (Olson *et al.*, 2021). Micronutrient malnutrition is often known as "hidden hunger", it poses a major threat to one's health.To fulfill these deficiencies, food manufacturers have started adding chemicals which consist of vitamins and minerals into the product during product manufacturing (Anonymous, 2021).

Incorporation of vitamins and minerals in commonly consumable foods to increase their nutritional value during processing is referred as food fortification. Food had been fortified with vitamins, minerals, and other micronutrients. When added to food, these chemicals don't have any discernible taste, texture or odour. It is a tried-and-true method for boosting diet, preventing and controlling micronutrient deficiencies,which is both secure and affordable (Olson *et al.*, 2021).The WHO defines fortification as the deliberate practice of increasing the content of one or more micronutrients, such as vitamins and minerals, in a food with the goals of improving the nutritional quality of supply food, providing public health benefit with minimal health risk, as well as aiding the restoration of the micronutrients lost during processing.

When food producers are legally obliged by government to fortify particular foods with specific micronutrients, then it is termed as Mandatory food fortification where as when food producer fortify specific food on its own in compliance with government regulations and standards, then it is termed as voluntary food fortification(WHO, 2018).Food enrichment and fortification are two different concepts where food enrichment is restoration of nutrients that are lost during processing and food fortification is incorporation of micronutrients to enhance the nutritional quality of product (Liyanage and Hettiarachchi, 2011).

Need for Foods Fortification

Large portion of the population in India has micronutrient deficiencies, which are typically brought on by a lack of iodine, vitamin A, vitamin B12, iron, folic acid, and other nutrients, resulting in goitre, night blindness, anaemia, and numerous birth abnormalities. Foods with added nutrients make it easier for people to consume what they need in diet (Anonymous, 2021). Nutrient losses occur frequently during the processing of food. Therefore, food fortification can enhance food nutrition.

As per National Family Health Survey, about 53.1% women in the reproductive age group and 58.4% children (6-59 months) are anemic (Kishore *et al.*, 2020).In order to combat deficits such vitamin and mineral shortages, iodine deficiency syndromes, iron deficiency, and anaemia, the WHO recommends widespread food fortification. U.S. food producers have been fortifying foods since the 1920s to prevent nutrition-related illnesses (Anonymous, 2021).

Factors on which food fortification depends are(Olson *et al.*, 2021):

- Dietary composition
- Production system
- Micronutrient deficiency existing
- Food processing capacity
- Available infrastructure

Vehicles for food fortification

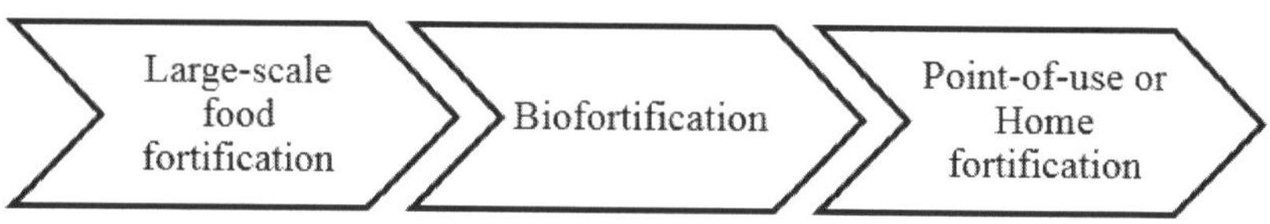

Large-Scale Food Fortification

Large-scale food fortification (LSFF) or Industrial fortification is incorporation of micronutrients in commonly consumed food for instance, flour, edible oil, salt, condiments, sugar etc during processing. In LSFF nutrients are normally added during post harvest processing. By boosting the nutritional value of staple foods that are often and extensively consumed by the population at risk of micronutrient deficiency, LSFF aims to prevent or reverse deficiencies in micronutrients. Large-scale food fortification programmes, which began in the early 20th century in developed nations, have contributed in the eradication of deficiency disorders in high-income nations, primarily in North America and Europe(Osendarp *et al.*, 2018).Due of their widespread and daily consumption, basic goods including flour, oil, salt, sugar, and condiments are frequently used as the vehicle for LSFF.

In the US, folic acid was made mandatory for the widespread fortification of improved cereal grain products in 1996, and it took full effect in 1998. The prevalence of NTDs was significantly lowered to 0.66 per 1000 pregnancies or fewer in just 5 years (Honein *et al.*, 2001). The widespread issue of childhood rickets was eradicated in Canada in 1965 when the requirement to add vitamin D to milk was implemented (CPHA, 2015).Programs for industrial fortification might be classified either as mandatory or voluntary. Mandatory fortification programs are common, especially in case of iodized salt and fortified flour.

Biofortification

Biofortification is the process of improving the nutritional value of food crops. It primarily focuses on raising provitamin A, zinc, iron etc in varied food crops agronomically or via plant breeding. Some of the biofortification projects are iron biofortification of rice, beans, maize and sweet potato; zinc biofortification of wheat, rice, beans, sweet potato and corn; and vitamin A biofortification of sweet potatoes, corn and cassava(Olson *et al.*, 2021). Biofortification done via genetic engineering simultaneous introduce multiple micronutrients in a single crop and increase level of micronutrient accumulation (Van Der Straeten *et al.*, 2020).

The simultaneous introduction of genes that increase micronutrient concentration, bioavailability and antinutrient content, which restricts the bioavailability of nutrients in plants, is also possible using transgenic techniques. Agronomic strategies, conventional plant breeding, and genetic engineering are key methods of biofortification.

Studies showing that consuming vitamin A biofortified orange fleshed sweet potatoes raised circulation of b-carotene and had a moderate effect on vitamin A status provided evidence for the effectiveness of biofortification. In Rwanda and Maharashtra, India, iron-biofortified beans and pearl millet enhanced haemoglobin and total body iron storage (Bouis and Saltzman, 2017). Due to biofortified provitamin-A maize, children in Zambia aged 5 to 7 with vitamin A insufficiency had their total body stores of vitamin A boosted, and their visual function had significantly improved (Gannon *et al.*, 2014). In Kenya, provitamin A in cassava was successful in improving the vitamin A status of pupils (Talsma *et al.*, 2016).

Point-of-Use Fortification or Home Fortification

Point-of-use fortification is also known as home fortification, it is the addition of micronutrient (i.e. vitamins and minerals)into the food that has been cooked and is ready to be eaten in other words when the micronutrient are not added to the foods during the processing but just before consumption at home or at schools or child-care facilities, it is called point-of-use fortification(De-Regil *et al.*, 2013). In 2012, the term "point-of-use" wasadopted by WHO. Point-of-use fortification of complementary foods with micronutrient powders (MNPs) arecrucial interventions for improving micronutrient intake in children aged 6-24 months(Olson *et al.*, 2021).

MNPs are single-dose packets consisting of numerous vitamins and minerals in powdered form that could be sprinkled on the food.To increase the nutritional content of food while retaining its flavour and colour, a powdered premix containing iron and possibly other vitamins and minerals is added either during or after cooking (De Regil *et al.*, 2017).When compared to no treatment or a placebo, home fortification with MNPs decreased anaemia by 31% and iron deficiency by 51% in newborns and young children, but had no impact on growth (Olson *et al.*, 2021).

Benefits of Fortification

According to FSSAI, Benefits of Fortification are:

- Benefit-to-cost ratio is high.
- Every 1 Rupee spent on fortification results in benefits of 9 Rupees to the economy.
- Overall cost of fortification is extremely low, with the exception of the initial investment in both equipment and premixes.
- Compared to the usual price variance, the price rise for fortified food is just about 1-2%.
- Other benefits are:
- Substantial portion of the population eats staple food, so nutrients are typically incorporated in them.
- For safe consumption, the micronutrients are added in a minimal amount and in accordance with the set guidelines.
- It does not encourage any changes in people's food habits or eating patterns.
- Addition of micronutrients does not affecttaste, aroma or texture of the food.

Fortified products

The World Health Organization (WHO) has been collaborating with numerous networks for food fortification at the regional, national, and local levels for many years. The WHO, has developed basic guidelines for the inclusion of essential nutrients into the food (Olson *et al.*, 2021).

Table 1: Studies conducted to examine role of food fortification

S.No.	Food Product	Micronutrient	Location	Outcome	References
1.	Rice	Iron (Micronized ground ferric pyrophosphate)	Bangalore, Karnataka	Increase iron store reduce the prevalence of iron deficiency	Moretti *et al.*, 2006
2.	Rice	Iron	Bangalore, Karnataka	Reduce prevalence of iron deficiency. There was a clear decrease in blood lead with iron repletion.	Zimmermann *et al.*, 2006

S.No.	Food Product	Micronutrient	Location	Outcome	References
3.	Khichdi	Iron, vitamin A, and folic acid	Maheshtala, South 24 Parganas district, West Bengal	Increase iron store and reduces the prevalence of iron deficiency and anemia	Varma *et al.*, 2007
4.	Salt	Dual fortification with iodine and iron	Anckaltaluk, Bangalore urban district, Karnataka	DFSs were efficacious in reducing the prevalence of anemia and iron deficiency in school-age children.	Andersson *et al.*, 2008
5.	Laddoo (cereal-legume snack)	Calcium and Vitamin D	Pune, India	Increase TBLH BMC of toddlers having habitually low dietary calcium intake.	Ekbote *et al.*, 2011
6.	Rice kernels	Iron (Micronized ferric pyrophosphate)	Ranga Reddy district, Andhra Pradesh	Improves iron stores and reduces iron deficiency	Radhika *et al.*, 2011
7.	Wheat	Iron (NaFeEDTA)	Bangalore, Karnataka	Improve Fe status. The drop in anemia prevalence was more pronounced.	Muthayya *et al.*, 2012
8.	Rice	Vitamin B_{12}	Bangalore, Karnataka	Small reduction in anemia prevalence improve Vitamin B_{12} status and physical performance	Thankachan *et al.*, 2012
9.	Milk	Vitamin D	NCR, India	Improved S.25(OH) D (serum 25 hydroxy vitamin D) level in children aged 10-14 years.	Khadgawat *et al.*, 2013
10.	Salt	Iodine and iron	Panighata tea estate, Darjeeling district, West Bengal	Improve hemoglobin, sTfR,ferritin and body iron concentrations in a population that ranged from moderately iron deficient anemic to iron sufficient.	Haas *et al.*, 2014

Wheat Flour

Wheat is a staple food for most of the Indians especially for those living in the wheat growing areas. People normally had wheat in the form of chapattis or rotis. So to improve the nutritional status of the large population, wheat flour is preferred for fortification. Wheat flour is fortified with iron, folic acid and vitamin B12which helps in fighting anaemia and formation of blood (Field *et al.*, 2021). Wheat flour fortification is very secure and effective means for improving public health.

Mandatory fortification of wheat flour was first introduced in 1942, and 85 countries have presently authorized its use. In North and South America, in order to lower the risk of birth defects, fortification of wheat flour with folic acid has been mandated (Osendarp *et al.*, 2018).

Rice

Rice is fortified to increase its nutritional value by adding specific vitamins and minerals during the postharvest process. In India, rice is a staple diet for 65% of the population, reaching even to the most vulnerable and underprivileged people. Micronutrient deficiencies are addressed through rice fortification in nations, with high per capita rice intake. The complexity of the supply chain, structure and capacity of the rice sector, legislative and regulatory environment, and the scale of the relevant programme, all affect the cost of fortification (Yusufali *et al.*, 2015).

Salt

Double Fortified Salt (DFS) is an innovative new fortified food product which provide little but crucial amounts of iodine and iron to human beings through their diet. A long-term solution to address iodine and iron deficiency may be to fortify salt with both iodine and iron. DFS formulations are typically designed to supply all the daily dietary requirements for iodine and circa 30% of daily requirements of iron (Drewnowski *et al.*, 2021). The creation of double fortified salt was pioneered by India's National Institute of Nutrition (NIN). DFS with encapsulated iron has been produced by the micronutrient initiative (Sivakumar and Nair, 2002). The usage of iodized salt has increased from 20% to 70% worldwide over the past few decades, making it the most popular form of fortification (UNICEF, 2008).

In India, every year 9 million pregnant women and 8 million newborns are at high risk of iodine deficiency disorders (IDD). IDD results in several disorders like goitre, intellectual disability, brain damage, cretinism, abortion, psychomotor defects, hypothyroidism, hearing and speech impairment etc. Most of the IDD's effects are invisible and incurable but they are also entirely avoidable. Most of the newborns in India are at high risk to IDD. IQ point of kids born in iodine deficient areas is 13.5 points lower than those in iodine sufficient areas (Pandav *et al.*, 2013). Iodine deficient soil is associated with IDDs. This therefore causes lack of iodine in crops cultivated in iodine-deficient soil, which in turn, results in a lack of iodine in cattle and human diets. Fortification of salt with iodine, can be used to correct this dietary iodine shortage. As salt is used everyday and by everyone, so, salt has been found to be a useful delivery system for iodine. In India, 91% of the population currently uses iodized salt (Pandav *et al.*, 2013).

Edible oil

Edible oils are fortified with vitamin A and D. This fortification with vitamin A and D is a viable technique to address micronutrient malnutrition. Oil that has been fortified offers roughly 25–30% of the recommended daily allowance (RDA) for vitamin A and D(Anonymous-1). Edible oils are becoming popular kind of fortification and until now 27 countries have made vitamin A fortification of oils mandatory (Osendarp *et al.*, 2018). In India, numerous micronutrient deficiencies are pervasive and continue to be serious public health issue, having a negative influence on everyone's health and productivity. Vitamin A insufficiency affects more than 57% of kids, and it can either be symptomatic or subclinical. A large percentage of expecting mothers and their newborns are deficient in vitamin D. According to the 2011 NSSO report on the Household Consumer Expenditure Survey in India, oil is consumed by all demographic groups, and as a result its daily consumption has climbed to 20-30g /person /day(Anonymous-1).

Milk

Calcium, high-quality proteins, and the fat-soluble vitamins (Vitamin A and D) are all abundant in milk. During milk fat removal process, vitamins A and D are also lost. Since it is simple to add back the vitamins that were lost, several nations have adopted mandatory provisions to do so. It is termed as replenishment, as the nutrients are reintroduced which were lost during processing.

In India, milk is fortified with vitamin A and vitamin D due to widespread deficiencies existing in the population. Since all demographic groups consume milk, fortifying it with specific micronutrients is a useful way to combat micronutrient deficiency. With a per capita availability of 407 grams/day and a production of 198.4 million tonnes (GOI, 2021), India is the world's greatest producer of milk. India's dairy industry has advanced from a state of scarcity to that of abundance.

Micro-nutrients Added

Fortification is the process of boosting the nutritional value of staple food like rice, wheat, oil, milk, and salt by adding essential vitamins and minerals including iron, iodine, zinc, and vitamins A and D. These nutrients might not be originally present in the food before processing or might have been lost during processing. The iodization of salt is the practice that is used most frequently worldwide.

Table 2: List of mandatory food fortification in India (Sirohi *et al.*, 2018)

Food Items	Nutrient	Level of fortification	Source nutrients
Salt	Iodine (manufacturer level)	Not less than 30 ppm on dry weight	Potassium iodate
	Iodine (distribution level)	Not <15 ppm on dryweight	Potassium iodate
	Iron	850-1100 ppm	Ferrous sulfate or ferrous fumarate
Oil	Vitamin A	25 IU/mg	Retinyl acetate, retinylpalmitate, and retinylpropionate
	Vitamin D	4.5 IU/mg	Cholecalciferol and ergocalciferol
Milk	Vitamin A	770 IU	Retinyl acetate, retinylpalmitate, and retinyl propionate
	Vitamin D	550 IU	Cholecalciferol and ergocalciferol
Vanaspati Atta	Synthetic Vitamin A	Not <25 IU per g	Retinyl acetate and retinylpalmitate
	Iron	20 mg	Sodium iron (III) ethylene diamine tetra acetate, trihydrate (sodium federate-Na Fe EDTA)
	Folic acid	1300 µg	Folic acid
	Vitamin B12	10 µg	Cyanocobalamin and hydroxycobalamin
Maida	Iron	60 mg	Ferrous citrate, ferrous lactate, ferrous sulfate,ferrous pyrophosphate, electrolytic iron, andferrous fumarate. Sodium iron (III) ethylene diamine tetra acetate,trihydrate (sodium federate-Na Fe EDTA)
	Folic acid	1300 µg	Folic acid
	Vitamin B12	10 µg	Cyanocobalamin and hydroxycobalamin
Rice	Iron	20 mg	Ferric pyrophosphate. Sodium Iron (III)ethylene diamine tetra acetate,trihydrate (sodium federate-Na Fe EDTA)
	Folic acid	1300 µg	Folic acid
	Vitamin B12	10 µg	Cyanocobalamin and hydroxycobalamin

- **Vitamin A:** Vitamin A is necessary for vision, lack of it can cause night blindness in children under the age of five and a high incidence of bitots spot. Fruits and vegetables contains beta, which our body cannot fully convert into retinol and are not absorbed by body. Vitamin A supplementation is the preferred approach in areas where the issue is widespread since it is the quickest way to enhance children's vitamin A level (Agrawal and Agrawal, 2013).

Large reductions in vitamin A deficiency have also been observed in El Salvador and Honduras, where fortification was paired with supplementation. Guatemala's sugar fortification programme has almost eliminated vitamin A deficiency. Nigeria, Morocco, Yemen, Bangladesh, and Pakistan are putting national plans to fortify cooking oils with vitamin A into effect because the poorer sections of the population in Africa and Asia do not consume as much sugar as in Latin America. Safety is a problem due to the great efficacy of vitamin A fortification, thus care must be taken not to over fortify (Mannar and Hurrell, 2018).

- **Vitamin D:** Vitamin D is typically not abundant in Indian cuisine. Rickets, development issues, bowlegs, knocked knees etc. are some of the problems due to vitamin D deficiency. Vitamin D absorption is ensured through exposure to sunlight, particularly between 11:00 am and 1:00 pm. One can get one-third of their daily vitamin D requirements via fortified milk and edible oils. 70-90% of people in our country are vitamin D deficient (Aparna *et al.*, 2018).

Many people rely on dietary vitamin D to maintain a sufficient status, especially the elderly and those living in northern latitudes. Vitamin D is naturally present in only few number of food at high amount. Consumption of vitamin D is much below the recommended dietary guideline in many nations, including Europe and North America. For the prevention of rickets and osteomalacia, lower cut-off values have been advocated, whereas considerably higher cut-off values have been suggested for the prevention of falls and fractures in the elderly. The method of transportation is particularly problematic and even in areas where fluid milk or margarine are intentionally or legally fortified, many people still get less vitamin D than advised because milk consumption is declining. In the US, yoghurt, butter, margarine, cheese, orange juice, and bread have all voluntarily been fortified in addition to milk (Mannarand Hurrell, 2018).

- **Iron:** The majority of the iron we require comes from meat, which is absorbed by our body to the tune of 50%. Vegetables only provide a small amount of intake and only 3% of absorption because of this, iron deficiency has become a significant issue in the nation. In addition to the iron in wheat flour and rice, iron insufficiency can be addressed by fortifying iodized salt (Double Fortified Salt). Iodized salt had a poor acceptance rate (40-60%) during 1960-2007, but after required legislation was passed in 2007, the acceptance rate has now rose to 92%. NIN produced salt with added iron in the 1970s.

Anaemia is common among both undernourished and nourished people. Anaemia was a major cause of maternal death in the 1970s. 50% reduction in anaemia by 2025 is attainable if the distribution of foods supplemented with iron is increased (Hasan *et al.*, 2022).Iron deficiency does not result in a condition that is easily recognizable, thus, it is challenging to demonstrate the effect and the level of iron fortification. Anaemia, poor pregnancy outcomes, poor work performance, and delayed brain development are all effects of iron deficiency that have a variety of underlying causes(Mannarand Hurrell, 2018).

A significant issue has been that, while the more insoluble, less bioavailable iron compounds are less effectively absorbed and produce frequent colour and flavour changes in some food vehicles. The main source of iron should have a lot of phytate, a powerful inhibitor of iron absorption (Gera *et al.*, 2012). Furthermore, Costa Rica recently showed that a national effort to fortify milk powder and maize flour with iron considerably decreased the prevalence of anaemia in mothers and children (Martorell *et al.*, 2015).

- **Iodine:** The fortification of salt with iodine has proven to be the most successful fortification method worldwide. Iodine is added to salt in a simplistic manufacturing process that only costs a few pennies per person per year. Iodized salt cost1 rupee, yet it yields benefits worth close to 80 rupees(Anonymous-2).Successful salt iodization has averted mental retardation, subclinical iodine deficiency diseases and goitre as well as decreased their incidence. It has also increased national production. In India, private sector produces 99% of iodized salt.

Through CSR programme, the private sector is essential in raising public awareness of food fortification. According to a poll, iodine salt is now available to 92% of the population (Yadav and Pandav, 2018). Iodized salt is available to a sizable section of the populace in more than 120 nations. As per UNICEF report 2015, around 80 million babies are shielded from the risk of mental impairment brought on by iodine deficiency every year, thanks to the iodization of around 76% of salt ingested globally.

Double fortified salt, which is produced in India, has the potential to be distributed to economically underprivileged populations worldwide through both private and public channels. Contrary to developed countries, low- and middle-income countries take time for fortification of frequently produced goods like salt in a numerous cottage-scale firms, which lengthens the time it takes for a fortification intervention to be effective(Mannarand Hurrell, 2018).

- **Folic acid:** The target population for folic acid fortification is not a large population with low folate intakes or a reported poor folate status, but rather young women with an augmented requirement for folate. According to FSSAI, fortification offers a high benefit to cost ratio andrice should be fortified with 75-125µg folic acid (Peña-Rosas *et al.*, 2019). Media is incredibly important in raising public awareness of the value of folic acid fortification.

Spina bifida and anencephaly are the results of NTDs, which happen when the neural tube fails to close early in pregnancy. Consuming folate prior to conception is necessary to have an impact on NTDs.The highest rates of neural tube abnormalities are found in several countries of Africa, South East Asia, and India (4.7-9 per 1000 in Punjab & Haryana). In the US, folic acid was added to flour in 1998. Folic acid fortification of wheat, maize, or rice is practiced in about 80 nations (Tablante *et al.*, 2019). Despite the fact that only 10% of preventable birth abnormalities are currently avoided, Bell and Oakley (2009) estimated that 27% of the world's population had access to folic acid fortified wheat. This is mostly because of insufficient coverage in low- and middle-income nations. High folate intakes may have negative health effects, and some worry that too much folic acid may hide the neurological damage and increased risk of memory loss brought on by anaemia which is the result of B12 insufficiency(Mannarand Hurrell, 2018).

- **Vitamin B_{12}:** Most bacteria and algae manufacture vitamin B12 with the aid of enzymes. Through assimilation of microbes into the food of animal origin, vitamin B12 (cobalamin) produced by microbes makes its way into the food chain of humans. These microorganisms that produce vitamin B12 are supported in their growth by intestinal fermentation and as a result, the vitamin is absorbed and incorporated into the animal tissues. The blood and neurological system are primarily affected by B12 deficiency (Green *et al.*, 2017). The classic signs of B12 insufficiency were initially identified in pernicious anaemia (Stabler, 2013). Rarely, it is brought on by a diet deficient in vitamin B12,instead, it is brought on by a deficiency of a naturally occurring stomach component that prevents vitamin B12 absorption.
- **Zinc:** Zinc act as a cofactor for metalloenzymes necessary for the normal operation of the immunological, gastrointestinal, dermatologic, neurological, and reproductive systems of the human body, zinc is a vital part of some chemical reactions in the body (MayoWilson *et al.*, 2014). 85% of the zinc in the body is located in muscle and bone tissue, 11% in skin and liver, and the rest is found in trace levels in other tissues like plasma (0.1%) (Das *et al.*, 2018). Daily zinc losses are primarily through the digestive system, with minimal losses through the kidneys. Although it can be found in various foods, whole-grain cereals, meat, fish, nuts, seeds, and legumes are the ones with the highest concentrations. Moreover, it was found that dairy products contributed more to zinc levels than veggies. Cereals are frequently chosen as food fortification targets. In countries like Indonesia (30 ppm in wheat flour), Jordan (20 ppm in wheat flour), Mexico (16 ppm in maize and wheat flours), and South Africa (15 ppm in maize and wheat flours), mass zinc fortification is the main strategy for increasing dietary consumption (Lim *et al.*, 2013).

Fortification in Government Programmes (FSSAI, 2017)

Table 3: Fortified staples inclusion in states government programmes

Programmes	Beneficiaries
Fortification in Mid-Day Meal Scheme	Children enrolled in government, local and government-aided schools up to 8th class.
Fortification in Integrated Child Development Services (ICDS)	Pregnant women, nursing moms, and kids between the ages of 6 months and 3 years receive Take Home Rations (THR). Kids between the ages of 3 and 6 receive a hot cooked meal.
Fortification in Public Distribution System	Beneficiaries are classified as AAY or PHH (Priority Households /Non AAY).
Other Schemes	
SABLA: A Scheme for Adolescent Girls	Girls in their adolescence (11 to 18 years old) for their holistic development.
Welfare Institutions and Hostels	Elderly for significantly enhancing their nutritional status.
Tea Plantation Workers	Vulnerable group of people to improve the nutrition status

Health Benefits of Fortified Foods

The benefits of fortified foods include:

- Fortified foodiscost-effective. Even after addition of micronutrients the product pricing is still reasonable and has better nutritious value.
- Eating fortified food helps in preventing diseases like rickets, anaemia etc. that are caused by nutrient deficiencies.
- They are useful during pregnancy since pregnant women require extra nutrients than usual and cannot get them from regular meals.
- Foods enriched with folic acid reduce the incidence of birth abnormalities in expectant women.
- Food that has been fortified helps people, especially elderly people in maintaining healthy levels of micronutrients, strong bones, aid in digestion, and prevent heart problems.
- Fortified foods provide the children with additional nutrients, which are crucial for their growth and development.

Limitations of Fortified Foods

- Foods that have been fortified contain excessive levels of carbohydrates, fats, sodium and other components that can cause issues including obesity.
- It should not be taken along with any medication prescribed by a doctor unless authorised by them.
- The poorer section of society, where the risk of micronutrient insufficiency is greatest, have not yet received fortified foods.

- It might enrich the meal with an excessive amount of vitamins and minerals, which could be hazardous to the consumer.
- Premixes and equipments both require a large upfront investment.

Conclusion

Fortification is one of the most scalable, sustainable and affordable strategies to combat hidden hunger. It is a research-backed tactics that enhance the overall quality of the diet and being a cost-effective solution, it has the potential to alleviate malnutrition on a global scale. When the deficiency rate is high and the target demographic is challenging to reach, fortification is exceptionally alluring, this is particularly true for folic acid and iodine. The best source of fortifying fuel is food. Food fortification has produced favourable outcomes in terms of social, economic and environmental aspects as well as the prevention and control of micronutrient deficiencies in vulnerable populations, particularly women and children.When the majority of the population faces an intolerably high danger of being deficient in a particular micronutrient, mass fortification is the appropriate course of action. Along with the primary fortification activities, community education and promotion programmes should be conducted to increase knowledge, acceptability and equity.

References

Agrawal, S. and Agrawal, P. 2013. Vitamin A supplementation among children in India: Does their socioeconomic status and the economic and social development status of their State of residence make a difference? International Journal of Medicine and Public Health, 3(1): 48-54.

Andersson, M., Thankachan, P., Muthayya, S., Goud, R.B., Kurpad, A.V., Hurrell, R.F. and Zimmermann, M.B. 2008. Dual fortification of salt with iodine and iron: a randomized, double-blind, controlled trial of micronized ferric pyrophosphate and encapsulated ferrous fumarate in southern India. The American Journal of Clinical Nutrition, 88(5): 1378-87.

Anonymous. 2021. What to know about fortified foods. Available on: https://www.webmd.com/diet/what-to-know-about-fortified-foods. Accessed on 24 June, 2022.

Anonymous-1.Available on: https://ffrc.fssai.gov.in/commodity?commodity=fortified-oil. Accessed on 24 June, 2022.

Anonymous-2.FSSAI. Available on: https://ffrc.fssai.gov.in/micronutrient#iron. Accessed on 24 June, 2022.

Aparna, P., Muthathal, S., Nongkynrih, B. and Gupta, S.K. 2018.Vitamin D deficiency in India.Journal of Family Medicine and Primary Care,7(2): 324-330.

Bell, K.N. and Oakley, G.P. 2009. Update on prevention of folic acid preventable spina bifida and anencephaly. Birth Defects Research Part A, 85(1): 102-107.

Bouis, H.E. and Saltzman, A. 2017.Improving nutrition through biofortification: a review of evidence from harvest plus, 2003 through 2016.Global Food Security, 12: 49-58.

CPHA. 2015. Food fortification with vitamins and minerals. Canadian Public Health Association.https://www.cpha.ca/food-fortification-vita mins-and-minerals.Accessed August 3, 2022.

Das, J.K., Khan, R.S. and Bhutta, Z.A. 2018. Food Fortification in a Globalized World. Zinc Fortification, Academic press, 213-219.

De-Regil, L.M., Jefferds, M.E.D. and Peña‐Rosas, J.P. 2017. Point‐of‐use fortification of foods with micronutrient powders containing iron in children of preschool and school‐age. Cochrane Database of Systematic Reviews, (11), Art. No.: CD009666.

De-Regil, L.M., Suchdev, P.S., Vist, G.E., Walleser, S. and Pena-Rosas, J.P. 2013. Home fortification of foods with multiple micronutrient powders for health and nutrition in children under two years of age (Review). Evidence-Based Child Health A Cochrane Review Journal, 8(1): 112-201.

Drewnowski, A., Garrett, G.S., Kansagra, R., Khan, N., Kupka, R., Kurpad, A.V., Mannar, V., Martorell, R. and Zimmermann, M.B. 2021. Key considerations for policymakers-iodized salt as a vehicle for iron fortification: current evidence, challenges, and knowledge gaps. The Journal of Nutrition, 151(Suppl 1): 64S-73S.

Ekbote, V.H., Khadilkar, A.V., Chiplonkar, S.A., Hanumante, N.M., Khadilkar, V.V. and Mughal, M.Z. 2011. A pilot randomized controlled trial of oral calcium and vitamin D supplementation using fortified laddoos in underprivileged Indian toddlers. European Journal of Clinical Nutrition, 65(4): 440-446.

Field, M.S., Mithra, P. and Peña-Rosas, J.P. 2021. Wheat flour fortification with iron and other micronutrients for reducing anaemia and improving iron status in populations.Cochrane Database of Systematic Reviews, (1): CD011302.

FSSAI. 2017. Large Scale Food Fortification in India. Available at: https://nutritionconnect.org/sites/default/files/uploads/resources/2019-04/Large%20Scale%20Food%20Fortification%20in%20India.%20The%20Journey%20So%20Far%20and%20the%20Road%20Ahead..pdf.Accessed on 25 June, 2022.

Gannon, B., Kaliwile, C., Arscott, S.A., Schmaelzle, S., Chileshe, J., Kalungwana, N., Mosonda, M., Pixley, K., Masi, C. and Tanumihardjo, S.A. 2014. Biofortified orange maize is as efficacious as a vitamin A supplement in Zambian children even in the presence of high liver reserves of vitamin A: a community-based, randomized placebo-controlled trial. The American Journal of Clinical Nutrition,100(6): 1541-50.

Gera, T., Sachdev, H.S. and Boy, E. 2012.Effect of iron-fortified foods on hematologic and biological outcomes: systematic review of randomized controlled trials. The American Journal of Clinical Nutrition, 96(2): 309-324.

GOI. 2021. Economic Survey 2020-21. Department of Economic Affairs. Available at: https://www.indiabudget.gov.in/economicsurvey/ebook_es2021/index.html#p=613. Accessed on 28 June, 2022.

Green, R., Allen, L.H., Bjorke-Monsen, A.L., Brito, A., Guéant, J.L., Miller, J.W., Molloy, A.M., Nexo, E., Stabler, S., Toh, B.H., Ueland, P. and Yajnik, C. 2017. Vitamin B12 deficiency.Nature Reviews Disease Primers,3: 17040.

Haas, J.D., Rahn, M., Venkatramanan, S., Marquis, G.S., Wenger, M.J., Murray-Kolb, L.E., Wesley, A.S. and Reinhart, G.A. 2014. Double-fortified salt is efficacious in improving indicators of iron deficiency in female Indian tea pickers. Journal of Nutrition, 144(6): 957–964.

Hasan, M.M., SoaresMagalhaes, R.J., Garnett, S.P., Fatima, Y., Tariqujjaman, M., Pervin, S., Ahmed, S. and Mamun, A.A. 2022. Anaemia in women of reproductive age in low- and middle-income countries: progress towards the 2025 global nutrition target. Bulletin of the World Health Organization, 100(3): 196-204.

Honein, M.A., Paulozzi, L.J., Mathews, T.J., Erickson, J.D. and Wong, L.Y. 2001. Impact of folic acid fortification of the US food supply on the occurrence of neural tube defects.JAMA, 285(23): 2981-2986.

Khadgawat, R., Marwaha, R.K., Garg, M.K., Ramot, R., Oberoi, A.K., Sreenivas, V., Gahlot, M., Mehan, N., Mathur, P. and Gupta, N. 2013. Impact of vitamin D fortified milk supplementation on vitamin D status of healthy school children aged 10-14 years. Osteoporos International, 24(8): 2335-43.

Kishore, S., Singh, M., Jain, B., Verma, N., Gawande, K., Kishore, S., Aggarwal, P. and Verma, S.K. 2020.A study to assess prevalence of anaemia among beneficiaries of AnaemiaMukt Bharat Campaign in Uttarakhand.Journal of Family Medicine and Primary Care, 9(3): 1691-1694.

Lim, K.H.C., Riddell, L.J., Nowson, C.A., Booth, A.O. and Szymlek-Cay, E.A. 2013. Iron and zinc nutrition in the economically developedworld: a review. Nutrients, 5: 3184-3211.

Liyanage, C. and Hettiarachchi, M. 2011. Food fortification.Ceylon Medical Journal, 56(3): 124-127.

Mannar, M.G.V. and Hurrell, R.F. 2018.Food Fortification: Past Experience, Current Status, and Potential for Globalization. Food Fortification in a Globalized World.Academic Press, 3-11.

Martorell, R., Ascencio, M., Tacsan, L., Alfaro, T., Young, M.F., Addo, O.Y., Dary, O. and Flores-Ayala, R. 2015. Effectiveness evaluation of the food fortification program of Costa Rica: impact on anemia prevalence and hemoglobin concentrations in women and children. The American Journal of Clinical Nutrition, 101(1): 210-217.

Mayo-Wilson, E., Junior, J.A., Imdad, A., Dean, S., Chan, X.H.S., Chan, E.S., Jaswal, A. and Bhutta, Z.A. 2014. Zinc supplementation for preventing mortality, morbidity, and growth failure in children aged 6 months to 12 years of age. Cochrane Database of Systematic Reviews, 1465-1858.

Moretti, D., Zimmermann, M.B., Muthayya, S., Thankachan, P., Lee, T.C., Kurpad, A.V. and Hurrell, R.F. 2006. Extruded rice fortified with micronized ground ferric pyrophosphate reduces iron deficiency in Indian schoolchildren: a double-blind randomized controlled trial. The American Journal of Clinical Nutrition,84(4): 822-9.

Muthayya, S., Thankachan, P., Hirve, S., Amalrajan, V., Thomas, T., Lubree, H., Agarwal, D., Srinivasan, K., Hurrell, R.F., Yajnik, C.S. and Kurpad, A.V. 2012. Iron fortification of whole wheat flour reduces iron deficiency and iron deficiency anemia and increases body iron stores in indian school-aged children. Journal of Nutrition, 142(11): 1997-2003.

Olson, R., Gavin-Smith, B., Ferraboschi, C. and Kraemer, K. 2021.Food fortification: the advantages, disadvantages and lessons from sight and life programs. Nutrients, 13(4): 1118.

Osendarp, S.J.M., Martínez, H., Garrett, G.S., Neufeld, L., De Regil, L., Vossenaar, M. and Darnton-Hill, I. 2018. Large-scale food fortification and biofortification in low- and middle-income countries: a review of programs, trends, challenges, and evidence gaps. Food and Nutrition Bulletin, 39(2): 315-331

Pandav, C.S., Yadav, K., Srivastava, R., Pandav, R. and Karmarkar, M.G. 2013. Iodine deficiency disorders (IDD) control in India. Indian Journal of Medical Research,138(3): 418-433.

Peña-Rosas, J.P., Mithra, P., Unnikrishnan, B., Kumar, N., De-Regil, L.M., Nair, N.S., Garcia-Casal, M.N. and Solon, J.A. 2019.Fortification of rice with vitamins and minerals for addressing micronutrient malnutrition.Cochrane Database of Systematic Reviews, (10): CD009902.

Radhika, M.S., Nair, K.M., Kumar, R.H., Rao, M.V., Ravinder, P., Reddy, C.G. and Brahmam, G.N. 2011. Micronized ferric pyrophosphate supplied through extruded rice kernels improves body iron stores in children: a double-blind, randomized, placebo-controlled midday meal feeding trial in Indian schoolchildren. The American Journal of Clinical Nutrition, 94(5): 1202-10.

Sirohi, A., Pundhir, A. and Ghosh, S. 2018. Food fortification: a nutritional management strategy in India. Innovare Journal of Food Science, 6(2): 1-8.

Sivakumar, B. and Nair, K.M. 2002. Double fortified salt at crossroads. Indian Journal of Pediatrics,69(7): 617-23.

Stabler, S.P. 2013.Clinical practice.Vitamin B12 deficiency.New England Journal of Medicine,368(2): 149-160.

Tablante, E.C., Pachón, H., Guetterman, H. and Finkelstein, J.L. 2019.Fortification of wheat and maize flour with folic acid for population health outcomes. Cochrane Database of Systematic Reviews, 7(3).

Talsma, E.F., Brouwer, I.D., Verhoef, H., Mbera, G.N., Mwangi, A.M., Demir, A.Y., Maziya-Dixon, B., Boy, E., Zimmermann, M.B. and Melse-Boonstra, A. 2016. Biofortified yellow cassava and vitamin A status of Kenyan children: a randomized controlled trial. The American Journal of Clinical Nutrition, 103(1): 258-67.

Thankachan, P., Rah, J.H., Thomas, T., Selvam, S., Amalrajan, V., Srinivasan, K., Steiger, G. and Kurpad, A.V. 2012.Multiple micronutrient-fortified rice affects physical performance and plasma vitamin B-12 and homocysteine concentrations of Indian school children. The Journal of Nutrition, 142(5): 846-52.

UNICEF. 2008. Sustainable Elimination of Iodine Deficiency: Progress since the 1990 World Summit for Children. United Nations Children's Fund, New York, USA.

Van Der Straeten, D., Bhullar, N.K., De Steur, H., Gruissem, W., MacKenzie, D., Pfeiffer, W., Qaim, M., Slamet-Loedin, I., Strobbe, S., Tohme, J., Trijatmiko, K., Vanderschuren, H., Van Montagu, M., Zhang, C. and Bouis, H. 2020. Multiplying the efficiency and impact of biofortification through metabolic engineering. Nature Communications, 11: 5203.

Varma, J.L., Das, S., Sankar, R., Mannar, M.G., Levinson, F.J. and Hamer, D.H. 2007. Community-level micronutrient fortification of a food supplement in India: a controlled trial in preschool children aged 36-66 mo. The American Journal of Clinical Nutrition, 85(4): 1127-33.

WHO. 2018. Food fortification Q&A. Available at: https://www.who.int/vietnam/news/feature-stories/detail/food-fortification-q-a. Accessed on 28 June, 2022.

Yadav, K. and Pandav, C.S. 2018. National iodine deficiency disorders control programme: current status and future strategy. Indian Journal of Medical Research, 148(5): 503-510.

Yusufali, R., Ghoos, K., Milani, P. and Rosenzweig, J. 2015. Scaling up rice fortification in Asia. Sight and life supplement. Sight and Life Magazine, 43-51.

Zimmermann, M.B., Muthayya, S., Moretti, D., Kurpad, A. and Hurrell, R.F. 2006. Iron fortification reduces blood lead levels in children in Bangalore, India. Pediatrics, 117(6): 2014-21.

Chapter 7

Role of Food Processing, Value Addition and Secondary Agriculture in Increasing Farm Profits

Neeraj Gupta, Fozia Hameed, Vikas Gupta and Monica Reshi*

Division of Post Harvest Management, Chatha, SKUAST-Jammu-180009
Basmati Rice Research Centre (BRRC), SKUAST-J, R. S. Pura, J&K-181102
****email: neeruguptapht@gmail.com***

Abstract

Secondary agricultural sector plays a vital role in increasing farmer's income due to its immense untapped potential. It adds value to the primary agricultural produce through valorization and waste utilization. Potential avenues of the secondary agriculture include nurseries, vermicompost, biofertilizers, food processing products, flavours, dyes etc. The waste obtained from the industrial processing of fruits, vegetables, cereals, medicinal and aromatic plants is rich in several bioactive compounds and can either be added as source of ADF (antioxidant dietary fiber) in the preparation of food products like therapeutic, functional and nutraceutical or utilized for the production of food-grade industrial products like bioethanol, citric acid, pectin, pigments, dyes rather being used as livestock feed. When properly explored, these industrial by-products could substantiate country's economic gain, reduce environmental pollution and increase farmer's earnings. This chapter mainly focuses on the potential of secondary agriculture, bio-processing and value addition in different avenues of secondary agriculture for increasing annual income of farmers and uplifting country's economy.

Keywords: Value addition, food processing, secondary agriculture, by-product utilization, farmer's profit

Introduction

Agricultural sector, the backbone of Indian economy constitutes the principal source of livelihood for ~60% of population in India (Yadav *et al.*, 2020). This sector is mainly divided into three categories viz primary, secondary and tertiary agriculture. The cultivation, growing and harvesting of any agricultural commodity is primary agriculture while as agro-based manufacturing comprises of secondary agriculture. Secondary agriculture is also defined as a processes and practices which adds value to the primary agricultural commodity (O'Shea *et al.*, 2012) and includes all food and non-food bio-resource based products for human and industrial use such as protein, oil and milk from soybean, oil from rice bran, starch and corn syrup from corn seeds, bio-fuels from lingo-cellulosic biomass (rice residues and sugarcane bagasse) and so on. Tertiary agriculture includes business related processes or selling of products. All three sectors are interdependent having their own functions and importance. In India, previous strategies for agricultural development has mainly focused on increasing farm output and improving nations food security so as to combat the food shortage that India faced during the mid 1960s. Adoption of such production strategies not only resulted in multiplying the food production by 3.7 times but has also made India food self sufficient. However, the strategy neither appreciated the need to raise the farmers profit nor addressed any direct measure to promote farmers welfare. Further, research studies have also shown that farmers' income does not always follow the increase in production output (Chand *et al.*, 2015). So in this milieu, understanding the importance of agricultural sector, the paradigm shift in focus on agriculture policies from production to income have evolved recently and the Government of India has set the target of doubling farmers' income by 2022 (NCAER, 2017).

As per ministry of food processing industries, India produced about 291.95 MTs of food grains during 2019-2020 which is higher by 6.74 MTs achieved during 2018-2019 (MOFPI, 2019), thus, succeeded in achieving the target of self-sufficiency in food-grain production. However, annually only less than 10% of its agri-output is processed; which accounts to 2% of fruits and vegetables, 6% of poultry, 21% of meat, 23% of marine and 35% of milk and most of the processing is primary processing (MOFPI, 2019). A major fraction of the India's total agri produce is lost due to post-harvest losses which accounts for about

US $1.5 billion (Rs 92,000 crores) annually (MOFPI, 2019) as a result of non-availability of post-harvest processing infrastructure like cold chains, proper transportation facilities, proper storage facilities, gaps in food processing, imperfect market information, limited market access and so on (Kumar and Kalita, 2017). Although, the core inefficiencies in the post-harvest value chain leading to post-harvest losses is shown in Fig. 1.

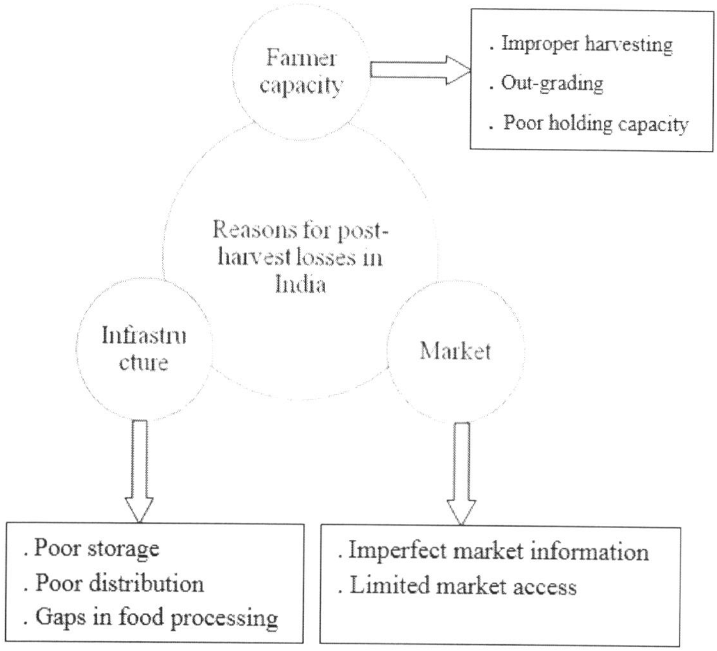

Fig. 1. Core inefficiencies in the post harvest value chain leading to post-harvest losses

Value addition of the agricultural produce, strengthening of the infrastructure in the food processing sector and usage of modern food processing technologies is gaining importance which not only helps in reducing post-harvest losses and farm distress but also contribute towards improving the income level of the farmers either by reducing food wastages or enhancing food as well as processed food exports. Also, series of policy initiatives taken by the GOI keeping in view the key issues and need of farmers include development of mega food parks, integrated cold chain and preservation infrastructure facilities, quality control laboratories and capacity building through establishment of National Institute of Food Technology, Entrepreneurship and Management (NIFTEM), Indian Institute of Crop Processing Technology (IICPT) and Center of Innovative and Applied Bioprocessing (CIAB) which mainly works on secondary agriculture and development of value added products through bio-processing. With this regard, the Central government in 2015 has also changed the name of Ministry of Agriculture to Ministry of Agriculture and Farmers Welfare.

Presently, the average annual income of the farmers in India is less than Rs. 1 lakh. The farm economy has gradually changed due to younger generation with education and expertise finding an opportunity to work in the non-farm sectors (Chengappa, 2013). Secondary agriculture, thus, plays a vital role in generating employment opportunities for growing younger unskilled, semi-skilled and skilled workforce, shifting disguised unemployment from traditional agriculture to food processing and agriculture export, thereby reducing poverty by contributing to the India's overall GDP. Further, value addition through secondary agricultural processing provide 2-4 times higher income to farmers as well as contribute to overall growth of the agro-processing sector. Researchers also paved new and alternative ways in creating new products and applications with value added approach at the cost of recycling secondary agricultural waste.

Food Processing Sector: Current Scenario

The food value chain is a sunrise sector that has witnessed a tremendous growth and development in the recent years because of the availability of modern tools and technologies to the farmers (Murthey and Yogesh, 2014). This sector contributes around 11% of agricultural value- addition and 9% of manufacturing value-addition, thus, serves a crucial backward and forward linkage between agriculture and industry: the two pillars of Indian economy (Shelly and Kaur, 2015). Strengthening the synergy is of critical importance not only in reducing the wastage of agri-output but also raise the value of the agricultural commodities by increasing shelf-life as well as by fortifying the agri-produce, thereby making nutritionally sufficient products, ensuring remunerative income to farmers on one hand and affordable prices to consumers on other hand (Singh *et al.*, 2012).

In India, the food processing sector is composed of two broad segments viz organized and unorganized segments. According to the ministry of food processing industries, unorganized sector contribute 75% of employment, 12% of output and 27% of value addition in the food processing sector while as organized sector accounts for 26% of employment, 90% output and 72% of value addition. Further, under the organized segment, three food commodities viz. grains and starch, oil and fats and sugar constitute 52.1% of the value of food produce and three major perishable commodities viz. meat, fish and fruits and vegetables contribute a negligible share of 7.8% in food output. In India, food processing sector is high growth and high profit sector because of its immense potential for value addition. The share of Indian food processing industry in country's total food market is 32 per cent. Its contribution in manufacturing Gross Domestic Product (GDP) is 14 per cent.

Fruits: Production, Processing and Value addition

Fruits play an important role in human health and nutrition due to their medicinal and aesthetic value. The dietary importance of fruits accounts for its richness in vitamins, minerals, fiber and phytochemicals (Slavin and Lloyd, 2012). Despite the nutritional security, the production of fruits helps in improving the country's economy by raising the farmers' income and generating employment opportunities. More recently, the huge demand for the high value fruits has increased owing to large population growth and changing lifestyle of the health conscious consumers (Yadav *et al.*, 2020). In India, horticulture sector has shown a voluminous growth over the last few years as a result of amazing weather conditions, investments through National Horticulture Mission (NHM) and a number of other programmes, thus, contributing a percentage share of around 33% towards agriculture (Kashish and Dawan, 2016). It is the second largest producer of fruits in the world after China with a production of 97.35 MTs for the year 2017-2018 with its projected score touching 98 MTs by the year 2020-2021. India is the leading producer of some fruits like banana, mango, papaya and citrus. However, India's contribution in the global export market of food products is only 2.31 per cent.

Further, out of the total fruit production only 2% is processed for value addition (MOFPI, 2019). From production to consumption, around 30-50% (worth thousand crore rupees) is lost as waste due to lack of post-harvest infrastructure and management (Rohm *et al.*, 2015). A high level of waste and by-products is generated as a result of seasonal glut or during transportation, storage and processing. Fruit waste or by-product consists of peel, pomace, pellets, trimmings, seeds, shells and bagasse and accounts for about 25-60% of the weight of fresh fruit (Coman *et al.*, 2019). Sometimes these wastes and by-products have higher nutritional and functional value than the final product due to their associated numerous bioactive compounds including proteins, micronutrients, dietary fibers and secondary metabolites like phenols, flavonoids, antioxidants, anthocyanins, tannins and saponins (Torres-Leon *et al.*, 2017). Most of these by-products generated are not utilized and end up in landfills which in turn can cause a negative impact on environment (Torres-Leon *et al.*, 2017). Valorization of the waste/by-products, their bioactive compounds and surplus fruits is the need of the hour so as to reduce the post-harvest losses and thus raise the earnings of the farmers and also generate various employment opportunities (Yadav *et al.*, 2020). Possible valorisation of by-products obtained from the main fruit processing indutries like apple, mango, banana and citrus are described below.

Valorization of Apple By-Products

In case of apple processing industry, the major by-product is apple pomace which accounts for about 25% of fresh apple weight (Perussello *et al.*, 2017). Worldwide production of apple pomace is about 3.0-4.2 ^{x} 106 MTs per year (Yadav *et al.*, 2020). After China and India (1000 K Tonnes), Brazil is the leading producer of apple pomace with a production of 800 K Tonnes per year followed by Germany (250 K Tonnes), Japan (160 K Tonnes), Iran (97 K Tonnes), US-

American (27 K Tonnes) and New Zealand (20 K Tonnes), respectively (Lyu *et al.*, 2020). Apple pomace consists of peel and flesh (95.5%), seeds (3-4%) and stem (1%). Usually producers discard tonnes of apple pomace annually at a cost to themselves. Alternatively, it is used as animal feed. Apple pomace is rich in nutrients such as micronutrients, macronutrients, dietary fiber, pectin and phenolic compounds (O'Shea *et al.*, 2012). Therefore, discard of apple pomace is a waste of resources. Valorization of the waste can minimize environmental pollution, raise the country's economy as well as can increase the farm profits. Numerous processing technologies have been used to produce various kinds of products like pectin, ethanol, flavor components, pigments, citric acid, lactic acid and phenolic compounds (Perussello *et al.*, 2017). Apple pomace as an ingredient can be utilized in a variety of food products like baked goods such as cakes, muffins, bread and biscuits, confectionary and sausages (Henriquez *et al.*, 2010). Addition of apple pomace not only increases the nutritive value of such products but also improves the health of the consumers because of its nutrient and phytochemical composition. Research studies have also revealed that addition of apple pomace in bakery products increased the dietary fiber content upto 40% of the developed product (Masoodi and Chauhan, 1998; Alongi *et al.*, 2019). Further, pectin extracted from the apple pomace contains methoxyl content (4-10%), anhydrouronic acid (63-65%), acetyl value (0.68%) and degree of esterification (33.44-49.35%) (Gazala *et al.*, 2017). Due to the functional properties of pectin like gelling, thickening and stabilizing abilities and water holding capacity, apple pomace as a substitute for pectin can be used in the formulation of confectionary and sugar based foods like jams, jellies etc (Shalini and Gupta, 2010). Hussein *et al.*, 2015 developed an apple pomace based jam made by fruit by-products like apple pomace, carrot peel, banana peels and mandarin peels. Results revealed from the study showed that apple pomace jam had higher phosphorous (220 mg/100 g), total phenolic (82.5 mg CAT/100 g) and total flavonoid content (30.1 mg CAT/100 g). Further, sensory analysis revealed that apple pomace jam had highest score for overall acceptability resulting in fruity flavor and satisfying appearance (Hussein *et al.*, 2015).

Valorization of Banana By-Products

Major by-product of banana fruit is peel which constitutes 30-40% of the whole fruit weight (Chueh *et al.*, 2019). Further, approximately 20-25 tonnes of pseudo-stem are generated from nearly thousand banana processing plants yielding 5% edible starch (Schieber *et al.*, 2001). Banana peel is usually discarded for animal feed or utilized for the extraction of pectin. Starch and pectin extracted from banana peel were evaluated for their potential application in food industry as gelling, thickening and stabilizing agents (Padam *et al.*, 2014). Research studies have highlighted the potential health benefits of the by-product and reported to contain high content of polyphenols and antioxidants (Mohapatra *et al.*, 2010). However, potential antioxidants present in banana peel include anthocyanin, carotenoids (α, β and xanthophylls), flavanal-3-ols, flavonols, catechin, epicatechin, epigallocatechin, procyanidin, tripenes, sterols, dopamine and L-dopa (Vu *et al.*, 2018). Biotechnological applications of banana peel include production of proteins, ethanol, α -amylase, hemi-celluloses and celluloses. Also, the anthocyanins present in the banana bracts could be used as a potential source of natural food colorant (Ove *et al.*, 2019). Futher, research studies have shown that bread supplemented with banana peel flour as a source of dietary fiber showed higher nutritional content and sensory acceptance when replaced with 10% banana peel flour (Eshak, 2016). Also, noodles developed from banana peel flour lower glycemic index and decrease the digestion time as a result of its high resistant starch content (Ramli *et al.*, 2009).

Valorization of Citrus By-Products

Citrus by-product is the waste obtained after the extraction of juice and consists of peels (60-65%) (albedo and flavedo), pulp (10-15%) (juice sac residue), rag (25-30%) (membranes and cores) and seeds (0-10%) (Mamma and Christakopoulos, 2014). This waste product accounts for about 60-70% of the original fruit weight. Global production of this solid residue is approximately 120 MTs annually while in India, the waste generated is about 7.8 million tonnes. Huge amount of the residue (usually dried) is either utilized for animal feeding or not as commonly, used as a source for pectin extraction (Lario *et al.*, 2004). However, the residual material is a good source of valuable compounds like dietary fiber, soluble sugars, amino acids, vitamins, oils and lipids, antioxidants, polyphenols, carotenoids and flavonoids (Zema *et al.*, 2018). The main flavonoids present are in particular flavanones and flavones like hesperidin, eriocitrin, naringin and narirutin (Schieber *et al.*, 2001). The use of citrus peel in food and pharmaceutical industry depicts its valuable composition. The peel is rich in essential oil- composed of D-limonene, α-terpinolene, α-pinene, citronella, β-citronellol and so on. In food industry, the essential oils are used as preservatives and flavoring agents and in cosmetic industry; the oil is used for the preparation of soaps, perfumes and other products (Chavan *et al.*, 2018). The growing concerns on the serious microbial spoilage and the interest in shifting from synthetic to natural antimicrobial agents suggest its utilization in the extraction of useful compounds (Shakir and Salih, 2015). Citrus residue can also be a source of natural food colorant (Abdullah and Razak, 2018). Further, research studies have reported that after the extraction of essential oils, the solid residue left is either used for the production of dietary fiber or cellulolytic fiber- a supporter of semi-solid fermentation (Rivas *et al.*, 2008).

Valorization of Mango By-Products

Mango, a popular tropical fruit is generally eaten raw or in the form of freshly squeezed juice or nectar. A major portion of the fruit is processed leaving behind tonnes of waste ranging from 40-60% which include peel (12-15%) and seed kernel (15-20%) (Ben-Othman *et al.*, 2020). Research studies had reported mango waste (peel and seed) to be a potential source of nutraceutical due to their high content of bioactive compounds such as carotenoids, phenolic acids, cathechins, tannins and dietary fiber, thus, making it a promising functional ingredient (O'Shea *et al.*, 2012). Due to their high nutrient and phytochemical composition, mango waste (peel and seed kernel) can be incorporated in a variety of food products, improving dramatically the nutritive value of the developed products without negating the quality. Mango peel and seed kernel powder was added to biscuits to improve the phenolic content. Results revealed that addition of peel and kernel powder enhance the flavor while increase the phenolic and antioxidant content (Ashoush and Gadallah, 2011). Further, mango by-products can also be used for the preparation of active packaging material in gelatin based films.

Vegetables: Production, Processing and Value Addition

India is a country which is blessed with diverse agro-climatic conditions, thus, making it possible to grow a wide variety of vegetables. Vegetables are rich sources of vitamins, minerals, dietary fibers, micronutrients, antioxidants and phytochemicals (Butnariu and Butu, 2014). Thus, the shifting of farmers from long-established cereal crop to vegetable farming has occurred as a result of high yield, short duration, economic viability, and employment generation both on farm as well as off-farm. India being the second largest producer of vegetables in the world produces 162 MTs from an area of 9.2 Mha having productivity of 17.3 T/ha (Rais and Sheoran, 2015). Production of vegetables is increasing faster than population resulting in market glut. Further, inadequate post-harvest management facility, non-availability of adequate low cost cold chain facilities and secondary vegetable processing has led to huge wastage accounting to the loss of Rs. 500 billion annually, thus, affecting the livelihood of farmers. Vegetables are generally less processed and usually eaten raw. Out of the total production, only 2-3% is processed leaving behind a considerable amount of waste. The waste generated from vegetable processing industry includes peel, seeds, pomace and skin. Research studies have found the waste to be a potential source of valuable compounds like vitamins, dietary fiber, oils, carotenoids, polyphenols etc. (Bandyopadhyay *et al.*, 2014). Valorization of the waste is the easiest way to utilize their nutritional potential. In India, value addition is limited to pickles, chutneys, preserves and candies at cottage level, which is highly decentralized and large units are located at small scale with unskilled labours. However, value addition in fruits and vegetable sector is expected to increase from 8 to 20% by 2025.

In the world's vegetable production market, India contributes a share of 14% annually (Yadav *et al.*, 2020). Around seventy different varieties of vegetables are cultivated in India but more focus has been given to vegetables like carrot, onion, tomato, potato, cauliflower and brinjal. Secondary processing of these vegetables generates enormous waste which has made India the leading producer of waste annually. About one-third of the waste in the form of seeds, skin, pomace and peels is lost worth crores of rupees at the secondary level. Keeping in view the nutrient composition, vegetable waste can be incorporated in a variety of food products so as to add value to the developed product. Numerous processing technologies have been used to produce bio-diesel, bio-gas and bio-fertilizer from the vegetable waste. The waste is also used as animal feed and as a source for color synthesis through solid-state fermentation.

Valorization of Carrot By-Products

Carrot is a highly palatable dietary fiber rich stable vegetable consumed widely. It is a vital source of phenolic compounds, vitamins minerals and carotenoids (Clementz *et al.*, 2019). Being seasonal in nature, carrots which meet the quality standards are generally processed in frozen, chopped or canned form. Remaining carrots are discarded as waste accounting about 25-30% (Kaur *et al.*, 2020). Further, the major product from carrot processing is carrot juice. The entire vegetable is usually pressed in a juice extracting machine to extract the juice from the vegetable, resulting in much waste. This waste is termed as carrot pomace. The juice yield is however reported to be only 60-70% and up to 30-40% is left-over pomace which consists of about 80% carotene (Schieber *et al.*, 2001). The waste is either used as cattle feed or end up in the municipality bins. Several attempts were made by the researchers at utilizing the carrot pomace as a source of dietary fiber for the development of various functional foods like baked goods (bread, cake and biscuits), extruded products (noodles, pasta), confectionary and functional beverages. Addition of hydrolyzed carrot pomace as a functional food ingredient in beverages was studied by Stoll *et al.*, 2003. Results concluded that carrot pomace can successfully be incorporated in the carrot juice. Cloud stability of the juice after bottling was found to be satisfactory. Furthermore, addition of pomace naturally increased the total phenolic content.

Valorization of Cauliflower By-Products

Cauliflower, the main Brassica vegetable is mostly included in soups and stews or eaten as such with the main meal. In industrial processing of cauliflower, only 30-40% is commercially utilized; remaining 60-70% is discarded as waste-constituting 50% leaves and 10-20% stem, thus having highest waste index (Femenia *et al.*, 1998). The by-products of cauliflower are rich in bioactive compounds and may be used as a potential functional ingredient to design novel functional foods (Stojceska *et al.*, 2008). These bioactive compounds include non-starch polysaccharides (Feminia *et al.*, 1998), flavonoids (Llorach *et al.*, 2003), glucosinolates (Cabello-Hurtado *et al.*, 2012) and hydroxyl-cinnamic acid (Llorach *et al.*, 2003). Utilization of cauliflower by-products helps in improving the nutritional content of a product. Stojceska *et al.*, 2008 explored the utilization of cauliflower trimmings in ready-to-eat extruded snacks and studied its effect on the textural and functional characteristics. The findings of the study concluded that cauliflower by-product can successfully be incorporated in extrudes as it increased the dietary fiber content, protein content and water absorption capacity. Sensory analysis revealed that ready-to-eat snacks containing up to 10% trimmings were acceptable. Cauliflower by-products can also be utilized as fat substitute in the production of beef sausages (Abul-Fadl, 2012).

Valorization of Onion By-Products

Onion is a highly valued vegetable used mostly for culinary purpose. But onions unlike other vegetables are many times under-estimated regarding their health benefits (O'Shea *et al.*, 2012). From the culinary point of view, onions are either prepared as pre-chopped and peeled, canned, pickled or in a powder or flake form (Nile *et al.*, 2018). Approximately 50 K Tonnes of onion by-product is generated annually (Roldan *et al.*, 2008). The wastes are in the form of skin, roots, fleshy scales and onions which does not meet the quality standards. Unfortunately, onion by-products can neither be utilized as animal feed due to its pungent odor, nor used as a fertilizer due to its phyto-pathogenic agents (Benitez *et al.*, 2011). However, the residues left after processing can effectively be valorized to obtain bioactive compounds of interest. Onion by-products have also been found to contain high amounts of antioxidants making it an efficient free-radical scavenging agent. The extracts from the onion by-products had been reported to contain ferulic acid, protocatechuic acid, anthocyanins, flavonols and organo-sulphur compounds (Ng *et al.*, 2000). Despite its nutraceutical potential, very less attention is placed on the onion waste unlike peel and pomace. Now-a-days attempts are being made by the researchers to find its application as potential functional food ingredient in the food, nutraceutical and pharmaceutical industries (Griffiths *et al.*, 2002). A research conducted by Roldan *et al.*, 2008 investigated onion by-product as a potential food ingredient. Researchers developed an onion by-product paste and found that pasteurization process helps in preserving the bioactive compounds, maintain a significant antioxidant activity and inhibit browning.

Valorization of Potato By-Products

Potato is a staple and the fourth largest produced edible food crop after rice, wheat and maize (Ben-Othman *et al.*, 2020). Being an important constituent in the human diet, potatoes are rich source of carbohydrates, vitamins, minerals, dietary fiber and phenolic compounds (Abu-Ghannam and Crowley, 2006). Potato is widely used in the food processing industries for the production of chips, French fries, cakes, croquettes, potato mash and other potato-based snacks (O'Shea *et al.*, 2012). Approximately 20-30% of potatoes are discarded annually as they fail to meet the quality standards. Further as a result of secondary processing, huge quantities of unavoidable waste are generated per year which generally is used for the production of bio-fuels or organic biofertilizers (Pathak *et al.*, 2017). Potato waste consists of mostly peel, a valuable source of nutrients which can be converted into various value added products. Aqueous extracts have shown that potato peel contain higher amounts of phenolic acids, especially chlorogenic, gallic, protocatechuic and caffeic acid than potato flesh (Fritsch *et al.*, 2017). Therefore, researchers are finding new ways of using this by-product. Addition of potato peel as a starch and fiber source could improve both the functional characteristics as well nutritional properties of the developed product. Sabeena Farvin *et al.*, 2012 used potato peel as a source of natural antioxidants to prevent lipid and protein oxidation in minced mackerel. Results of the findings concluded that potato peel can successfully be employed to retard oxidation process. In another study, Kaack *et al.*, 2006 explored the utilization of potato peel in the preparation of wheat bread. Both the hardness and gumminess of the resulting bread were decreased due to solubilized potato peel fibre. The results also showed an increase in the dietary fiber content from 10.8% in the control to 17.5% in the potato peel enriched bread. Sensory analysis also revealed that bread with 12% potato peel fiber was acceptable in terms of color, texture and taste.

Valorization of Tomato By-Products

Globally, tomato is the second most popular vegetable crop produced after potato (Bugatti *et al.*, 2019). Nutritionally, tomatoes contain significant valuable bioactive compounds like carotenoids, especially lycopene, vitamins, mostly vitamin C and E and other phenols (Mirabella *et al.*, 2013). Each year several million tones of tomatoes are produced, consumed and processed. The products that are prepared from the tomatoes include sauces, juices, pastes, ketchups, etc. generating large amounts of wastes. The waste obtained from the tomato processing industry is generally divided into two categories 1) tomato peel, by-product obtained from the canning industry and represents up to 3-5% of the whole tomato and 2) tomato pomace, mixture of seeds, skin and pulp and accounts for 50% on dry weight basis (Al Khraisat, 2015). Annually around 5.4-9.0 ^{x} 106 tonnes of tomato pomace is produced worldwide (Lu *et al.*, 2019). Such waste being a rich source of high organic content molecules is extremely important for the food industry. Tomato waste, if not properly managed is prone to spoilage due to its high moisture content and nutrient richness which not only increases the disposal burden on one hand but also results in the considerable loss of potentially valuable nutrients on the other hand (Bugatti *et al.*, 2019). Such valuable and cheap waste from the tomato processing industry can be utilized for the production of usable resources such as dietary fiber, lycopene, edible films and tomato seed oil (Lu *et al.*, 2019). Phytochemical screening of tomato pomace showed that the by-product contain significant polyphenolic content having a potential antioxidant activity (Ilahy *et al.*, 2011). Therefore, such value added bioactive compounds can be used as potential antioxidants for the formulation of functional foods like extruded snacks, baked goods, fermented sausages and hamburgers or serve as additives in food system for shelf-life enhancement (O'Shea *et al.*, 2012). Due to its functional properties, tomato pomace can be added to tomato ketchups as thickener replacing hydrocolloids, thereby, improving the color and texture of the product (Farahnaky *et al.*, 2008).

Cereals, Pulses and Oilseeds: Production, Processing and Value Addition

Cereals

Cereal grains are the primary human food resources commonly cultivated worldwide. They are regarded as the vital source of essential nutrients such as macronutrients (carbohydrates, proteins) as well as micronutrients (vitamins, minerals and fiber) (Papageorgiou and Skendi, 2018). In developing countries like India, approximately 60% of energy is derived from the cereals alone while in case of the developed countries; it is only 30 per cent. Cereal crops like wheat, maize and rice constitute a lion share of about 87-90% in total Indian food basket (Singh *et al.*, 2017). World cereal production in 2016 reached ~2600 MTs; whereas coarse cereal (barley, oat, millet and sorghum) production reached 1330.02 million tonnes. Global cereal production is however expected to reach 3053 MTs in 2028. India is the second largest producer of cereals in the world. In India, more than 272 million metric tonnes of cereals were estimated to be produced at the end of the year 2020. The huge demand for cereals in the global market is creating an excellent environment for the export of Indian cereal products, thus, making India also the largest exporter of cereal products worldwide (Khatkar *et al.*, 2016). India exports cereals worth Rs. 47,287.12 crore during the year 2019-20.

Cereal grains for human consumption are generally milled to eliminate bran and germ so as to meet the consumer quality standards, leaving behind grains scant of key nutrients (El Mekawy *et al.*, 2013). The scientific approach towards the novel processing technologies like extrusion, popping, milling, steam flaking and pelleting for value addition of major cereals could pave a way for further progress (Yadav *et al.*, 2020). Apart from this, the economic value of the cereal grains can also be enhanced by the development of RTS-snacks, parboiled rice, extruded and puffed products (Yadav *et al.*, 2020). Cereal processing, an important part of the food production chain leave a large quantity of by-products throughout the entire food processing (Papageorgiou and Skendi, 2018). These by-products are mainly composed of germ, pericarp, husk, spent grains, parts of endosperm, aleurone, testa, and bran obtained from dry and wet milling, malting and pearling and contain significant valuable compounds such as vitamins, minerals, dietary fiber and phenolics, beneficial for human health (Saini *et al.*, 2019). Currently these by-products either end up as waste or used as animal feed or fertilizers. Researchers now-a-days are finding ways to provide novel, sustainable and innovative solutions for the utilization of cereal by-products and grains which does not meet the consumer sensory expectations. In that respect, these promising materials can be valorized by applying various biotechnological techniques, thus leading to numerous economic and environmental advantages as well as important opportunities towards new product development (NPD) in the food, chemical, packaging, nutraceutical (dietary supplements and food additives), cosmetic and pharmaceutical industries (Papageorgiou and Skendi, 2018).

Valorization of Wheat and Wheat By-Products

Wheat is a largely cultivated and consumed crop globally. Out of the total wheat grain production in the world (731 MTs), Indian accounts for about 99.70 MTs annually. Thus, comprises of about 12% of the world's production of cereals. In India, the major wheat producing states include Uttar Pradesh (27.52 MTs), Punjab (15.17 MTs) and Haryana (10.5 MTs) (Singh *et al.*, 2017). Though development in the production technologies had resulted in increased wheat grain production, but still a post-harvest loss of up to 16 million metric tons persists (Yadav *et al.*, 2020). In terms of quantity produced, wheat market in India comprises of wheat flour, maida and sooji, of which wheat flour is mostly consumed due to its taste, hygiene and nutritive value (Gandhi and Koshy, 2006). Further, in baking industry demand for wheat flour is 30%, maida 55% and multi-grain flour 5 per cent. Processing of wheat generate by-products such as bran and germ, among which only 10% of wheat bran is exploited for valorization due to its richness in fiber, micronutrients and bioactive compounds (Huang *et al.*, 2014). Researchers have recently analyzed the composition of wheat bran as potential functional ingredient (Papageorgiou and Skendi, 2018). Addition of wheat bran in food products could dramatically improve the nutritional composition and perhaps the health of the consumer. Wheat bran has been used by researchers in a variety of products such as cereal based baked goods, fried and other ready-to-eat foods (Hossain *et al.*, 2013). Another by-product obtained from the wheat processing is germ which is a significant oil source. Wheat germ oil is reported to contain phytosterols and octacosanol, when incorporated in cookies reduced low-density lipoproteins in rats fed with cookies (Arshad *et al.*, 2007). However, the literature available at present on wheat germ oil and its effects on products is lacking but what is available suggests that its addition to food products would be beneficial due to its high content of antioxidants.

Valorization of Rice and Rice By-Products

Rice, a staple food crop is the king of Indian food basket (Singh *et al.*, 2015). The livelihood of around 70% of the South Asian population is dependent on rice (Yadav *et al.*, 2020). With the adaption of advanced production technologies, an impressive growth in rice production as well as productivity was witnessed in India post-independence. The annual production of rice in India is approximately 110 MTs accounting for about 22.3% of global production. However the estimated increase in the rice production by 2025 is 130 million tonnes (Singh *et al.*, 2017). Rice grains are separated into their constituent parts such as husk (20%), bran (8%), germ (2%) and white rice (70%) by the process of milling. Post-harvest losses in rice occurs throughout the entire food chain, but the loss is higher at farm level (75%) compared to storage (3.5%). Rice straw-a milled by-product is either used as fuel or bio-fertilizer. Rice straw is a good source of linocellulosic materials. However, the main by-product of rice is rice polishing-a significant source of thiamine and crude fat. Broken rice, another by-product of rice is either used for the preparation of ready-to-eat products like gluten-free rice bread, idli and dosa or processed into fortified analogues which not only result in crop exploitation but also raise economics of farmers (Yadav *et al.*, 2020). Rice bran is another by-product of rice milling with high oil content. It is also rich in fiber, b-complex vitamins, proteins, amino-acids and phytochemicals. Thus, rice bran can be explored as potential functional food ingredient in cooked or baked food products (Sharif *et al.*, 2013). Also, rice bran can be used as a coating material to provide permeability barrier in case of fruits, vegetables and candies (Zou and Yang, 2019). It is better to stabilize rice bran before incorporating in a food product to avoid lipid oxidation and deterioration of quality and organoleptic characteristics (Skendi *et al.*, 2020). Recently, it has been used in micro/nano encapsulate matrices for probiotics showing high encapsulation efficiency and enhanced protection to simulated gastrointestinal tract and storage conditions (Raddatz *et al.*, 2020; Rocha, 2016). In addition, it was successfully added in yoghurts in order to increase its antioxidant activity, causing a decrease in syneresis and viscosity values (Demirci *et al.*, 2017). Rice husk is composed of cellulose (34.4%), hemicellulose (24.3%), lignin (19.2%), ash (18.85%) and other trace elements (3.25%). The ash of rice husk can be used as refining agent in case of vegetable oils as it acts as a good absorbent (Zou and Yang, 2019). Rice husk and straw can also be used as a source for the production of bio-ethanol. Further, rice straw residues are used to produce bio-char-capable of reducing CO2 emissions (Yadav *et al.*, 2020).

Valorization of Maize and Maize By-Products

After rice and wheat, maize-the marvels performer is the third major crop produced in India (Singh *et al.*, 2017). Currently, maize is cultivated in about 10.2 mha in India, accounting for ~9% of total food grain production (Khatkar *et al.*, 2016). Maize is important to India as 15 million Indian farmers are engaged in maize cultivation. It qualifies as a potential crop for doubling farmer's income. The consumption of maize has increased at a CAGR of 11% in last five years. Today, maize is a source of more than 3500 products including specialized maize like quality protein maize (QPM). Despite the production strength, around 5-7% of the maize is lost as a result of improper shelling, drying and storage. Processing of maize includes both wet and dry milling. The dry milling is done to obtain endosperm-used as meals, flour or grits and by-products such as fiber and germ-used for the production of oil where as wet milling results in the formation of desired products such as starch and germ and by-products including fiber, protein, and gluten meal (Yadav *et al.*, 2020). Further the maize cake obtained after the extraction of oil is a rich source of dietary fiber and thus can be used for the fortification of cakes. Maize is a multi-faceted crop used for the industrial production of sweeteners, beverages, corn oil and alcohol. Maize is mostly cultivated for the production of starch but now-a-days researchers have found its potential use in the production of edible films for food packaging

(Sucheta *et al.*, 2019). As renewable resource, maize is an excellent source for the production of bioethanol and other biofuels such as biodiesel and biogas. A part from this, maize processing waste water contains arabino-xylans-a water soluble material which possesses the gelling ability, thus, could find potential market in the food industry over the use of other gums.

Valorization of Pulses and Pulse By-Products

Pulses are the staple cereal grains in the diet of vegetarians in India as they are the potential source of proteins, amino-acids, vitamins and minerals. They contain 20-22% proteins, which is twice the protein in wheat and thrice that in rice. India is the largest producer, importer and consumer of pulses in the world contributing 25% of the global production (Yadav *et al.*, 2020). A wide variety of pulses are grown worldwide. However, the major pulses grown in India are bengal gram, chickpea, lentil, cowpea, faba beans, moong bean, urd bean and pigeon pea; out of these, chickpea and pigeon pea share maximum per cent production. In India, the pulses are mainly grown on semi-arid areas, thus fluctuating rainfall and insect infestation during storage results in higher losses (Ahlawat *et al.*, 2016). Further, milling losses occur as a result of unskilled labour. Keeping in view the key issues and need of farmers, initiatives taken by some of the institutes such as CSIR-Central Food Technological Research Institute (CFTRI), Mysore; Central Institute of Agricultural Engineering (CIAE), Bhopal; Central Institute of Post Harvest Engineering and Technology (CIPHET), Ludhiana; and Indian Agricultural Research Institute (IARI, New Delhi include development of small scale milling equipments (Tiwari and Shivhare, 2016). Compared to traditional milling, development in processing technologies has increased the availability of dal to 70-75 per cent. Milling of pulses resulted in a number of by-products such as brokens, germ meal and husk. Husk is used as dry fodder for livestock, as lingo-cellulosic material for production of biofuel and as fertilizer. Brokens are used for the preparation of ready-to-eat foods like snacks and instant food products. By-products from pulses have also been used for novel product development like bean curd, gluten-free bread and fermented sausages of lupin (Singhal *et al.*, 2016).

Valorization of Oilseeds

Annual oilseed production in India is 30.06 MTs from an area of 26.67 mha, accounting for 10% of the total oilseed production worldwide. Out of it, 6-7% is limited to vegetable oil and 7% protein meal. Major edible oilseeds produced in India are soybean, mustard, groundnut, rapeseed, sunflower, sesame, safflower and niger (Yadav *et al.*, 2020). Of the total oilseed production in India, groundnut, soybean and mustard share a maximum percentage of 88 per cent. Most of the oilseed meal is used as feed for livestock or as a source for biodiesel production. Oilseeds can also be used as lubricants in industries or as hydraulic oil due to their high stability which is mostly achieved as a result of genetic modifications or trans-esterification. Mustard seeds can be used as weedicide while as sunflower meal can be used as fertilizer due to its high nitrogenous protein (Abiodun, 2017).

Medicinal and Aromatic Plants: Production, Processing and Value Addition

Medicinal and aromatic plants are the herbaceous plants that are mostly used in cosmetics, pharmaceuticals and cooking. More recently, these plants have found wide application in food processing as a result of their antioxidant potential. Indian herbal industry is progressively moving towards global competitive market with a market size of approximately US$ 72 billion (Yadav *et al.*, 2020). However, global trade of medicinal and aromatic plants is shown in Table 1. India being the second leading producer of pharmaceutical plants holds only 4% compared to China, which holds 36.5% of this market. Further, India contributes only 0.5% towards the global trade of medicinal and aromatic plants. Adoption of novel production technologies has resulted in increased production of medicinal and herbal plants. Global production of medicinal and aromatic plants has increased to 10,22,500 tonnes in the year 2015-2016 as compared to 2,02,000 tonnes in the year 2005-2006 (Charan, 2013). In India, Madhya Pradesh is the largest producer of herbal plants accounting 44% of production followed by Rajasthan with 19 per cent. Apart from production, Rajasthan stands first with 56% area under cultivation as compared to 25% in Uttar Pradesh. Further, Karnataka is also identified as potential cultivar of these plants with highest production of Ashwagandha contributing 61.65 per cent (Mohapatra *et al.*, 2018). In case of aromatic plants, lemon grass and mint are the major herbs produced in India contributing 95% share in total area and production. Globally, the demand for essential oils has also increased from 174 tonnes in 2013 to 245 tonnes by the year 2020 (Govindsamy *et al.*, 2013). Traditionally, around 90% of the herbal plants were used in the manufacture of Ayurvedic, Siddha, Unani and homeopathy system of medicine. Now-a-days, these herbaceous plants have found their importance in every field. Presently, there are around 11 CSIR institutes experimenting on the medicinal and aromatic plants for their commercial use in the preparation of functional foods, herbal cosmetics, incense sticks, nutraceutical, etc.

Table 1. Worldwide trade of medicinal and aromatic plant and their value added products

Type	Particulars	Trade
Aromatic plants (250 types of essential oils)	Essential oils	US$ 1.2 billion
	Flavours and fragrances	US$ 20 billion
	Cosmetics	US$ 150 billion

Type	Particulars	Trade
Medicinal plants (400 species (75-100 in large quantities)	Plant based drugs	US$ 170 billion 2050: US$ 005 trillion
	Herbal teas, drug adjuncts	US$ 10 billion
	Nutraceuticals	US$ 80 billion
	Functional foods	US$ 57 billion

Valorization of Medicinal Plants

Medicinal plants possess pharmacological properties such as anticancer, anti-fungal, anti-microbial, antioxidant etc. and thus are used for the treatment of diseases like cancer, Hepatitis, AIDS, Alzheimer's disease (Jivad and Rabiei, 2014). Since time immemorial, various pharmaceutical molecules were produced from the medicinal plants on industrial scale. These include artimisinin, deseridine, ectoposide, ginkgolides, lectinam, paclitaxel, reseinnamine, reserpine, teniposide, vinblastin and vincristine (Dhar *et al.*, 2016). Serpentine, a pharmaceutical used for lowering blood pressure is isolated from Rauwolfia serpentine root. Nathapodytes nimmoniana is enriched with camptothecin and hence, used for treating cervical cancer particularly by Japanese. Medicinal plants are also therapeutic in nature, thus, used in the preparation of some herbal beverages including herbal juices, sharbats, herbal beer and herbal drinks (nannari, sugandhipala, aloe, noni) (Rao *et al.*, 2009). Herbal chocolates, sweets and candies are also being marketed internationally. Medicinal plants are also used for the preparation of functional foods like diabetic flour.

As herbal sweeteners

Changing lifestyle and health consciousness has resulted in the usage of low calorie sweeteners. Compared to sucrose and other sweeteners, Stevia, an herbal sweetener has high nutritional value and can thus be used as substitute of sucrose. Another natural herb Paraguay-non-calorie sweetener is 100-300 times sweeter than cane sugar (Koubaa *et al.*, 2015). Steviol glycoside-a compound responsible for its sweetness is considered GRAS for diabetic and obese people. Paraguay can be used in the food products such as fruit juices, ice-creams, soft drinks, baked goods, candies, chewing gum and as table top sweetener for tea and coffee. It is a used as a therapeutic agent in products like toothpastes, mouth washes, food supplements and diabetic foods (Yadav *et al.*, 2020).

Valorization of Aromatic Plants

With time numerous technologies have been developed for the isolation of aroma molecules from aromatic plants. These techniques include steam distillation, fractional distillation, super critical fluid extraction, expression, solvent extraction, etc. Value addition of aromatic plants results in the formation of essential oils, aroma chemicals, gums, perfumed water, oleoresins and hydrosol (Rao *et al.*, 2009). Some of the value added products from aromatic plants are depicted in Table 2. Essential oils are widely used in fragrance industry, food and beverages, cosmetics, homecare and healthcare (Shaaban *et al.*, 2012). Herbs and essential oils can also be used as flavoring and coloring agents in exotic dishes and local cuisine. Oils of thyme, lemongrass and rosemary not only add flavor but also improve the organoleptic characteristics of a food product. Lemon oil is used in air purifiers, laundry detergents, soaps and floor cleaners because of its anti-bacterial and deodorizing properties. Citronella oil is used as natural insect repellent and pesticide, thus could be used in many medicated formulations.

Table 2. Value added products from aromatic plants

Name	Value added products
Ambrette (*Abelmoschus moschatus*)	Liqueurs, vermouth and bitters, non-alcoholic beverages, ice cream, candy, baked products
Basil (Tulsi) (*Ocimum basilicum/ sanctum*)	Liqueurs, sauces, condiments, non-alcoholic, beverages, candy, vinegar, canned meats, ice cream, gelatin and puddings, baked products
Davana (*Artemisia pallens*)	Cakes, pastries, beverages, baked products
Hops (*Humulus lupulus*)	Beer, flavouring bitters, tobacco, tea
Lemongrass (*Cymbopogon flexuosus*)	Herbal tea, flavoured chicken, confectionery, products
Marjoram (*Origanum marjorana/ Marjorana hortensis*)	Salad dressings, vermouths, bitters, icecream, sauces, canned meats, non-alcoholic beverages, baked goods, candy, condiments
Mint (*Mentha arvensis/ piperita*)	Candy, chewing gum, beverages, paan, paan masala, bitter chocolatesicecream, meats, baked goods, gelatin and puddings, icings, toppings, non-alcoholic beverages, jellies
Patchouli (*Pogostemon patchouli*)	Alcoholic beverages, candy, non-alcoholic, beverages, frozen dairy desserts, baked goods, gelatin, meat and meat products
Rose (*Rosa species)*	Gulkand, sweets, food flavouring Rosemary (*Rosmarinus officinalis*): Salad dressings, herbal tea, sauces, meat, condiments, other food products
Sweet flag (*Acorus calamus*) Thyme (*Thymus vulgaris*) Vanilla (*Vanilla planifolia*)	Liquors, beer, gin, vinegar, confectionery, salad dressings, herbal tea, liquors, Icecream, baked goods, non-alcoholic beverages, candy, icings, syrups, toppings, liqueurs, confectionery

Conclusion

India being an agrarian economy has now transformed from agriculture-based economy to products and service based economy. Primary agriculture is although important but only a limited fraction can be extracted from the agricultural produce. Secondary agriculture can add 2-3 fold value to the primary agriculture and thus can allow farmers to get better returns from their harvest, improve rural and urban economy as well as create millions of new employment opportunities. However, the value addition of primary agriculture is extremely low, processing only 2% of the horticultural produce leaving major portion as waste. Adoption of novel technologies can improve the value of agricultural produce both at the production and processing levels. The waste or by-product generated from food processing can be explored through value addition for more valuable products like biofuel, biodegradable commodities like plates, industrial chemicals, pigments, coloring and flavoring agents, natural additives and extraction of bioactive compounds. The waste or by-product utilization could be another source for increasing farmer's income, reduce environmental pollution and strengthen primary economy. Further, research methodology and processing technology used for value addition to secondary agriculture should be economical, ecological and sustainable. Both research institutions (ICAR, CSIR and DBT) and agro-industries should work in collaboration for the upliftment of this sector.

References

Abdullah, S.R. and Razaq, R.K. 2018. Extracting carotenoids pigments from citrus peel and studying their functional properties. Journal of Tikrit University of Agricultural Sciences, 18:146-163.

Abiodun, O.A. 2017. The role of oilseed crops in human diet and industrial use. In: Ahmad, P. (eds.) Oilseed crops: Yield and adaptations under environmental stress. Wiley, Hoboken, pp. 249-263.

Abu-Ghannam, N. and Crowley, H. 2006. The effect of low temperature blanching on the texture of whole processed new potatoes. Journal of Food Engineering, 74 (3): 335-344.

Abul-Fadl, M.M. 2012. Nutritional and chemical evaluation of white cauliflower by-products flour and the effect of its addition on beef sausage quality. Journal of Applied Sciences Research, 8 (2): 693-704.

Ahlawat, I.P.S., Sharma, P. and Singh, U. 2016. Production, demand and import of pulses in India. Indian Journal of Agronomy, 61 (4): 33-41.

Al Khraisat, A.A.A. 2015. Evaluation of metabolizable nutrient values of Tomato Pomace (TP) in Nile tilapia (Oreo chromisniloticus). International Journal of Fisheries and Aquatic Sciences, 2: 164-167.

Alongi, M., Melchior, S. and Anese, M. 2019. Reducing the glycemic index of short dough biscuits by using apple pomace as a functional ingredient. Leben-Wissens Technology, 100: 300-305.

Arshad, M.U., Anjum, F.M. and Zahoor, T. 2007. Nutritional assessment of cookies supplemented with defatted wheat germ. Food Chemistry, 102 (1): 123-128.

Ashoush, I.S. and Gadallah, M.G. 2011. Utilization of mango peels and seed kernels powders as sources of phytochemicals in biscuit. World Journal of Dairy and Food Sciences, 6 (1): 35-42.

Bandyopadhyay, K., Chakraborty, C. and Bhattacharyya, S. 2014. Fortification of mango peel and kernel powder in cookies formulation. Journal of Academics and Industrial Research, 2: 661-664.

Benítez, V., Mollá, E., Martín-Cabrejas, M., Aquilera, A., López-Andréu, F. and Cools, K. 2011. Characterization of industrial onion wastes (Allium cepa L.): Dietary fibre and bioactive compounds. Plant Foods for Human Nutrition (Formerly Qualitas Plantarum), 66 (1): 48-57.

Ben-Othman, S., Jõudu, I. and Bhat, V. 2020. Bioactives from agri food wastes: Present Insight and Future Challenges. Molecules, 25: 510; doi:10.3390/molecules25030510.

Bugatti, V., Brachi, P., Viscusi, G. and Gorrasi, G. 2019. Valorization of tomato processing residues through the production of active bio-composites for packaging applications. Frontiers in Materials, 36 (6):1-10.

Butnariu, M. and Butu, A. 2014. Chemical composition of vegetables and their products. Handbook of Food Chemistry, 1-49.

Cabello-Hurtado, F., Gicquel, M., and Esnault, M.A. 2012. Evaluation of the antioxidant potential of cauliflower (Brassica oleracea) from a glucosinolate content perspective. Food Chemistry, 132 (2): 1003-1009.

Chand, R., Saxena, R. and Rana, S. 2015. Estimates and Analysis of Farm Income in India, 1983-84 to 2011-12, Economic and Political Weekly, 1 (32).

Charan, C. 2013. Production of herbal and medicinal plant: aninnovative effort towards sustainable development (a case study of Bihar). Global Journal of Management Business Studies, 3 (2): 145-152.

Chavan, P., Singh, A.K. and Kaur, G. 2018. Recent progress in the utilization of industrial waste and by-product of citrus fruit-A review. Journal of Food Process Engineering, 1-10.doi: 10.1111/jfpe.12895.

Chengappa, P.G. 2013. Secondary Agriculture: A Driver for Growth of Primary Agriculture in India. Indian Journal of Agricultural Economics, Vol.68, No.1, Jan.-March 2013.

Cheuh, C.C., Lin, L.J., Lin, W.C., Huang, S.H., Jan, M.S., Chang, S.C., Chung, W.S. and Lee, T.T. 2019. Antioxidant capacity of banana peel and its modulation of Nrf2-ARE associated gene expression in broiler chicken. Indian Journal of Animal Sciences, 18 (1): 1394-1403.

Clementza, A.,Torresia, P.A., Mollib, J.S., Cardell, D., Mammarella, E. and Yoria. J.C. 2019. Novel method for valorization of by-products from carrot discards. LWT- Food Science and Technology, 100: 374-380.

Coman, V., Teleky, B., Mitrea, L. Martău, G.A., Szabo, K., Călinoiu, L. and Vodnar, D.C. 2019. Bioactive potential of fruit and vegetable waste. Advances in Food and Nutrition Research, https://doi.org/10.1016/bs.afnr.2019.07.001

Demirci, T., Akta, S.K., Sözeri, D. Öztürk, H.I. and Akın, N. 2017. Rice bran improves probiotic viability in yoghurt andprovide added antioxidative benefits. Journal of Functional Foods, 36: 396-403.

Dhar, R.A., Shahnawaz, M. and Qazi, P.H. 2016. General overview of medicinal plants: a review. Journal of Phytopharmacology, 6 (6): 349-355.

El Mekawy, A., Diels, L., De Wever, H. and Pant, D. 2013. Valorization of cereal based bio-refinery by-products:Reality and Expectations. Environmental Science and Technology, dx.doi.org/10.1021/es402395g.

Eshak, N.S. 2016. Sensory evaluation and nutritional value of balady flat bread supplemented with banana peels as a natural source of dietary fiber. Annals of Agricultural Science, 61 (2): 229-235.

Farahnaky, A., Abbasi, A., Jamalian, J., and Mesbahi, G. 2008. The use of tomato pulp powder as a thickening agent in the formulation of tomato ketchup. Journal of Texture Studies, 39 (2): 169-182.

Femenia, A., Robertson, J.A., Waldron, K.W. and Selvendran, R.R. 1998. Cauliflower (Brassica oleracea L), globe artichoke (Cynara scolymus) and chicory witloo (Cichorium intybus) processing by-products as sources of dietary fibre. Journal of the Science of Food and Agriculture, 77 (4): 511-518.

Fritsch, C., Staebler, A.,Happel, A., Márquez, M.A.C., Aguiló-Aguayo, I., Abadias, M., Gallur, M., Cigognini, I.M.,Montanari, A., López, M.J., Suárez-Estrella, F.,Brunton, N., Luengo, E., Sisti, L., Ferri, M. and Belotti, G. 2017. Processing, valorization and application of bio-waste derived compounds from potato, tomato, olive and cereals: A Review. Sustainability, 9: 1492.

Gandhi, V.P. and Koshy, A. 2006. Wheat marketing and its efficiency inIndia. Paper presented at Conference of the International Association of Agricultural Economist (IAAE), Gold Coast, Australia, 12-18 August, 2006.

Gazala, K., Masoodi, F.A., Masarat, H.D., Rayees, B. and Shoib, M.W. 2017. Extraction and characterization of pectin from two apple juice concentrate processing plants. International Food Research Journal, 24 (2): 594-598.

Govindsamy, R., Arumugam, S. and Simon, J.E. 2013. An assessment of the essential oil and aromatic plant industry with a focus on Africa. In: Juliani, H.R. and Simon, J.E. (eds) African natural plant products volII: discoveries and challenges in chemistry, health and nutrition, ACS symposium series. ACS Publications, pp 289-321 https://doi.org/10.1021/bk-2013-1127.ch018.

Griffiths, G., Trueman, L., Crowther, T., Thomas, B. and Smith, B. 2002. Onions-A global benefit to health. Phytotherapy Research, 16 (7): 603-615.

Henriquez, C., Speisky, H., Chiffelle, I., Valenzuela, T., Araya, M. and Simpson, R., 2010. Development of an ingredient containing apple peel, as a source of polyphenols and dietary fiber. Journal of Food Science, 75 (6): 172-181.

Hossain, K., Ulven, C., Glover, K., Ghavami, F., Simsek, S., Alamri, M.S., Kumar, A. and Mergoum, M. 2013. Inter-dependence of cultivar and environment on fibre composition in wheat bran. Australian Journal of Crop Science, 7 (4): 525-531.

Huang, Q., Shi, C.X., Su, Y.B., Liu, Z.Y., Li, D.F., Liu, L., Huang, C.F., Piao, X.S. and Lai, C.H. 2014. Prediction of the digestible and metabolizable energy content of wheat millingby-products for growing pigs from chemical composition. Animal Feed Science and Technology, 196: 107-116.

Hussein, A.M.S., Kamil, M.M., Hegazy, N.A., Mahmoud, K.F. and Ibrahim, M.A. 2015. Utilization of some fruits and vegetables by-products to produce high dietary fiber jam. Food Science and Quality Management, 37: 39-45.

Ilahy, R., Hdider, C., Lenucci, M.S., Tlili, I. and Dalessandro, G. 2011. Phytochemical composition and antioxidant activity of high-lycopene tomato (Solanumlycopersicum L.) cultivars grown in Southern Italy. Scientia Horticulturae, 127 (3): 255-261.

Jivad, N. and Rabiei, Z. 2014. A review study on medicinal plants used in the treatment of learning and memory impairments. Asian Pacific Journal of Tropical Biomedicines, 4: 780-789.

Kaack, K., Pedersen, L., Laerke, H.N. and Meyer, A. 2006. New potato fibre for improvement of texture and colour of wheat bread. European Food Research and Technology, 224 (2): 199-207.

Kashish, A. and Dhawan, V. 2016. A study on production and trade performance of fruits in India. Agricultural Research Journal, 54 (1): 108.

Kaur, G.J., Kumar, D., Orsat, V. and Singh, A. 2020. Assessment of carrot rejects and wastes for food product development and as a biofuel. Biomass Conversion and Biorefinery, https://doi.org/10.1007/s13399-020-01096-z.

Khatkar, B.S., Chaudhary, N. and Dangi, P. 2016. Production and consumption of grains: India. Encyclopedia of Food Grains, 1, pp. 367-373.

Koubaa, M., Rosello-Soto, E. and Zlabur, J. 2015. Current and new in sights in the sustainable and green recovery of nutritionally valuable compounds from Stevia rebaudiana Bertoni. Journal of Agricultural Food Chemistry, 63: 6835-6846.

Kumar, D. and Kalita, P. 2017. Reducing post-harvest losses during storage of grain crops to strengthen food security in developing countries. Foods, 6 (8) doi:10.3390/foods6010008

Lario, Y., Sendra, E., Garcia-Perez, J., Fuentes, C., Sayas-Barbera, E., Fernandez-Lopez, J. 2004. Preparation of high dietary fiber powder from lemon juice byproducts. Innovative Food Science and Emerging Technologies, 5: 113-117.

Llorach, R., Espín, J. C., Tomás-Barberán, F. A., &Ferreres, F. 2003. Valorization of cauliflower(Brassica oleracea L. var. botrytis) by-products as a source of antioxidantphenolics. Journal of Agricultural and Food Chemistry, 51(8): 2181-2187.

Lua, Z., Wanga, J., Gaoa, R., Yea, F. and Zhao, G. 2019. Sustainable valorisation of tomato pomace: A comprehensive review. Trends in Food Science and Technology, 86: 172-187.

Lyu, F., Luiz, S.F., Azeredo, D.R.P. and Cruz, A.D. 2020. Apple pomace as a functional and healthy ingredient in food products: A Review. Processes, 8:319; doi:10.3390/pr8030319

Mamma D, Christakopoulos P. 2014. Biotransformation of citrus byproducts into value added products. Waste Biomass Valori, 5: 529-549.

Masoodi, F.A. and Chauhan, G.S. 1998. Use of apple pomace as a source of dietary fiber in wheat bread. Journal of Food Processing and Preservation, 22: 255-263.

Mirabella, N., Castellani, V. and Sala, S. 2013.Current options for the valorization of food manufacturing waste: a review, Journal of Cleaner Production, doi: 10.1016/ j.jclepro.2013.10.051. S0959-6526(13) 00744-0

MOFPI, 2019. National food processing policy, 2019.

Mohapatra, D., Mishra. S. and Sutar, N. 2010. Banana and its by-product utilisation: an overview. Journal of Scientific and Industrial Research, 69: 323-329.

Mohapatra, U., Rudrapur, S., Hiremath, D.B. and Mohapatra, S. 2018. Medicinal and aromatic plants sector in Karnataka: an economic perspective and SWOT analysis. Journal of Pharmacognosy and Phytochemistry, (Special Issue) 3: 232-235.

Murhty, T.M.S. and Yogesh, M.S. 2014. An overview of Food Processing Industry in India-Challenges and Opportunities. Online International Interdisciplinary Research Journal, 4 (5).

NCAER, 2017. The 2017 Rabi season and doubling farmers income. https://www.ncear.org/event.

Ng, A., Parker, M.L., Parr, A.J., Saunders, P.K., Smith, A.C., and Waldron, K.W. 2000. Physicochemical characteristics of onion (Allium cepa L.) tissues. Journal of Agricultural and Food Chemistry, 48 (11): 5612-5617.

Nile, A., Nilea, S.H., Kimb, D.H., Keumc, Y.S., Seokb, P.G. and Sharma, K. 2018. Valorization of onion solid waste and their flavonols for assessment of cyto-toxicity, enzyme inhibitory and antioxidant activities. Food and Chemical Toxicology, https://doi.org/10.1016/j.fct.2018.02.056.

O'Shea, N., Ek, Arendt and Gallagher, E. 2012. Dietary fibre and phyto-chemical characteristics of fruit and vegetable by-products and their recent applications as novel ingredients in foodproducts. Innovative Food Science and Emerging Technology, 16: 1-10.

Ove, T.A., Kamal, M.M., Nasim, S.M.N.I., Momin, M.M.I. and Mondal, S.C. 2019. Extraction and quantification of anthocyanin from banana bracts using different pH and solvent concentration. International Journal of Food and Nutrition Sciences, 4 (2): 60-64.

Padam, B.S., Tin, H.S., Chye, F.Y. and Abdullah, M.I. 2014. Banana by products: an underutilized renewable food mass with great potential. Journal of Food Science and Technology, 51 (12): 3527-3545.

Papageorgiou, M. and Skendi, A. 2018. Introduction to cereal processing and by-products. Sustainable recovery and reutilization of cereal processing by-products, doi.org/10.1016/B978-0-08-102162-0.00001-0.

Pathak, P.D., Puranik, N.M., Jambhulkar, S.J., Mandavgane, S.A. and Kulkarni, B.D. 2017. Valorization of potato peel: a biorefinery approach. Critical Reviews in Biotechnology, 1-44.

Perussello, C.A., Zhang, Z., Marzocchella, A, and Tiwari, B.K. 2017. Valorization of Apple Pomace by Extraction of Valuable Compounds. Comprehensive Reviews in Food Science and Food Safety, 16: 776-796.

Raddatz, G.C., Da Fonseca, D.S.B., Poletto, G., Jacob-Lopes, E., Cichoski, A.J., Muller, E.I. Flores, E.M.M., de Bona da-Silva, C. and Ragagnin de Menezes, C. 2020 Influence of the prebiotics hi-maize, inulin and rice branon the viability of pectin microparticles containing Lactobacillus acidophilus LA-5 obtained by internalgelation/emulsification. Powder Technology, 362: 409-415.

Rais, M. and Sheoran, A. 2015. Scope of supply chain management in fruits and vegetables in India. Journal of Food Processing and Technology, 6 (3): 1-7.

Ramli, S., Abbas F.M., Yong, Y.S. and Easa, A.M. 2009. Utilization of banana peel as a functional ingredient in yellow noodle. As. Journal of Food and Agricultural Industries, 2 (3): 302-310.

Rao, B.R., Sastry, K.P., Kumar, R., Rajput, D.K., Patel, R.P. and Reddy, L.P.A. 2009. Value added products from medicinal and aromatic plants. national level seminar on medicinal & aromatic plants and value added products. Medicinal and Aromatic Plants and Value Added Products 9-10 January, 2009.

Rivas, B., Torrado, A., Torre, P. and Converti, A. 2008. Submerging citric acid fermentation on orange peel atohydrolysate. Journal of Agricultural and Food Chemistry, 56 (7), 2380-87.

Rocha, J.M. 2016.Microencapsulation of probiotic bacteria. Ecronicon Microbiology, 3: 480-481.

Rohm, H., Brennan, C. Turner, C., Günther, E., Campbell, G., Hernando, C., Struck, S. and Kontogiorgos, V. 2015. Foods, 4: 690-697.

Rolda N.E., Sanchez-Moreno, C., Ancos, B. and Cano, M.P. 2008. Characterisation of onion (Allium cepa L.) by-products asfood ingredients with antioxidant and antibrowning properties. Food Chemistry, 108: 907-916.

Sabeena Farvin, K.H., Grejsen, H.D. and Jacobsen, C. 2012. Potato peel extract as a natural antioxidantin chilled storage of minced horse mackerel (Trachurus trachurus): Effect on lipid and proteinoxidation. Food Chemistry, 131: 843-851.

Saini, A., Panwar, D., Panesar, P.S. and Bera, M.B. 2019. Bioactive compounds from cereal and pulse processing byproducts and their potential health benefits. Australian Journal of Nutrition and Metabolism, 6 (2): 1068.

Schieber, A., Stintzing, F.C. and Carle, R. 2001. By-products of plant food processing as a source of functional compounds-recent developments. Trends in Food Science and Technology, 12: 401-413.

Shaaban, H.A.E., El-Ghorab, A.H. and Shibamoto, T. 2012. Bioactivity of essential oils and their volatile aroma components: Review. Journal of Essential Oil and Research, 24: 203-212.

Shakir, I.K. and Salih, S.J. 2015. Extraction of essential oils from citrus by-products using microwave steam distillation. Journal of Chemical Petroleum Engineering, 16 (3): 11-22.

Shalini, R. and Gupta, D.K. 2010. Utilization of pomace from apple processing industries: A Review. Journal of Food Science and Technology, 47: 365-371.

Sharif, M.K., Butt, M.S., Anjum, F.M. and Khan, S.H. 2013. Rice bran: a novel functional ingredient. Critical Reveiw of Food Science and Nutrition, 54 (6): 807-816.

Shelly, M. and Kaur, K. 2015. Impacts of food processing industry on economic growth, FDI and exports of India. Pacific Review Business International, 7 (12): 63-72.

Singh, A.K., Singh, A.K., Kumar, K., Prakash, V., Sundaram, P.K. and Yadav, S.K. 2017. Indian cereals saga: standpoint and way forward. Journal of Agricultural Search, 4 (1): 1-9.

Singh, L., Singh, P., Kotru, R., Singh, K.N., Singh, P., Lone, B.A. and Qayoom, S.2015. Economising fertility in temperate rice through green manuring under Kashmir valley condition. Journal of Agricultural Search, 2 (1): 14-19.

Singh, S.P., Tegegne, F. and Ekenem, E. 2012. The Food Processing Industry in India: Challenges and Opportunities. Journal of Food Distribution Research, 43 (1).

Singhal, A., Karaca, A. C., Tyler, R. and Nickerson, M. 2016. Pulse proteins:from processing to structure-function relationships. https://cdn.intechopen.com/pdfs/50953.pdf.

Skendi, A., Zinoviadou, K.G., Maria Papageorgiou, M. and João M. 2020. Rocha advances on the valorisation and functionalization of by-products and wastes from cereal-based processing industry. Foods, 9: 1243; doi:10.3390/foods9091243.

Slavin, J.L. and Lloyd, B. 2012. Health benefits of fruits and vegetables. Advances in nutrition, 3: 506-516.

Stojceska, V., Ainsworth, P., Plunkett, A., Ibanoglu, E. and Ibanoglu, S. 2008. Cauliflower by-products as a new source of dietary fibre, antioxidants and proteins in cereal based ready-to-eat expanded snacks. Journal of Food Engineering, 87 (4): 554-563.

Stoll, T., Schweiggert, U., Schieber, A. and Carle, R. 2003. Application of hydrolyzed carrot pomace as a functional food ingredient to beverages. Journal of Food Agriculture and Environment, 1: 88-92.

Sucheta, Rai, S.K., Chaturvedi, K. and Yadav, S.K. 2019. Evaluation of structural integrity and functionality of commercial pectin based edible films incorporated with corn flour, beetroot, orange peel, muesli and rice flour. Food Hydrocoll, 91: 127-135.

Tiwari, A.K. and Shivhare, A.K. 2016 Pulses in India: retrospect and prospects. Government of India, Ministry of Agriculture and Farmer's Welfare, Directorate of Pulses Development.

Torres-Leon, C., Rojas, R., Serna-Cock, L., Belmares-Cerda, R. and Aguilar, C.N. 2017. Extraction of antioxidants from mango seed kernel: optimization assisted by microwave. Food Bio-production Process, 105: 188-196.

Vu, H.T., Scarlett, C.J. and Vuong, Q.V. 2018. Phenolic compounds within banana peel and their potential uses: A review. Journal of Functional Foods, 40: 238-248. https://doi.org/10.1016/j.jff.2017.11.006.

Yadav, S.K., Kauldhar, B.S., Sandhu, P.P., Thakur, K. and Sucheta, Sharma, T.R. 2020. Retrospect and prospects of secondary agriculture and bioprocessing. Journal of Plant Biochemistry and Biotechnology, 29 (1): 1-14.

Zema, D.A., Calabrò, P.S., Folino, A., Tamburino, V. and Zappia, G.S.M. 2018. Valorization of citrus processing waste: A review. Waste Management, 80: 252-273.

Zou, Y. and Yang, T. 2019. Rice husk, rice husk ash and their applications. In: Cheong L-Z, Xu X (eds) Rice bran and rice bran oil. Academic Press and AOCS Press, Urbana.

Chapter 8

Blockchain Technology: Benefits, Constraints and Future Perspective in the Food Sector

Seerat Gupta, Neeraj Gupta, Daman Preet Kour and Ankita Choudhary*

Division of Post Harvest Management, Sher-e-Kashmir University of Agricultural Sciences and Technology, Chatha, Jammu-180009, J&K, India.
**Corresponding author: seeratgupta03@gmail.com*

Abstract

A potential technology with significant ramifications to serve the food business is blockchain technology. A globally accessible database containing entries in the form of encrypted blocks, or a public ledger that keeps track of finished transactions or digital events that have been shared among those who participated and potentially accessible at any point in the future, is recognized as a blockchain. The reliable and decentralized capability of the blockchain is usually employed for world-wide financial systems, however it's relatively simple to expand to various sorts of contracts and processes, including tracking the entire worldwide supply network. Furthermore, by incorporating the authoritative source of the claim (such as the certification body) in the blockchain to validate the claim's authenticity and assured the business clients as well as the end consumers, blockchain may strengthen the integrity of credence claims that include ethically produced, organic or faith-based claims for instance kosher or halal.

Despite the evident advantages of blockchain and the potential upgrades to increase openness, dependability and correctness of information, significant obstacles to its widespread adoption still exist. Blockchain might enhance traceability, facilitate much more efficient recalls, as well as minimize the risk of fraudulent products along with other forms of illegitimate transactions. However, several aspects relating to the many blockchain knots computers constantly linked to the blockchain networks having the responsibility of storing and disseminating an updated copy of each block in a food supply chain results being of utmost significance. These technologies seem to have a lot of potential and are very promising, with ample scope for applications across a variety of industries. However, they are still in their infancy and are challenging to implement because of their complexity.

Introduction

The globalization of the agri-food industry has made it more complicated, with many interconnected ecosystems ranging from primary production to foodservice and retail (Wee *et al.*, 2016). Every country's economy depends on access to food, which is also a basic human right. For the health, productivity and sustainable growth of economies, having accessibility to safe, nourishing and economical food also serves as crucial (Campoy-Munoz *et al.*, 2017).

Blockchain in the agri-food sector can boost reliability, streamline processes, eliminate unwanted middlemen from the supply chain, as well as boost consumers confidence in accessible food items, among other benefits (Papa, 2017). These days, the technologies comparatively low transaction costs, transparency and quick implementations may aid with food security. The resilient and decentralized functionality of the blockchain is typically utilized by financial institutions worldwide but it is simple to extend it to contracts and operations like tracking the worldwide supply network (Galvez *et al.*, 2018). A blockchain is a decentralized database that stores information in the form of encrypted "blocks" (smaller datasets), also known as a public ledger, which may be used to track all past and present digital transactions and occurrences. Information can be withdrawn after it has been added to the public ledger and each transaction has been authorized by the majority of system users. Farmers, processors, traders, wholesalers, retailers and consumers are just a few of the various parties involved in the agri-food sector who often want high-quality, reliable goods with the most information possible (Verbeke, 2005).

Blockchain has several significant benefits for the agri-food industry. Authorized blockchains can help to accomplish crucial goals in the battle against the cyber security risk to sufficiently secure consumer data and maintain the integrity of the entire industrial system as cyber threats systematically progress into growing technologies. Due to well-known vulnerabilities (such as food safety, food quality and food fraud) as well as the rising price of food commodities, the agri-food industry is one of the few sectors that is thought to get a significant amount of public attention. FSC management must therefore adopt an effective strategy regarding its regulation and risk minimization of incorrect or unauthorized practices (Ting *et al.*, 2014). An expensive product safety recall might be necessary for ensuring (Maloni and Brown, 2006) FSC effectiveness despite taking into consideration consumer concerns regarding food quality, safety, nutrition and fraud. Negligence to effectively reduce risk might result in a variety of food frauds or low quality, unsafe products being released onto the market. With this technology, it is possible to ensure that food actually comply with the information on their labels regarding their place of origin, method of production, its processing technology, composition etc (Gerbig *et al.*, 2017; Danezis *et al.*, 2016). Blockchain technology is undoubtedly a promising solution for reducing uncertainty and thereby reducing some friction in FSCs.

Blockchain technology has been referred as an imminent game-changer during the past couple of years as it has grown in significance and utility across various industries which include the manufacturing and finance sectors (Yadav *et al.*, 2020). Supply chains can be made more efficient by this technology's major features, which include privacy, security, smart contracts (Christidis and Devetsikiotis, 2016) and traceability.

Blockchain

Blockchain is described as a digital, decentralized and distributed ledger in which transactions are logged and added in chronological order with the goal of creating permanent and tamperproof records (Treiblmaier, 2018). It can be considered as a fusion of numerous technologies, tools, along with methodologies suitable for specific problem or commercial use case (Rejeb *et al.*, 2018). Users are able to exchange resources and value directly owing to a blockchain, which eradicates the requirement for third parties to act as middlemen in transactions (Alammary *et al.*, 2019). The arrangement of multiple PCs operates as a decentralized ledger for all data and transactions. Blockchain is one of the most rapidly developing and emerging technologies that seeks to completely increase transparency within a particular supply chain by enabling simple, safe tracing or backtracking (Antonucci *et al.*, 2019). The blockchain system stores data in blocks that are disseminated among peers, with alterations executed using a consensus technique (Pence, 2019).

Blockchains comprise a series of "nodes" installed on a communication network which permits accessibility to a few standard communication protocols. A "block" is created whenever a transaction takes place and each block is linked to the one prior to it and the one after it in order to create an irreversible chain where all of the blocks consist of identical transactions. This linkage is collectively referred to as a "blockchain," and it is this chain of transactions that gives the name "blockchain" its meaning. A chain of trust is established as frequently as a new record is validated and included in the blockchain by creating numerous copies in a decentralized manner (Saberi *et al.*, 2019). In particular, when integrated with smart contracts, blockchains promise transparent, immutable, shareable and secure platforms that can enable unique business solutions (Andoni *et al.*, 2019). By eliminating central management, blockchains and distributed ledger technology (DLT) serve mainly to promote distributed transactions (Tian, 2016).

The blockchain systems have various characteristics that distinguish them apart from conventional digital ledgers or relational databases, where databases and database systems come in a broad variety of forms and designs. Blockchain systems are often dispersed across a network of computers, making it impossible for a single computer to control them. All users who are a part of the blockchain network share in the ownership of the transactions that take place within a blockchain. Consumers will have access to a wider range of detailed information on the origin of the food, the type of ingredients, the cultivation and processing techniques, as well as other details pertaining to the social conditions of the employees, thanks to blockchain-enabled FSCs (Behnke and Janssen, 2020). One of the primary perks of a blockchain is that its transactions can be determined all the way back to the beginning of the blockchain, allowing for the provision of information about an asset on the blockchain and the explanation of how this asset has grown and evolved as time passed (Lansiti and Lakhani, 2017).

Types of Blockchain

A blockchain is a specific kind of database that keeps a digital log of the past transactions. Based on the credibility of being a validator whose capabilities are open or subject to limitations by authorizing entities, blockchain platforms can be roughly classified into two distinct categories:

1) Permissionless: Public blockchain regulations built on 'proof of work' consensus algorithms are accessible for downloads and have no restrictions in all respects. As a consequence of this, anyone can access the public blockchain technology, install a public node on their personal device and commence validating network transactions in addition to performing the consensus process (the process of establishing brand-new blocks which are subsequently included to the blockchain) despite of acquiring any permission (Dinh *et al.*, 2017). Bitcoin, Ethereum, Monero, Dash, Litecoin and Dogecoin are a few examples of public blockchains.

These blockchain networks are ungoverned and anyone can validate the transactions. Thus all individuals have access to explore information stored on the blockchain platform. Despite being extremely transparent, this blockchain platforms perform inadequately. A public block explorer, for example, makes it possible for anybody to read the transactions that are recorded on the blockchain.

These transactions are frequently pseudo-anonymous or anonymous. By means of disintermediation, public blockchain technology provides the capability to upend established business paradigms. A public ledger may be validated by any network user, making it entirely distributed and censorship-resistant. Participation in the consensus process is open to all users that sign up for the public blockchain system (Dong *et al.*, 2018). The scalability of these open systems is inadequate (Dinh *et al.*, 2017) and work is being done to scale blockchains.

2) Permissioned: Blockchain platforms in this category include private and consortium. An individual or collection of entities may own and manage these blockchain platforms. The governing entities grant both the right for validating transactions and the authority to read data. With improved performance and scalability, transparency declines. Scalability is still constrained, though (Dinh *et al.*, 2017). A group of people, organizations or reliable authority often oversees consortium or federated blockchains. In essence, only a limited number of nodes are permitted to join the blockchain network. This is one of the primary distinctions between it and a public blockchain, in which anyone with Internet access can take part in the process of confirming transactions and developing new blocks.

Higher scalability and increased transaction privacy make consortium blockchains faster. These blockchain variants are frequently utilized in the financial industry (Gatteschi *et al.*, 2018). It has a multi-center (partial decentralization) structure, meaning that the participating system's member nodes are pre-configured to comply with a set of requirements, enabling the identification and verification of each node's identity. The regulatory authority of these nodes is simply set by the alliance chain (Gatteschi *et al.*, 2018). R3 (Banks), B3i (Insurance) and EWF (Energy) are a few examples of consortium blockchain. Consortial blockchains can be successfully used to lower transaction costs, eliminate data duplication, replace outdated systems, streamline document processing and build full compliance processes.

Due to its complete management by just one organization, a private blockchain is viewed as a centrally managed network. Browsing the blockchain may be either (partially) open to the general public or just available to a small group of people who have been asked to join the network or have been allowed access. The spectrum of participants and transaction-validating members is set by the central controller. The method of controlling access may differ; for instance, current participants may invite potential new participants, a regulatory body may provide a licence to join, or a group of participants may make such decisions. By establishing teams and members who can internally verify transactions, private blockchains serve a way to benefit from blockchain technology. Members who manage the blockchain are susceptible to security breaches, comparable to a centralized system, in contrary to public blockchains.

Scalability, state compliance with data privacy laws and other regulatory challenges can all be addressed by private blockchains. MONAX and Multichain (Dong *et al.*, 2018) are instances of private blockchains. Private chains are widespread yet have a low degree of decentralization.

Benefits of Blockchain Technology in Food Supply Chains

Input supply, post-harvest handling, manufacturing, processing, transportation, distribution, marketing and retailing are all included in food supply chains, as are agricultural upstream and downstream sectors. There is currently a lack of available data that makes it impossible to verify the origin, traceability and danger connected with the entire supply chain of agri-food products from production to consumer markets (Fortuna and Risso, 2019). Blockchain has been identified as a potentially ground-breaking technology in the agri-food industry with the ability to completely transform a variety of operations and (digital) business operations (Woodside *et al.*, 2017). With the help of this technology, FSC stakeholders can improve vital features like end-to-end traceability and enable quick product safety recalls (George *et al.*, 2019). The provision of a reliable, auditable and easily accessible record of food items' entire lifecycle, from beginning of production to retail, could also alleviate tension amongst FSC stakeholders (Babich and Hilary, 2019).

The supply chain might determine the provenance, current location and present situation of any food product at any stage with the use of blockchain, thereby providing a common information platform for all participants. By empowering businesses to create accurate supply-demand forecasting models and to have complete visibility into the supply chain, blockchain contributes to boost efficiency in the supply chain in this approach. Blockchain can promote the acceptance of a business model based on the circular economy by keeping track of the total amount of food wasted and salvaged across the entire food supply chain, which will reduce product loss and enhance profits. Digital data are stored in a decentralized and immutable manner using blockchain.

The origin of items cannot always be ascertained by current tracking systems and neither can the accuracy and transparency of data about food quality and safety (Kim and Laskowski, 2018). The initial purchase can be recorded on a blockchain to permanently record the product's validity and ownership of the certificate can be transferred in an efficient manner that can be governed by smart contracts. The connection between the physical object and the related authentication certificate is ensured by integrating an IoT device as a sensor and connecting it to the blockchain (Hughes *et al.*, 2019). By affixing an RFID tag with a specific ID, this type of transparency, dependability and visibility enables to identify the source of products and ensures their validity. As individuals of the supply chain are unable to alter the data recorded in the system, blockchain technology also lowers the danger of manipulation.

For ensuring the authenticity and high quality of the fresh basil used in its pesto, Barilla is adopting blockchain technology. In an attempt to minimize the possibility of receiving counterfeit goods, Nestlé monitors the export flows of its chocolates (Baci Perugina) (Kerner, 2018). The massive online retailer Alibaba has introduced the "Food Trust Framework" programme with a similar goal in mind, enabling Chinese users of Tmall Global which is an Alibaba-owned marketplace, to track the delivery of food from Australia and New Zealand (Austrade, 2018). Walmart uses IBM Blockchain to keep tabs on approximately 25 products coming from multiple vendors, including fresh fruits, leafy greens, dairy, meat, chicken etc (Chang *et al.*, 2019). Apart from tracing a product's origin, Walmart seeks to address food safety concerns by pinpointing the origin of an outbreak of a foodborne illness, coordinating more efficient recalls of products that may be contaminated and consider retaining more information to promote sustainable supply chain practices (Smith, 2018).

Secure data sharing among the supply chain participants offers enhanced surveillance and hence promotes food safety by reducing the chance of supply chain members unwittingly transfer counterfeit foods. Brands can manage potentially hazardous food fraud thanks to the full transparency of data, which also enables them to rapidly and pro-actively determine the source of contamination threats. This is done without causing any supply chain disruptions. Blockchain technology might be incorporated into food supply networks to assist sustainable practices. By utilizing blockchain technology, stakeholders may be able to swiftly notice changes and modify supply and demand by exchanging information about all phases of food production and consumption. Additionally, it might make it possible to share information with food banks with the objective to distribute leftovers and cut down on food waste.

There is a lot of potential for biodegradable microsensors to be used in combination with smart contracts and to register activities on blockchains. It might result in the measurement of the product's temperature, humidity and other crucial quality variables. Real-time monitoring of transactional information, ambient conditions, inventory as well as quality standards along the supply chain is crucial for extending the safety, quality and shelf life of perishable products, which is critical for fresh food in particular (Sander *et al.*, 2018). These microsensors are currently quite expensive, but their price might decline significantly in the future, enabling widespread use and creating a desirable relation between food products & the IoT (Schlaefli, 2017).

The production, processing and distribution of goods could all be trusted in a food supply chain that is based on blockchain technology. The majority of buyers are looking for assurances about the things they buy apart from their provenance. By automatically digitizing and permitting the sharing of audits, certificates, along with other pertinent documents (Tse *et al.*, 2018) on sustainable practices, blockchain elevates customer confidence in the security, quality and authenticity of the produce which they are purchasing.

Traceability

Food traceability is defined by Aung and Chang (2014) as a part of logistics management that gathers, stores and transmits sufficient data about a food, feed, food-producing animal or substance at all points in the food supply chain so that the product can be tracked both upward and downward, as well as examined for safety and quality control as needed. Although food organizations require traceability to develop customer trust and maintain food quality and safety requirements (Dutta *et al.*, 2020), implementing traceability is challenging and costly when food products involve diverse ingredients from distinct nations (Swaffield *et al.*, 2019). Blockchain can be employed for monitoring food sources around the world by creating origin chains which are safe, visible and adaptable to evolving regulations. (Dutta *et al.*, 2020). According to Bumblauskas *et al.* (2020), blockchain could be adopted to supplement food traceability in both novel and existing supply chains as the technology is a remedy to siloed FSC activities and can be incorporated into preexisting systems and technologies without involving significant process or behavioural modifications by FSC stakeholders (Keogh *et al.*, 2023).

The blockchain system makes all agri-food information in the supply chain transparent and open, permitting logistics enterprises to set up real-time tracking for agri-food products, supervision authorities carry out traceability management and responsibility investigations regarding defective products, and thus consumers to obtain full product details throughout the entire agri-food supply chain, all of this assist to establish a healthy market environment.

Traceable inventory may be monitored from its raw material source to its distribution to its end user via smart contracts that can enhance supply chain traceability. This can be accomplished through the application of a serial number, RFID tag, or smart sensors. Smart sensors can provide data pertaining to product quality (particularly for perishable commodities) in combined with location and environmental information. Real-time product status information will boost the capacity of determining more accurate and rapid choices. Delays can be minimized as a result, and the entire supply chain can stay nimble (Yadav and Singh, 2019). The confidence that blockchain offers on the traceability of food boosts customer satisfaction, increases their willingness to make payments for traceable food items and raise awareness concerning food items' certified, secure and sustainable origins (Tiscini *et al.*, 2020).

Trust and transparency

Consumer trust is a vital component in maintaining a market for food items in a business-to-consumer (B2C) context, particularly for organic and specialty products that fetch high and premium prices (Insch and Cuthbert, 2018; Carolan, 2006). Transparency is a prerequisite for increasing customer trust in food goods since it lowers uncertainty and increases information visibility across the entire FSC procedures (Ab Rashid and Bojei, 2019). The establishment of long-term relationships is aided by trust between trade parties (DuPuis and Gillon, 2009; Keogh *et al.*, 2023). Transparency is a vital quality for FSCs and a management priority because of the increased focus on issues relating to the environment, society, governance (also known as ESG) and growing consumer expectations (Faisal, 2015). Transparency is typically defined in the FSC context as the level of coverage from upstream to downstream FSC stakeholders, as well as the visibility of information to downstream members (Sporleder and Goldsmith, 2002). Trust and transparency are strategic management concerns in FSC operations since they are often fast-paced, sensitive to climate-related variables and some items are perishable (Opara, 2003). Although increasing confidence and openness in FSCs is recognized as a challenging task, new technologies such as blockchain can be used to increase the transparency of food production systems (Astill *et al.*, 2019).

Agri-food companies may additionally benefit from blockchains transparency capabilities to boost monitoring of FSC procedures (Dos Santos *et al.*, 2019), impart real-time transparency (Khanna *et al.*, 2020), as well as strengthen trust in the FSC. By enabling greater transparency, blockchain technology can enhance collaboration, synchronize processes, ease friction and reduces FSC expenses (Kittipanya-ngam and Tan, 2020). Smart contracts may additionally be employed to develop and retain electronic means of identification, enabling transparent authentication among supply chain partners. This method makes it simple for parties to confirm that other parties have the necessary licences and authorizations to carry out their obligations. Furthermore, blockchain technology might be utilized to assess trustworthiness and manage credibility (Antonucci *et al.*, 2019). By lessening conflict, it can strengthen the bonds between trade partners (Dutta *et al.*, 2020). A blockchain consortium has been determined to be the most widely recognized and feasible model for commercial use (Kayikci *et al.*, 2020), as it has the potential to bring compatible organizations together in the FSC and offers an unprecedented degree of trust and transparency that relies on one's interpretation of the truth. The development of a blockchain-based model for food traceability, transparency of supply chains and audibility, for instance, was the result of a partnership among Walmart, IBM, and Beijing Tsinghua University (Galvez *et al.*, 2018). Food organizations can minimize the risk of product recalls, boost confidence in FSC procedures and enhance satisfaction among consumers by boosting FSC transparency using blockchain (Stranieri *et al.*, 2021).

Data management and security

The widespread implementation of new technologies is pushing FSCs towards a data-driven, digital supply chain environment (Kamble *et al.*, 2019). There are multiple obstacles brought on by the increased accessibility of knowledge from food preparation and distribution to final consumers. Food loss and waste, might results from inadequate and inaccurate management of information systems (Zhong *et al.*, 2017). The deployment of cutting-edge technology like IoT results in the production of tremendous amounts of information that might detrimentally affect agri-food organizations' capacity to manage and utilize the information collected by IoT sensors across the FSC (Keogh *et al.*, 2023). Thus, data management and FSC system integration are essential components of the modern corporate environment (Gessner *et al.*, 2007). Blockchain has developed as a technology with prospective to boost the effectiveness of operations, handling of information, responsiveness and smart contract management to strengthen FSC data management and security (Dutta *et al.*, 2020).

According to Kamble *et al.* (2020), blockchain driven data management systems in FSCs may significantly assist the food industry by managing the land and use documentation, purchasing information and use of agricultural machinery, seeds and agriculture chemicals, as well as labour inputs, traceability and monetary transactions throughout the entire FSC. Numerous types of agri-food data can be included into a traceability system via RFTD in conjunction with intelligent machinery. This process reduces the number of errors brought on by human factors significantly as it doesn't involve any manual intervention. Additionally, the use of blockchain technology makes it impossible for any of the system's participants to alter agri-food data, thereby enhancing the product's safety and quality. Moreover, the integration of blockchain with IoT offers various advantages related to data management and analytics, making it an imminent facilitator for efficient communication and data quality (Casino *et al.*, 2020, 2019). Strong data security, privacy and openness are only a few benefits of the FSC's integration of blockchain and big data analytics (Dutta *et al.*, 2020; Liu *et al.*, 2020).

Food trade

The ease of international trading in food items has been greatly aided by globalization (Alonso *et al.*, 2020). The application of blockchain and smart contracts in food trade worldwide can manage strategic collaborations by enabling reliable ownership transfer, cross-country trades and data sharing without the need for middlemen (Dutta *et al.*, 2020). Blockchain fosters honest interactions among FSC stakeholders, democratises the supply-chain procedures and boosts transparency in food trade activities. Blockchain is a decentralized economic infrastructure. Along with facilitating quicker transactions, blockchain technology in the food industry also helps to combat price gouging and payments that are retroactive (Kayikci *et al.*, 2020). Thus, the advent of blockchain brings about the creation of an authentic record for trade data and information, permitting food organizations to swiftly reconcile invoices and boost transparency as well as efficacy in transactions (Keogh *et al.*, 2023). The upholding of essential sanitary and phytosanitary requirements, enhancement of sustainability and alleviation of food-related contamination and safety concerns all have been anticipated to be made feasible by blockchain, according to Kamble *et al.* (2019).

Blockchain might assist the food supply chains in the lodging sector by attaining the outcomes at each one of the three pillars of sustainable growth (Stroumpoulis *et al.*, 2022).

- Blockchain technology can guarantee information integrity, which implies that information cannot be changed at any point without the consent of the authorized partners. This is important since it supports the social pillar (Saberi *et al.*, 2019). The traceability and accessibility that blockchain offer permits hospitality businesses to stay accountable for the status of food items throughout the whole supply chain. Food safety might therefore be maintained while preventing food fraud.
- Regarding the environmental component, blockchain offers in-depth transaction monitoring and product tracking, dropping rework and recall (Saberi *et al.*, 2019), as an outcome limits the energy usage and environmental pollution.
- Lastly, by breaking up food supply networks into smaller chains featuring lower transaction costs, fewer delays and better management, blockchain technology might strengthen the economic pillars of each party involved. It could additionally lead to better transaction security (Xu *et al.*, 2019), avoiding extra expenditures. Additionally, by providing all relevant information, hospitality businesses may be able to select partners that are more compatible in their desires and trustworthy (Mathew, 2018). In food supply chains, partners will be able to keep track of all transactions in order to regulate and improve the integrity of all dealings between suppliers, retailers, customers and other partners (Shih *et al.*, 2019).

Blockchain technology holds a tendency to make substantial improvements in key performance indicators for the supply chain. Data on product production, processing, transmission and sales are not readily available for traditional food supply chains. Accessibility of data which was initially kept in centralized databases that were internal to supply chain participants relies on subsequent interaction with other parties or the development of information-sharing systems that are overseen by a third-party organization (Montecchi *et al.*, 2019).

Some Supply-Chain Applications of Blockchain Technology in the Agro-Food Industry

Products	Initiative/ Organization	Purposes	References
Mangoes	Walmart, Kroger, IBM	Food traceability	Kamilaris *et al.*, 2018; Olsen *et al.*, 2019
Fruits	FruitChains	Public, immutable, ordered ledger of records	Galvez *et al.*, 2018
Grains	AgriDigital	Financial	Kamilaris *et al.*, 2018

Products	Initiative/ Organization	Purposes	References
Agri-food	AgriOpenData	Allow quality and digital identity to be certified	Galvez *et al.*, 2018; Olsen *et al.*, 2019
	Supply Chain Traceability System for China Based on RFID and Blockchain Technology	Trusted information throughout the agri-food supply chain	
Fresh food	Ripe	Enabling data transparency and transfer from farm to fork	Galvez *et al.*, 2018
Soybean	HSBC and Cargill; ING and Louis Dreyfus Co.	Authenticate products and get rid of the "paper trail" of checks at each step of the supply chain	Hochfelder, 2018; Olsen *et al.*, 2019
Coffee	FairChain coffee: Bext360 in partnership with Moyee Coffee	Traceability, transparency of the value added	Ciaian, 2018
Wine	Chainvine, Winecoin	Boost efficiency, income, accountability and security	Galvez *et al.*, 2018; Ciaian, 2018
Beer	Downstream	Food traceability	Kamilaris *et al.*, 2018; Olsen *et al.*, 2019
Chicken	Gogochicken; Grass Roots Farmers Cooperative; OriginTrail; ZhongAn	Urban consumers' concerns about food safety and food traceability	Kamilaris *et al.*, 2018; Ciaian, 2018
Fish	Provenance	Auditable system	Galvez *et al.*, 2018
Beef	"Paddock to plate" project, BeefLedger; JD.com	Food traceability	Kamilaris *et al.*, 2018

Constraints of Blockchain Technology in Food Supply Chains

According to Wang *et al.* (2019), blockchain technology has the potential to revolutionize supply chain planning, management and operations. There are a number of barriers preventing its wider implementation (Hackius and Petersen, 2017).

- The implementation of blockchain technology necessitates considerable modifications to the way businesses carry out everyday tasks. The supply chain's participants would have to use cutting-edge data collection techniques while also being compelled to integrate with other technologies and tools like WSN (Wireless Sensor Network), sensor-RFID (Radio Frequency Identification) equipment, GIS (Geographic Information System), GPS (Global Positioning System), cloud computing services and smart contracts (Chang *et al.*, 2019; Tian, 2016). The ability of blockchains and IoT devices to interact with one another still has to be utilized entirely.

- Emerging blockchain technology, together with the corresponding hardware and required system updates, continues to be expensive. For instance, the cost of using RFID tags to enhance traceability systems is substantially more than that of using standard barcodes, which may limit their use. The adoption of cutting-edge technology also necessitates the development of technological skills and expertise. The broad adoption of modern technology in the food supply chain might be disrupted by all of these variables, particularly for small and medium-sized businesses and operators who are suffering from the negative consequences of the technological disparity (Saurabh and Dey, 2020).
- Given the requirement to secure the storage and synchronization of a growing quantity of information to be recorded, scalability of operations is developing as an additional barrier owing to the current restrictions on the capacity of each blocks that are appended to the blockchain. The implementation of this technology may be hampered by the regulatory uncertainties brought on by the current complicated web of rules and codes in different nations, especially those pertaining to ownership, data protection and security. To effectively regulate the automated, international and denationalized nature of blockchains and smart contracts, there needs for better standardization.
- Everyone involved in the food supply network are jointly responsible for supplying accurate information about food products on a blockchain-based information exchange platform (Musso and Risso, 2012). The system may thus simply assure the authenticity of the provided information, not its veracity. The introduction of QR codes or RFID tags on the packaged product generates an additional issue since, by its own, neither of these technologies provides a guarantee that the product will accurately match the information on the label.
- On a global scale, developing countries may suffer with a shortage of qualified labour, inadequate infrastructure (such as roads, refrigerators and warehouses) and technological obstacles at agricultural levels and in the field of information and communication technologies (ICT) (Kamble and Raut, 2019). The required intrinsic emphasis and governance on the food product as well as the extrinsic emphasis creates and sustain FSCs that assures the integrity of food items in transit, designing global FSCs is a challenging endeavour (Ting *et al.*, 2014).

Future Research Directions for Managing the Obstacles Associated with Employing Blockchain Technology in the Agri-Food Value Chain

Lack of storage space is a major obstacle to the use of blockchain technology in the agri-food value chain. Though some researchers have broadly investigated a number of methods that could increase the storage capacity, for instance, storing some data on-chain and some data off-chain (Xu *et al.*, 2019), creating a multi-layer blockchain design with distinct purposes (Hamida *et al.*, 2017) and developing a new algorithm that include deep learning concept (Chen, 2018). It is easy to determine that additional processing power needs to be gathered when applying distinct solution approaches. Future research should therefore look towards appropriate data storage solutions that don't require more computing power. A new approach to resolving the problem of storage involves specifically integrating the artificial intelligence or deep learning concepts with existing algorithms, such as the ant colony algorithm, in order to maximize computational power efficiency (Zhao *et al.*, 2019). Additionally, feasible and analytical comparisons of adopting these possible remedies in the overall context of the agri-food value chain must be undertaken.

As a novel technology, blockchain does not yet have any laws or rules to abide by and no established method of integrating its consequences into normative settings. Although the use of blockchain technology in the agri-food value chain is still in its early stages, it can be anticipated that subsequent initiatives will eventually be carried out by different agri-food organizations (Ge *et al.*, 2017). Consequently, additional investigation should be made from theoretical and empirical aspects to analyze the implications of blockchain technology across the agri-food value chain, particularly to examine its impact in the world's agri-food trade, and then develop matching regulations based on the research findings.

The majority researchers relied on a single blockchain design when examining how to resolve privacy leakage problems and they explored a variety of techniques to maintain user privacy, notably Stealth Addresses, Homomorphic encryption, Ring signatures, Pedersen Commitments and Zero-Knowledge-Proof. A double blockchain design, namely one for the public and the other for those who engage in the agri-food value chain, was only partially explored by academics in terms of how to ensure privacy (Zhao *et al.*, 2019). As a result, future research should look into how to create and assess double chain topologies for blockchain. Additionally, run tests to see how combining the aforementioned data privacy plans can maximize the effectiveness of the data privacy measures.

Several authors have suggested various consensus algorithms, including the Proof-of-Stake and delegated Proof-of-Stake algorithms, in relation to the high cost issue. Although these two methods have a number of benefits, such as speeding up computation and using fewer nodes for block validation, they also have certain drawbacks (Zhao *et al.*, 2019). In an empirical investigation including the agri-food value chain, no one has, however, performed a comparison analysis with both of these approaches. Future scholars can systematically examine the influence and relative significance of various mechanisms utilizing empirical and analytical techniques in this field.

Finally, despite our concern over the lack of capabilities, we were unable to locate numerous resources that provided us with straightforward solutions to this problem. Two learning objectives were suggested (Ge *et al.*, 2017) to address the issue of a skills gap that were (Boyaci *et al.*, 2014) to identify how to utilize blockchain technology and (Ge *et al.*, 2017) to acquire a general understanding of blockchain technology.

Conclusion

Food safety is emerging as a more pressing concern as a result of the growing traceability requirements, the weaknesses in the food supply chain, the risk of perishable products and the drive to improve operational efficiency. The quality and freshness of food products can be better regulated and maintained with the help of well coordinated and interoperable FSCs. Blockchain technology is one of today's rising and developing technologies since it is upending every industry and has the potential to bring about many revolutionary changes in the agricultural sector by enhancing the supply chain and guaranteeing the integrity, security and traceability of data. By cutting out the intermediary in the supply chain management process, blockchain offers an effective and transparent method. Blockchain technology's provides the agriculture system with strong security features. All participants in the supply chain can have access to information, monitor actions in real-time and track food goods due to blockchain technology as well as associated equipment. Through greater transparency, cross-border traceability and information sharing, it can guarantee the correctness, dependability and validity of information regarding food product quality, sustainability and safety.

Moreover, these technologies seem very promising and full of tremendous potential, demonstrating versatility for applications in a variety of areas, but they are still developing and challenging to put into practice due to their complexity. Through more precise data collection and data exchange procedures, food organizations can use technology to create more effective food traceability systems. By making food supply chains more transparent, it can be used to track food-related problems from the beginning of their creation to the last stage (consumption) and prevent food wastage, fraud, as well as any other unethical practice or action. Additionally, it may offer customers and consumers better knowledge about the production and delivery of food, empowering them to make decisions which are more considerate of the environment. In order to promote the efficient application of technology in the food sector and its extensive adoption by the FSC, further awareness programmes must be conducted.

References

Ab Rashid, N. and Bojei, J. 2019. The relationship between halal traceability system adoption and environmental factors on halal food supply chain integrity in Malaysia. *Journal of Islamic Marketing*, 11(1): 117-142.

Alammary, A., Alhazmi, S., Almasri, M. and Gillani, S. 2019. Blockchain-based applications in education: A systematic review. *Applied Sciences*, 9(12): 2400.

Alonso, R.S., Sitton-Candanedo, I., Garcia, O., Prieto, J. and Rodriguez-Gonzalez, S. 2020. An intelligent Edge-IoT platform for monitoring livestock and crops in a dairy farming scenario. *Ad Hoc Networks*, 98(3): 102047.

Andoni, M., Robu, V., Flynn, D., Abram, S., Geach, D., Jenkins, D., McCallum, P. and Peacock, A. 2019. Blockchain technology in the energy sector: A systematic review of challenges and opportunities. *Renewable and Sustainable Energy Reviews*, 100: 143-74.

Antonucci, F., Figorilli, S., Costa, C., Pallottino, F., Raso, L. and Menesatt, P. 2019. A review on blockchain applications in the agri-food sector. *Journal of the Science of Food and Agriculture*, 99(14): 6129-6138.

Astill, J., Dara, R. A., Campbell, M., Farber, J.M., Fraser, E.D.G., Sharif, S. and Yada, R.Y. 2019. Transparency in food supply chains: A review of enabling technology solutions. *Trends in Food Science and Technology*, 91: 240-247.

Aung, M.M. and Chang, Y.S. 2014. Traceability in a food supply chain: Safety and quality perspectives. *Food Control*, 39: 172-184.

Austrade. 2018. Exporting food and beverage to china: a guide for australian business, Australia. Available at: http://files.chinagoabroad.com/Public/uploads/content/files/201811/food-and-bev-for-China-guide.pdf. Accessed on 2 July, 2023.

Babich, V. and Hilary, G. 2019. Blockchain and other distributed ledger technologies in operations. *Foundations and Trends® in Technology, Information and Operations Management*, 12(2-3): 152-172.

Behnke, K. and Janssen, M.F.W.H.A. 2020. Boundary conditions for traceability in food supply chains using blockchain technology. *International Journal of Information Management*, 52: 101969.

Boyaci, I.H., Temiz, H.T., Uysal, R.S., Velioglu, H.M., Yadegari, R.J. and Rishkan, M.M. 2014. A novel method for discrimination of beef and horsemeat using Raman spectroscopy, *Food Chemistry*, 148(2014): 37-41.

Bumblauskas, D., Mann, A., Dugan, B. and Rittmer, J. 2020. A blockchain use case in food distribution: Do you know where your food has been? *International Journal of Information Management*, 52: 102008.

Campoy-Munoz, P., Cardenete, M.A. and Delgado, M.C. 2017. Economic impact assessment of food waste reduction on European countries through social accounting matrices. *Resources, Conservation and Recycling*, 122: 202-209.

Carolan, M.S. 2006. Social change and the adoption and adaptation of knowledge claims: whose truth do you trust in regard to sustainable agriculture? *Agriculture and Human Values*, 23(3): 325-339.

Casino, F., Kanakaris, V., Dasaklis, T.K., Moschuris, S. and Rachaniotis, N.P. 2019. Modeling food supply chain traceability based on blockchain technology. In Y. F. Ivanov D. Dolgui A. (Ed.), *IFAC-PapersOnLine*, 52(13): 2728-2733.

Casino, F., Kanakaris, V., Dasaklis, T.K., Moschuris, S., Stachtiaris, S., Pagoni, M. and Rachaniotis, N.P. 2020. Blockchain-based food supply chain traceability: a case study in the dairy sector. *International Journal of Production Research*, 59(19): 5758-5770.

Chang, S.E., Chen, Y.C. and Lu, M.F. 2019. Supply chain re-engineering using blockchain technology: a case of smart contract based tracking process, *Technological Forecasting and Social Change*, 144: 1-11.

Chen, R.Y. 2018. A traceability chain algorithm for artificial neural networks using T-S fuzzy cognitive maps in blockchain, *Future Generation Computer System*, 80(2018): 198-210.

Christidis, K. and Devetsikiotis, M. 2016. Blockchains and smart contracts for the internet of things, *IEEE Access*, 4: 2292-2303.

Ciaian, P. 2018. Blockchain technology and market transparency. Available at: https://ec.europa.eu/info/sites/info/files/law/consultation/mt-workshop-blockchain-technology-and-mt_ciaian_en.pdf. Accessed on June 24, 2023.

Danezis, G.P., Tsagkaris, A.S., Camin, F., Brusic, V. and Georgiou, C.A. 2016. Food authentication: techniques, trends and emerging approaches. *TrAC Trends Analytical Chemistry*, 85: 123-132. https://doi.org/10.1016/j.trac.2016.02.026

Dinh, T.T.A., Wang, J., Chen, G., Liu, R., Ooi, B.C. and Tan, K.L. 2017. Blockbench: a framework for analyzing private blockchains. In *Proceedings of the 2017 ACM International Conference on Management of Data*. New York, NY, USA: Association for Computing Machinery. pp 1085–1100.

Dong, Z., Luo, F. and Liang, G. 2018. Blockchain: A secure, decentralized, trusted cyber infrastructure solution for future energy systems. *Journal of Modern Power Systems and Clean Energy*, 6(5): 958-67.

Dos Santos, R.B., Torrisi, N.M., Yamada, E.R.K. and Pantoni, R.P. 2019. IGR token-raw material and ingredient certification of recipe based foods using smart contracts. *Informatics*, 6(1): 11.

DuPuis, E.M. and Gillon, S. 2009. Alternative modes of governance: organic as civic engagement. *Agriculture and Human Values*, 26(1): 43-56.

Dutta, P., Choi, T.M., Somani, S. and Butala, R. 2020. Blockchain technology in supply chain operations: Applications, challenges and research opportunities. *Transportation Research Part E: Logistics and Transportation Review*, 142: 102067.

Faisal, M.N. 2015. A study of inhibitors to transparency in red meat supply chains in Gulf cooperation council (GCC) countries. *Business Process Management Journal*, 21(6): 1299-1318.

Fortuna, F. and Risso, M. 2019. Blockchain technology in the food industry, *Symphonya, Emerging Issues in Management*, (2): 151-159.

Galvez, J.F., Mejuto, J.C. and Simal-Gandara, J. 2018. Future challenges on the use of blockchain for food traceability analysis. *TrAC Trends in Analytical Chemistry*, 107: 222-232.

Gatteschi, V., Lamberti, F., Demartini, C., Pranteda, C. and Santamaria, V. 2018. Blockchain and smart contracts for insurance: Is the technology mature enough? Future Internet 10(2): 20. doi: 10.3390/fi10020020.

Ge, L., Brewster, C., Spek, J., Smeenk, A., Top, J., Diepen, F.V., Klaase, B., Graumans, C. and De Ruyter de Wildt, M. 2017. *Blockchain for agriculture and food: Findings from the pilot study*. (Wageningen Economic Research report; No. 2017-112). Wageningen Economic Research. Available at: https://doi.org/10.18174/426747. Accessed on 3 July, 2023.

George, R.V., Harsh, H.O., Ray, P. and Babu, A.K. 2019. Food quality traceability prototype for restaurants using blockchain and food quality data index. *Journal of Cleaner Production*, 240: 118021.

Gerbig, S., Neese, S., Penner, A., Spengler, B. and Schulz, S. 2017. Real-time food authentication using a miniature mass spectrometer. *Analytical Chemistry*, 89: 10717-10725.

Gessner, D.G.H., Volonino, D.L. and Fish, L.A. 2007. One-up, one-back ERM in the food supply chain. *Information Systems Management*, 24(3): 213-222.

Hackius N. and Petersen M. 2017. *Blockchain in Logistics and Supply Chain: Trick or Treat?*, In: Kersten, Wolfgang Blecker, Thorsten Ringle, Christian M. (Ed.): Digitalization in Supply Chain Management and Logistics: Smart and Digital Solutions for an Industry 4.0 Environment. Proceedings of the Hamburg International Conference of Logistics (HICL), 23, epubli GmbH, Berlin. pp 3-18.

Hamida, E.B., Brousmiche, K.L., Levard, H. and Thea, E. 2017. Blockchain for Enterprise: Overview, Opportunities and Challenges. In: *The Thirteenth International Conference on Wireless and Mobile Communications (ICWMC 2017)*, Nice, France.

Hochfelder, B. 2018. HSBC, Cargill successfully complete blockchain trade-finance transaction. Available at: https://www.supplychaindive.com/news/hsbc-cargill-blockchain-pilot/523554/. Accessed on June 30, 2023.

Hughes, A., Park, A., Kietzman, J. and Brown, C. 2019. Beyond Bitcoin: What blockchain and distributed ledger technologies mean for firms. *Business Horizons*, 62(3): 273-281.

Insch, A. and Cuthbert, R. 2018. Why country of origin still matters in food retailing: implications for promotion management research. *Journal of Promotion Management*, 24(3): 363-375.

Kamble, S.S. and Raut, R.D. 2019. Evaluating the factors considered for procurement of raw material in food supply chain using Delphi-AHP methodology - a case study of potato chips processing company in India. *International Journal of Productivity and Quality Management*, 26(2): 176-189.

Kamble, S.S., Gunasekaran, A. and Gawankar, S.A. 2019. Achieving sustainable performance in a data-driven agriculture supply chain: A review for research and applications. *International Journal of Production Economics*, 219: 179-194.

Kamble, S.S., Gunasekaran, A. and Sharma, R. 2020. Modeling the blockchain enabled traceability in agriculture supply chain. *International Journal of Information Management*, 52: 101967.

Kamilaris, A., Fonts, A. and Prenafeta-Boldu, F.X. 2018. The rise of the blockchain technology in agriculture and food supply chain. Available at: https://www.researchgate.net/profile/Andreas_Kamilaris/publication/327534824_The_Rise_of_the_Blockchain_Technology_in_Agriculture_and_Food_Supply_Chain/links/5b93d0ada6fdccfd5428b0f2/The-Rise-of-the-Blockchain-Technology-in-Agriculture-and-Food-Supply-Cha. Accessed on June 25, 2023.

Kayikci, Y., Subramanian, N., Dora, M. and Bhatia, M.S. 2020. Food supply chain in the era of Industry 4.0: blockchain technology implementation opportunities and impediments from the perspective of people, process, performance, and technology. *Production Planning and Control*. 1-21.

Keogh, J.G., Rejeb, A., Khan, N. and Zaid-Kaylani, K. 2023. *Blockchain: an enabler for safe food in global supply networks*, Editor(s): Michael E. Knowles, Lucia E. Anelich, Alan R. Boobis, BertPopping, Present Knowledge in Food Safety, Chapter 68, Academic Press, pp 1045-1066.

Kerner, M. 2018. Simplifying blockchain app development with azure blockchain workbench, microsotf azure. Available at: http://azure.microsoft.com/it-it/blog/simplifying-blockchain-app-development-with-azure-blockchain-workbench-2/. Accessed on 5 July, 2023.

Khanna, T., Nand, P. and Bali, V. 2020. Permissioned blockchain model for end-to-end trackability in supply chain management. *International Journal of e-Collaboration*, 16(1): 45-58.

Kim, H.M. and Laskowski, M. 2018. Toward an ontology-driven blockchain design for supply-chain provenance. *Intelligent Systems in Accounting, Finance and Management*, 25(1): 18-27.

Kittipanya-ngam, P. and Tan, K.H. 2020. A framework for food supply chain digitalization: lessons from Thailand. *Production Planning and Control*, 31(2-3): 158-172.

Lansiti M. and Lakhani, R.K. 2017. The truth about blockchain. Available at: https://hbr.org/2017/01/the-truth-about-blockchain. Accessed on 30 June, 2023.

Liu, P., Long, Y., Song, H.C. and He, Y.D. 2020. Investment decision and coordination of green agri-food supply chain considering information service based on blockchain and big data. *Journal of Cleaner Production*, 277(7674): 123646.

Maloni, M.J. and Brown, M.E. 2006. Corporate Social Responsibility in the Supply Chain: An Application in the Food Industry. *Journal of Business Ethics*, 68(1): 35-52.

Mathew, S. 2018. Blockchain is a game changer in supply chain. Available at: https://supplychaingamechanger.com/blockchain-in-the-supply-chain-is-a-game-changer/. Accessed on 23 June, 2023.

Montecchi, M., Plangger, K. and Etter, M. 2019. It's Real, Trust Me! Establishing Supply Chain Provenance Using Blockchain. *Business Horizons*, 62(3): 283-293.

Musso, F. and Risso, M. 2012. ICT Innovation for Buyer-Seller Relationships in International Supply Chains, in Indian Institute of Science, Bangalore. In: *Driving the Economy through Innovation and Entrepreneurship. Emerging Agenda for Technology Management*. Bangalore: Springer India.

Olsen, P., Borit, M. and Syed, S. 2019. Applications, limitations, costs, and benefits related to the use of blockchain technology in the food industry. *Nofima*. 6-38.

Opara, L.U. 2003. Traceability in agriculture and food supply chain: A review of basic concepts, technological implications, and future prospects. *Food, Agriculture and Environment*. 1(1): 101-106.

Papa, S.F. 2017. Use of blockchain technology in agribusiness: transparency and monitoring in agricultural trade. In: *International Conference on Management Science and Management Innovation (MSMI 2017)*. Atlantis Press, 2017.

Pence, H.E. 2019. Blockchain: will better data security change chemical education? *Journal of Chemical Education*, 97(7): 1847-1854.

Rejeb, A., Sűle, E. and Keogh, J.G. 2018. Exploring new technologies in procurement. *Transport and Logistics: The International Journal*, 18(45): 76-86.

Saberi, S., Kouhizadeh, M., Sarkis, J. and Shen, L. 2019. Blockchain technology and its relationships to sustainable supply chain management. *International Journal of Production Research*, 57(7): 2117-2135.

Sander, F., Semeijn, J. and Mahr, D. 2018. The acceptance of blockchain technology in meat traceability and transparency. *British Food Journal*, 120(9): 2066-2079.

Saurabh, S. and Dey, K. 2020. Blockchain technology adoption, architecture, and sustainable agrifood supply chains. *Journal of Cleaner Production*, 284: 1-13.

Schlaefli, S. 2017. Biodegradable microsensors for food monitoring. Available at: https://phys.org/news/2017-09-biodegradable-microsensors-food.html. Accessed on 28 June, 2023.

Shih, D.H., Lu, K.C., Shih, Y.T. and Shih, P.Y. 2019. A simulated organic vegetable production and marketing environment by using Ethereum, *Electronics*, 8(11): 1341.

Smith, M. 2018. In Wake of Romaine E. coli Scare, Walmart Deploys Blockchain to Track Leafy Greens, *Walmart Communications*. Available at: https://corporate.walmart.com/. Accessed on 29 June, 2023.

Sporleder, T.L. and Goldsmith, P.D. 2002. Network embeddedness in the food supply chain and firm strategy on signaling quality. In: *Paradoxes in food chains and networks Proceedings of the Fifth International Conference on Chain and Network Management in Agribusiness and the Food Industry*, Noordwijk, Netherlands. pp 720-728.

Stranieri, S., Riccardi, F., Meuwissen, M.P.M. and Soregaroli, C. 2021. Exploring the impact of blockchain on the performance of agri-food supply chains. *Food Control*, 119: 107495.

Stroumpoulis, A., Kopanaki, E. and Oikonomou, M. 2022. The impact of blockchain technology on food waste management in the hospitality industry. *Entrenova - Enterprise Research Innovation*, 7(1): 419-428.

Swaffield, S.R., Corry, R.C., Opdam, P., McWilliam, W. and Primdahl, J. 2019. Connecting business with the agricultural landscape: business strategies for sustainable rural development. *Business Strategy and the Environment*, 28(7): 1357-1369.

Tian, F. 2016. An agri-food supply chain traceability system for China based on RFID and blockchain technology. In: *2016 13th international conference on service systems and service management*. Doi: 10.1109/ICSSSM.2016.7538424.

Ting, S.L., Tse, Y.K., Ho, G.T.S., Chung, S.H. and Pang, G. 2014. Mining logistics data to assure the quality in a sustainable food supply chain: A case in the red wine industry. *International Journal of Production Economics*, 152: 200-209.

Tiscini, R., Testarmata, S., Ciaburri, M. and Ferrari, E. 2020. The blockchain as a sustainable business model innovation. *Management Decision*, 58(8): 1621-1642.

Treiblmaier, H. 2018. The impact of the blockchain on the supply chain: A theory-based research framework and a call for action. *Supply Chain Management: An International Journal*, 23(6): 545-559.

Tse, D., Zhang, B., Yang, Y., Cheng, C. and Mu, H. 2018. Blockchain application in food supply information security. In: *IEEE International Conference on Industrial Engineering an Engineering Management*. pp 1357-1361.

Verbeke, W. 2005. Agriculture and the food industry in the information age. *European review of agricultural economics*, 32(3): 347-368.

Wang, Y., Singgih, M., Wang, J. and Rit, M. 2019. Making sense of blockchain technology: how will it transform supply chains? *International Journal of Production Economics*, 211: 221-236.

Wee, H.M., Budiman, S.D., Su, L.C., Chang, M. and Chen, R. 2016. Responsible supply chain management – an analysis of Taiwanese gutter oil scandal using the theory of constraint. *International Journal of Logistics Research and Applications*, 19(5): 380-394.

Woodside, J.M., Augustine Jr., F.K. and Giberson, W. 2017. Blockchain technology adoption status and strategies. *Journal of International Technology and Information Management*, 26(2): 65-93.

Xu, X., Weber, I. and Staples, M. 2019. Blockchain in software architecture. In: *Architecture for blockchain applications*, Springer, Cham, pp 83-92.

Yadav, S. and Singh, S.P. 2020. Blockchain critical success factors for sustainable supply chain, *Resources, Conservation and Recycling*, 152: 104505.

Yadav, V.S. and Singh, A.R. 2019. Use of blockchain to solve select issues of Indian farmers. In: *Proceedings of the 1st International Conference on Advances in Mechanical Engineering and Nanotechnology (ICAMEN 2019)*, 2148(1), AIP.

Zhao, G., Liu, S., Lopez, C., Lu, H., Elgueta, S., Chen, H. and Boshkoska, B.M. 2019. Blockchain technology in agri-food value chain management: A synthesis of applications, challenges and future research directions. *Computers in Industry*, 109(2019): 83-99.

Zhong, D.R.Y., Tan, P.K. and Bhaskaran, P.G. 2017. Data-driven food supply chain management and systems. *Industrial Management and Data Systems*, 117(9): 1779-1781.

Chapter 9

Climate Smart Water Management Strategies for Sustainable Fruit Production of the North-West Himalayas

Vijay Kumar[1], Shahid Ahamad[2], Rakesh Kumar[1], Vijay Khajuria[3], Sanjay Khajuria[4], Shalini Khajuria[4], Saurav Gupta[4], Neerja Sharma[4], Abhay Kumar Sinha[4], and Brinder Singh[5]

[1]Rainfed Research Sub-station for Sub-tropical fruits, Raya, [2]Directorate of Research, [3]Farming System Research Centre, Chatha, [4]Krishi Vegan Kendra, Samba, [5]Advanced Centre for Rainfed Agriculture
Sher-e- Kashmir University of Agricultural Sciences and Technology, Jammu, (J&K), India
**Corresponding author E-mail: vijaykumar.1144@yahoo.com*

Abstract

Universal water management in rainfed horticulture is becoming more and more complex to overcome the predictable water scarcity stress. In adding to this, challenges of global warming and climate change would have to be met through the sensible application of water in agriculture or horticulture through climate-smart water technologies. Horticulture is an important sector in India and many developing countries, providing huge employment opportunities to rural populations, and supporting them to achieve food and nutritional security goals. The challenges of increasing food production and improving rural livelihoods, while safeguarding critical water resources for sustainable use through adaptive measures for effective water management, particularly in rainfed regions. These are the upsetting threats to the sustainable horticulture system and require to be dealt through scientific inventions and interventions. There is necessity for employment of current and advance techniques such as micro-catchment, water harvesting, *in-situ* and *ex-situ* conservation techniques and maintaining economic and ecological balance of water resources. Keeping in view of this present study is undertaken on "Climate Smart Water Management Strategies for Sustainable horticulture".

Key words: Climate Change, climate-smart water technologies, water saving technologies, rainfed condition

Introduction

Climate-smart agriculture (CSA) is frequently distinct as "agriculture that sustainably increases productivity, enhances resilience (adaptation), reduces or removes greenhouse gases where possible and enhances achievement of national food security and development goals" (FAO, 2013). Our agricultural practices should be focused in such a way so that our soils and plants act as carbon sinks and absorb carbon dioxide (CO_2) from the atmosphere. Climate smart agriculture differs from existing approaches because it focuses on principles of increased productivity and sustainability along with explicitly addressing adaptation and mitigation challenges while working towards food security for all. Climate change is one of the major challenges in India as well as the world. The world is also facing today which has direct bearing food security of the nation. Water resources are exhausting rapidly either due to natural causes or by faulty practices of human beings. Climate change will aggravate the scarcity of water for horticulture/ agriculture and human use, especially in the rainfed areas under *kandi* belt. Water scarcity is referring as a lack of sufficient water, or not having access to safe water supplies. Water is a pressing need in many areas of the world. That scarcity is spreading as water is needed to grow and process food, create energy, and serve industry for a continually growing population. Climate plays an important role in several ways; it is causing severe events like floods and droughts over the entire world and accelerating water management problems. It is also dependable for human lives causing damage of properties of water. In the 21st century, climate change has emerged as a significant challenge for horticulture/ agriculture, freshwater resources and the food security of the people. Hence, it is crucial to bring out the advance technologies and management practices to manage water resources under the changing climate to make horticulture system sustainable to meet the food demand for growing population. The maximum potential increases in crop yield in rainfed areas where in many of the world's poor live and where managing water is the key to such increases (Molden, 2007). Rainwater management techniques are the best option for fruit crops for sustainable. Rainwater harvesting (RWH) is based on the principle of depriving (naturally or artificially) section of the land of its share of rainfall and adding it to another section, where it can be used beneficially. Therefore, rainwater harvesting is the process of concentrating rainfall as runoff from a larger area for its productive use in a smaller target area (Oweis *et al.*, 2001). The collected runoff can be applied either directly to an horticulture or agricultural field for crop production or be stored in some kind of storage facility for domestic use and as well as supplemental irrigation. Rainwater harvesting can be classified as: micro- catchment, macro-catchment and rooftop. In the field of fruit crops water harvesting technologies several structural such as trenches/terraces, bunding, half moon basin, full moon basin, cup and plate basin with infiltration pits, have featured prominently among others as a strategy for overcome the water shortages associated with droughts in rainfed fruit systems. The *ex-situ* techniques of water harvesting i.e. pond, well and tank etc for mitigate the water shortage in the orchards. The aim of water management technologies like *ex-situ* or *in-situ* water harvesting as well as cultural practices in this case is to collect runoff water from areas of surplus and make it available, where and when there is water shortage as well as conserve the water in soil. This outcome in this areas increase in water availability by impeding and trapping surface runoff, maximizing water runoff storage or trapping and harvesting sub- The utmost importance of seeking additional technical assistance to these strategy, must be emphasized. Land managers, farmers and villagers should consult the appropriate experts for relevant information on cropland production. The scope of these plans includes soil water management and the use of stored rain water for fruit proposes uses as a strategy to adapt and mitigate climate change.

In-Situ Moisture Conservation Techniques

Mulching

Water is the most significant natural resource in rainfed areas for the economic development of the country. In rainfed areas, the rainfall is the little and erratic rainfall, productivity has been severely reduce due to near to the ground moisture content in soil profile during flowering and fruit development. Mulching is the defined the covering soil, which makes soil more favorable for plant growth and nourishment. Several kinds of materials are used for mulching, such as kitchen scraps, hay, straw, leaves, grass clippings, shredded bark, sugar bagasse, animal manure, sawdust, shells, whole bark nuggets and wool. Natural materials for mulching include straw, leaves, hay, farm residues, compost, bark, saw dust and even wool, while the inorganic mulches include plastic mulch, organic sheet mulch, rock and gravel. Organic sheet mulch has the advantage of being biodegradable. It play an important role in improves the soil aeration creates recovered biological activates and thus to make a consequent beneficial consequence on the soil fertility. Several reimbursement of organic mulches include lessening in the evaporation losses from the soil surface, improved water use efficiency and chemical control of weeds, which produce environmental hazards and have residual effects on crops.

According to a field study conducted in Rainfed Research substation for subtropical fruits Raya in experimental farm in Jammu, various organic and inorganic materials are used for moisture content and yield. The inorganic mulch i.e. black polyethylene mulch was retained maximum soil moisture content and yield followed by farmyard manure (FYM) and branker (*Adhotada vassica*). These findings suggest that black polyethylene mulch could be used effectively for mulching to conserve soil moisture and increase crop productivity (Kumar *et al.* 2015) in figure -1.

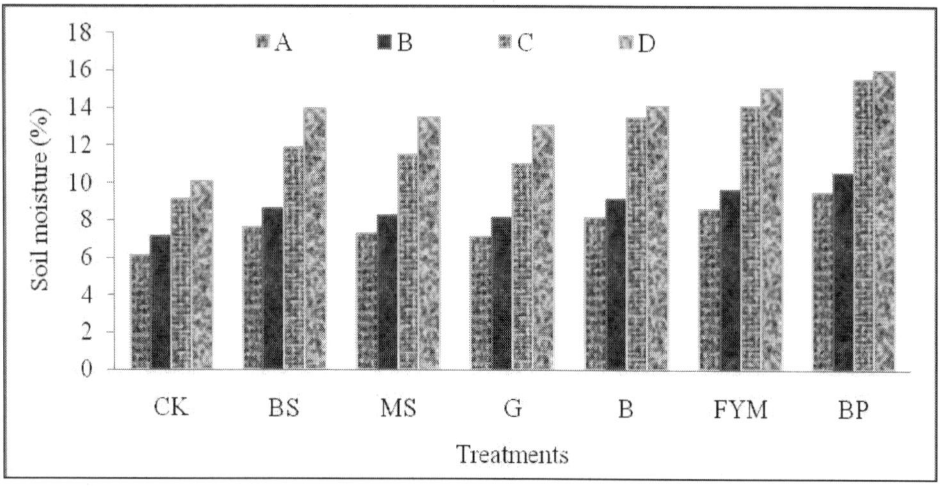

A= 80 DAM (2009-10), B= 80 DAM (2010-11), C= 110 DAM (2009-10) and D= 110 DAM (2010-11) Non-mulching (CK), bajra straw (BS), maize straw (MS), grasses (G), branker (B) (*Adhotada vassica*), farmyard manure (FYM) and black polyethylene (BP) different treatments

Fig. 1 Effect of different mulches on soil moisture (%) in Eureka lemon

Black polyethylene mulch controls of weeds and moisture more resourcefully and a mulch with a drip shows even enhanced results compared to an organic mulch. Biodegradable plastic mulches offer an alternative to polyethylene mulches, which is desirable, because they can reduce non-recyclable waste, conserve resources, and decrease environmental pollution as well as an expensive option to the farmers.

Fig. 2 Application of different types of mulch and hydrogel

The enhancement of soil moisture and nutrient status of various kinds of mulches through organics mulches materials like branker (*Adhotada vassica*) in locally available and biodegradable nature mulch augment the soil moisture content in soil for 18% and available nutrients status of nitrogen, phosphorus and potassium content in soil for 8.83 % Nitrogen, 10.81 % Phosphorus and 8.44% Potassium respectively higher than control in the standing crop of Eureka lemon. Rainfed areas also having narrow population of domestic animals, so availability of farmyard manure is also limited. The locally available farmyard manure may not be enough for need the requirement of the mulching materials.

Pusa hydrogel

Pusa hydrogel is a cross-linked polymer with a hydrophilic group which has the capacity to absorb huge quantities of water without dissolving in water. The gel was developed by the Indian Agricultural Research Institute (IARI), New Delhi and has been in use since 2012. It is non-toxic and biodegradable nature and the soil moisture and increases the plant yield by 10 - 25 per cent. Hydrogel (Super absorbent polymer) is a water retaining, cross-linked hydrophilic, biodegradable amorphous polymer which can absorb and retain water at least 400 times of its original weight and make at least 95 per cent of stored water available for crop absorption. The gel also increases growth rate and reduces water evaporation. When polymer is mixed with the soil, it forms an amorphous gelatinous mass on hydration and is capable of absorption and desorption over long period of time, hence acts as a slow release source of water in soil. The hydrogel particles may be taken as "miniature water reservoir" in the soil and water will be removed from these reservoirs upon the root demand through osmotic pressure difference. The large quantities of water retained by the polymer provide extra available water to crops which facilitates better crop growth. More available water in the soil also means less frequent irrigation and reduces irrigation frequency in horticulture crops. The hydrogel was able to retain available water for the plants up to fifteen days after irrigation and considerably for longer time increasing the soil moisture content due to water stress period of Kinnow mandarin. The recent study of Kumar *et. al.* (2022) in the application of trickle irrigation and Pusa hydrogel with black polyethylene mulch in Kinnow mandarin in experimental farm Raya, Jammu under various treatment of hydrogel which is one of the best result viz. 90 g hydrogel with much and three drays intervals water supply under rainfed conditions (Fig.-2)

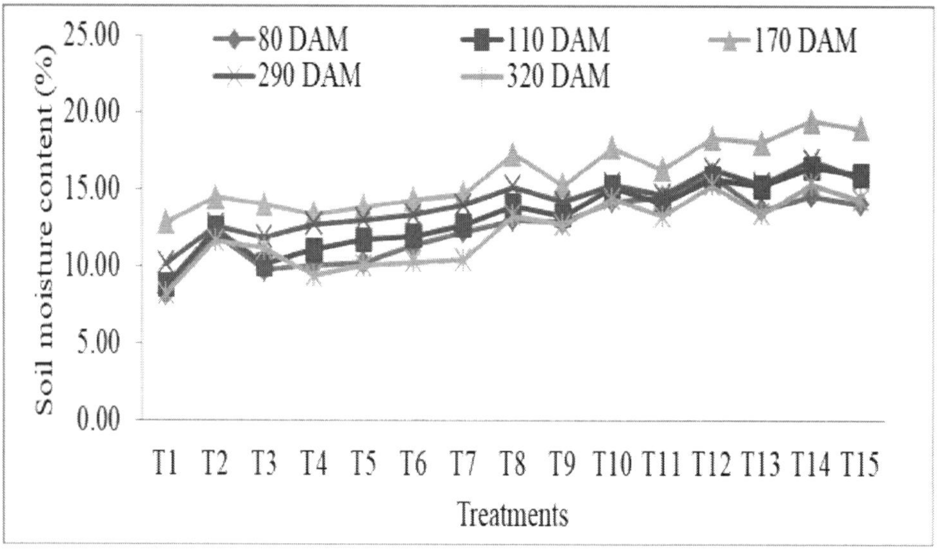

Fig.3: Soil moisture content (%) during different growth stages of Kinnow mandarin at 0-20 cm soil layer

The excellent water absorbency and water retention by hydrogel may prove especially practical in agriculture. It performs its wetting/drying cycles over a longer period of time, maintaining its very high water swelling and releasing capacity against soil pressure. The irrigation facilities is not available in rainfed areas is much more need for improvement of better crop growth, development and fruit quality attributes in horticulture crops. Its cultivation may be made more profitable by decreasing the demand of water and the frequency of irrigation.

Bunding

Bund is refer to small earthen barriers are constructed on horticulture and agriculture lands with slopes range from 1 to 6 per cent. They categorize the well-organized length of slope and consequently reduce enlarge in the velocity of runoff flow to stay away from rill and gully agreement. The bunds are generally two types i.e. graded bund and contour bund The graded bund is efficient in medium to high rainfall areas- having rainfall of 600 mm with moderately slope less than 10 percent of soil, deposition of eroded soil and raise the soil moisture content. This technique is reducing the speed of run-off water, for *in- situ* conservation of rain water and to reduce the soil erosion. The contour bund is refer in the comparatively narrow based embankments are made along the contour and across the slope. This barrier built across the slope (on the contours) for safe diversion of excess runoff and protects eroded soil. They are appropriate slopes of 1-7 % and comparatively low rainfall areas- having annual rainfall of fewer than 600 mm; particularly in the areas having light textured soil.

Trenches

The trench is referring to two meter length, 0.50 meter width and 0.50 m depth at one meter distance from the main trunk. These basins are suggested for fruit trees of the orchards and other plantation crops.

Bench terraces

The bench terraces are defined as the flat beds or platforms having suitable vertical drops across slope. The height of the riser should not be more than one meter and width of bench terrace depends on the degree of slope. The bench terraces are three types i.e. leveled and table top, outward sloping and inward sloping and it depends upon the nature of soil, climate and the crop.

Contour trenches

Contour trenches is refer as dig trenches along the contour and these trenches break the length of the slope and reduce the speed of run-off leading to condensed scouring capacity. These trenches techniques on mildly hill slope take away than 20 per cent as well as graded soils and also improve soil moisture. Hence the runoff generated in hilly areas which may differ from 20% to 40 % of rainfall, have to be managed doubtfully, rainfall can be retained by perennial vegetation followed by surface retention. *In-situ* moisture conservation practices are need to rainwater/ run-off management practice which are socially and efficiently viable as the levelling, field bunding, contour bunding, graded bunding, crescent bunding terracing and circular basin in rainfed orchards operation. According to filed study under temperate in hilly areas of Kumar and Kumar (2015) the management practices of circular basin, crescent bund and locally available mulches are used for the treatments. The higher soil moisture content in circular basin with 5 % slope and locally available mulch are 3.95 %, 4.51 % and 8.47 % as compared to circular basin with 5 % slope, crescent bunding and control (without practices) treatments were statistically significant in apple cultivars of fruit crops (Fig.3).

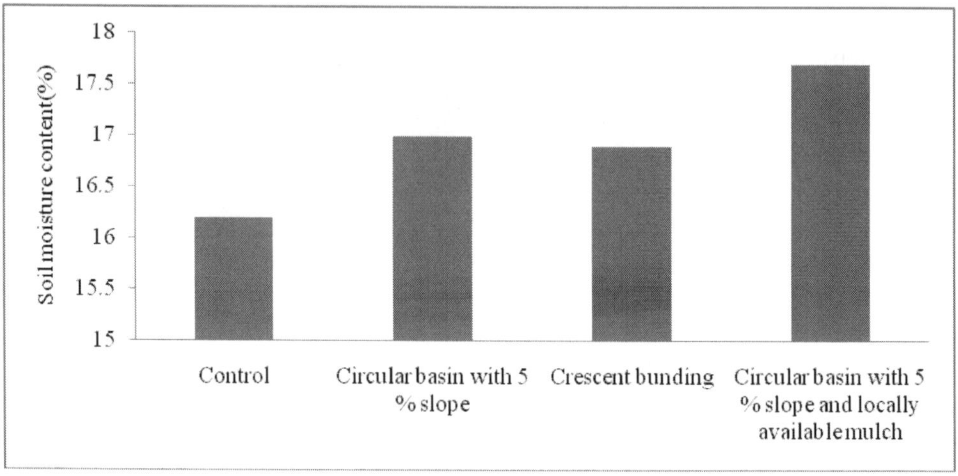

Fig. 4: Effect of *in-situ* water harvesting and mulching on soil moisture content in apple

Micro-Catchments Systems

Half moon systems

The half-moon systems/basin is defined as the semi-circular basin/beds of suitable diameter with the shape approaching a half moon. These basins are suggested for fruit trees or other plantation crops on steep slopes. The horticulture crops almost in rainfed/*kandi belt* areas. The half-moon basins are conserving in soil moisture content of fruit crops under rainfed areas.

Full moon

The full moon system is defined as circular basin at one meter distance from the main trunk. These basins are optional for fruit trees of the orchards and other plantation crops. The full moon basin is conserving the more moisture retention as compared to half moon basin, Cup and plate basins.

Cup and plate

The cup and plate systems are defined as the circular pit of 0.50 meter with and 0.50 meter depth around the tree at one meter distance from the main trunk. These basins are recommended for fruit trees of the orchards and other plantation crops.

Fig. 5 : *In-situ* moisture conservation techniques at RRSS, Raya, SKUAST-J

Ex-Situ Moisture Conservation Techniques

Water harvesting tank

The water harvesting term was firstly used in Australia by H.J. Geddes, to denote the collection and storage of any farm water either runoff or creek flow for irrigation use. This nomenclature has since then used very commonly, for expressing the meaning as collection of rainfall and all other water amounted to the same things. On the other hand, water harvesting is referring as the collection of runoff and its use for the irrigation of crops, trees and pastures and for livestock consumption during the dry spell. The fruit crops are in fact dependent on rainfall and are being practiced in varied agro-ecological and socio-economic environment under rainfed conditions. The majority of the fruit crops in rainfed systems include high rainfall Shiwaliks region of kandi belt but due undulating topography. Mainly of these villages face water insufficiency during summer's season and water table is very low approximate 500-550 meter below the ground level. This region agriculture is only dependent on rainfall, where the kind and amount of rainfall is such that if the vegetation is disturbed, large-scale erosion could ensue due to this diversity, rainfed horticulture offers diverse opportunities as well as challenges for the improvement of livelihood and sustainability of eco-system. On-farm, water harvesting through farm ponds on individual holding holds promise in rainfed areas to provide at least on critical life saving irrigation during dry season. The agriculture/horticulture crops are the main occupation of the people of kandi region of union territory of Jammu and Kashmir. The harvested rainwater in reservoirs can serve as a source of life saving irrigation to the crops and also have many indirect advantages like flood control, recharging of ground water, reclaiming land below the embankment and improvement of ecology etc. The present study was conducted in Rainfed Research Sub-station for Sub-tropical Fruits Raya, SKUAST-Jammu having water harvesting tank 20 x 20 m size is sufficient for nursery, shade net house approximate three kanals areas for producing 2500 plants every years and distribute the farmer communities.

Fig. 6 Water harvesting tank and ponds

Earth dams and water ponds

Earth dams are refer as the semi-circular or curved banks of earth, 3-4 meters high and 100 meters in length. Water ponds or pans are naturally occurring or excavated water storage structures without a constructed wall/dam. The tank should have a high depth to surface ratio to store maximum water behind the smallest possible dam. The most excellent catchment area would be a moderately steep and rocky landscape with no erosion and the dam should be placed in gentle sloping land in a wide shallow channel or broad hopelessness. It is chosen that these can be built by use manual labour and animal tracking.

Roof Top Harvesting

Roof water harvesting is an innovative upcoming technology used for collecting and storing of rainwater from rooftops, frequently in tanks. For the collection of water, a channel or drain pipe called Gutter is used along the edge of the rooftop to divert the flow. Gutter has a gentle slope towards the downpipe, which is intended for the free progression of water to the collection tank. This may be made up of wood, metallic sheet or any low-cost locally available material. In family rooftop rainwater harvesting systems, rainwater from the house roof is collected in a storage vessel or tank for use during the periods of scarcity. Generally this system is designed to support the drinking and cooking needs of the family at the doorstep. This water can be used for summer/kharif season for supply of supplementary irrigation for horticulture/agriculture crops. This water is the purpose of higher crops quality improvement. The roofs of buildings including greenhouses are, in many locations, a very valuable resource being used mainly for drinking and domestic purposes.

Summary and Conclusions

This chapter with different techniques in in-situ and ex-situ rainwater harvesting methods adopted in rainfed region. *In-situ* and ex-situ moisture conservation techniques for rainfed horticulture are the best opinion for improving the fruit yield and quality attributes. These practices of full moon half moon, bunding and trenches also improve the moisture conservation as well as yield and fruit quality attributes. Ex-situ moisture conservation of water storage, farm ponds and roof top technology would serve the purpose of climate resilience in rainfed horticulture and found more profitable in sandy loams soils. The socio- economic outline situation for rural people varies strongly from those in ancient times. Hence, to maintain the cost of orchard establishment and management proper intercropping of horticulture agri-horticulture and perennial fruit cropping systems should be recognized and promoted in the recurrently under rainfed region.

Acknowledgement

All the authors are thankful to the supporting staff of Rainfed Research Sub-station for Sub-tropical fruits, Raya, Sher -e- Kashmir University of Agricultural Sciences and Technology - Jammu for cultural practices of preparation of basins.

References

FAO.2013. Climate-Smart Agriculture Sourcebook. Rome.

Kumar, M. and Kumar V. (2015). Effect of different *in-situ* water harvesting techniques and low cost mulching in apple (cv *Mollies Delicious*) under temperate conditions. *Environment and ecology*, 33(4A): 1766-1769.

Kumar, V., Bhat, A.K., Sharma, V., Gupta, N., Sohan, .P and Singh, V.B. (2015). Effect of Different Mulches on Soil Moisture, Growth and Yield of Eureka Lemon (*Citrus limon* Burm) Under Rainfed Condition. *Indian Journal of Dryland Agricultural Research & Development*,30(1): 83-88.

Kumar, V., Kumar, R., Chand, G., Kumar, J., Singh, B., Dhotra, B., Khajuria, S., Gupta, S., Sharma, N., Sinha, A.K., Raina, V. and Ahamad, S. (2022). Influence of trickle irrigation and hydrogel on Kinnow mandarin (*Citrus reticulata* Blanco) under rainfed condition. *Agricultural Mechanization in Asia*, 53 (7): 9007-23.

Kumar, V., Sharma, .V, Gupta, N., Kumar, R, Singh, B., Kumar, M. and Singh, V.B.(2018). Soil mulches able to mitigate soil water deficiency impacts on Eureka lemon (*Citrus limon* Burm) production in rainfed environments in the Shivaliks of Jammu. *International Journal of Chemical Studies*, 6(3):2578-2584

Molden D (ed) 2007 Water for Food, Water for Life: A Comprehensive Assessment of Water Management in Agriculture (London: Earthscan) (Colombo: IWMI).

Oweis, T., Prinz, D. and Haehum., A.2001. Water harvesting: indigenous knowledge for the future of the drier environments. ICARDA, Aleppo, Syria. 40 pp.

Chapter 10

Potential Impact of Climate-Smart Agriculture on Food Security

*Jagmohan Singh and Rafeeya Shams**

Division of Food Science and Technology, Sher-e-Kashmir University of Agricultural Sciences and Technology of Jammu, India
Corresponding email id: rafiya.shams@gmail.com

1. Introduction

Agriculture production and food systems are altered by climate change, which creates uncertainty and vulnerability concerns for farmers and decision-makers (Vermeulen *et al.*, 2013). There is no question that the effects of climate change present significant challenges to global food security, and this is expected to get worse over the next few years due to pressures on livelihoods like population growth, economic development, urbanisation, and the frequent occurrence of natural hazards like extreme heat, drought, and flooding, among other things. According to estimates, 9 billion people would have worsened living conditions by 2050, with hunger and poverty taking the lead and making it very difficult to put food on the table (World Bank, 2015; FAO, 2013). In order to create agricultural systems that will improve and expand food production at all scales, from global to small, many international institutions are collaborating with the World Bank and the Food and Agricultural Organization (FAO) groups. The adoption of climate-smart agriculture (CSA) has since received support from a number of organisations, including those involved in research, policymaking, and investments, as well as from the corporate, public, and civil society sectors

(Taylor, 2018; Lipper *et al.*, 2014). In order to assist farmers adapt to climate change, climate smart agriculture is a collection of micro-level soil and water conservation strategies including planting and agroforestry. The current efficacy of CSA techniques and, in certain circumstances, farmers' willingness to embrace them in areas like Ethiopia, Peru, and Malawi have both been demonstrated in a number of recent articles (Amadu *et al.*, 2020; Marenya *et al.*, 2020).

By (a) sustainably boosting agricultural productivity and incomes, (b) adapting to climate change and strengthening resilience, and (c) lowering and/or eliminating GHG emissions, when practicable, the CSA strategy helps to achieve food and livelihood security as well as other developmental goals (FAO, 2010, 2013). Broadly speaking, CSA focuses on establishing resilient food production systems that can ensure the food and livelihood security of farming communities under climate change and variability. It also integrates climate change into the planning and execution of sustainable agricultural initiatives (Lipper*et al.*, 2014; Vermeulen, Campbell, & Ingram 2012). As a foundation for reorienting policy in response to climate change, it highlights synergies and trade-offs among food security, adaptation, and mitigation (Lipper *et al.*, 2014). Its goal is to clearly incorporate climate change as a key component in the identification and operationalization of sustainable agricultural growth. Naturally, for CSA to succeed, a comprehensive strategy open to unique local circumstances is needed (FAO, 2014). The CSA approach (a) addresses adaptation and increases resilience to climatic shocks; (b) views climate change mitigation as a potential co-benefit; (c) is location-specific and knowledge-intensive; (d) identifies integrated options that create synergies and reduce trade-offs; (e) identifies adoption barriers and offers appropriate solutions; (f) strengthens livelihoods by enhancing access to services, knowledge, and resources; and (g) integrates climate financing (FAO, 2014).

2. Need for Effective Implementation of CSA

The following four areas require immediate action from public, private, and civil society stakeholders: (1) developing evidence and assessment tools; (2) bolstering national and local institutions; (3) creating coordinated and evidence-based policies; and (4) increasing funding and its effectiveness (Lipper *et al.*, 2014).

The evidence base currently available is insufficient to assist efficient decision-making and is largely unavailable to national and local decision-makers. Many studies addressing the effects of climate change on agriculture have spatial and temporal scales that are inappropriate for planning at the national and local levels. This is due to the lack of information on future changes in climate variability at such scales and their effects on agriculture (Thornton *et al.*, 2014), which may be much more significant for local communities than long-term trends in climate variables. Additionally, there are technical problems associated with the downscaling of models to scales that are more appropriate for decision support (IPCC, 2012). The creation and use of problem-oriented approaches to adaptation planning offer enormous promise for determining effective responses to uncertainty (Vermeulen *et al.*, 2012). Exploiting synergies across global, regional, and local studies (Challinor *et al.*, 2014). Tools are required for assessing the potential for adaptation and mitigation of various policies and technologies at all scales, including local, national, and international levels. These tools should also cover the effects of extreme events and slow-onset changes on agriculture and food security, as well as ways to strengthen the resilience of agriculture and food systems and determine the benefits and costs of reducing emission growth. The effectiveness of research efforts at the national and international levels can be considerably improved by using landscape methodologies (Sayer *et al.*, 2013), as well as examination of the possibilities in current foresight and scenario programmes (Herrero *et al.*, 2014; Vervoort *et al.*, 2014). Finding obstacles to the adoption of agricultural techniques that respond to climate change and ways to overcome them, with an emphasis on the most vulnerable, such as smallholder producers, women, the poor, and marginalised groups, is another significant evidence base deficit. Although they are adjusting to the changing climate, farmers don't always implement practises that could be helpful (Arslan *et al.*, 2013; McCarthy *et al.*, 2011). Strong studies are especially needed to better understand what works where and why in various agro-ecologies and farming systems, making it easier to define what qualifies as "climate smartness" in various biophysical and socioeconomic circumstances. Strengthening national and local institutions to boost adaptive capability through improving people's access to resources, particularly information, is CSA's second priority action area. Although it has long been a focus of agricultural development initiatives, institutional development has had varying degrees of effectiveness due to poor design or financing (Kirsten, 2009). According to empirical evidence, public support is needed in four key areas to supplement private efforts: (1) extension and information sharing, especially regarding the use of evidence to adapt practises to local conditions; (2) coordinated efforts where practises have positive spillover effects, such as lowering flood risks, preventing pest outbreaks, or preserving biodiversity; and (3) comprehensive risk-management strategies for managing extreme weather events that have a wide-ranging impact (Arslan *et al.*, 2013).

Farmers may need access to particular inputs, such tree seedlings, seeds, or fertilisers, in order to use CSA practises. The absence of these inputs prevents widespread adoption. An essential component of productivity and effective resource utilisation is timely access to fertiliser, although this is frequently absent (Duflo *et al.*, 2011). These problems can be solved by using novel input delivery techniques, particularly those that make use of ICTs. Building supportive legislative and regulatory frameworks through improved coordination of agriculture, climate change/environmental, and food system policies is the third priority action area for CSA. In order to create an enabling policy environment, all relevant ministries must engage in dialogue to address trade-offs, gaps, and overlaps. It is crucial to coordinate national climate change instruments, such as national adaption programmes (NAPs), nationally applicable mitigation activities (NAMAs), and climate change investment plans, as well as national agricultural policies, strategies, and investment plans. A priority for 18 of the 44 nations organising NAMAs is agricultural activity (Lipper *et al.*, 2014).

3. Smart Practices and Technologies

Building the resilience of Indian agriculture to deal with climate change and variability is crucial for the food and livelihood security of farmers in general, and small and marginal farmers in particular, in the nation. This is because climate change and rising climatic variability are most likely to exacerbate the problem of food security by exerting environmental pressure on agricultural systems. In light of the World Bank's (2013) prediction that total crop production will increase by 60% by 2050 absent climate change, but only by 12% in the event of climate change under a 2°C warming by the 2050s, CSA gains considerable significance. In addition, compared to a scenario without climate change, the nation will need to import twice as many food grains to meet per capita calorie consumption. For the country's crop output to be sustainable, stable, and to provide rural communities with a stable source of food and income, it is essential to adapt to such climatic fluctuations (Ghosh,2019).

The weather-index-based crop insurance scheme (WBCIS), which was introduced in the nation in 2007 as an alternative to the yield-index-based National Agricultural Insurance Scheme, has been encouraging farmers to invest in their crops, raise agricultural productivity, and mitigate climate change, thereby boosting the resilience of smallholder production systems. The number of farmers covered by the programme expanded from 1,000 in 2003–2004 to over 12 million in 2011–2012, and both the number of payouts and insured farmers have grown significantly over time. By lowering farmers' risk when investing in farm implements and inputs like better seeds and fertilisers, this insurance programme has the potential to greatly increase food output (Neate, 2013). The Pradhan Mantri Fasal Bima Yojana, a crop insurance programme, has been highlighted in the Economic Survey 2017-2018 (Govt. of India, 2018) as a means of quickly determining losses and compensating farmers.

3.1. Smart Farming, Productivity and Food Security

When everyone, at all times, has physical, social, and economic access to enough, safe, and nutritious food that satisfies their dietary needs and food preferences for an active and healthy life, this is referred to as a condition of food security (FAO, 2001, p.49). Therefore, to ensure that there is enough food for everyone, it is necessary that (a) food be available in sufficient quantities, taking into account domestic production, imports, and national stocks; (b) household livelihoods be sufficient to ensure that people have access to food supplies; and (c) the available supplies must satisfy each individual's dietary and health needs (FAO, 2001). The main requirement is to increase food grain production because food security depends on the availability of food grains. We assess the effectiveness of various climate-smart agricultural practises and technologies in ensuring food security by enhancing agricultural productivity and improving the livelihoods and incomes of farm households in India against the backdrop of declining growth rates of yield and output of food grains in India since the early 1990s. Given that they account for more than 80% of farmers, contribute more than 50% of total agricultural output, cultivate 44% of the world's arable land, and ensure the livelihood and food security of millions of people, small landholders, with an average landholding size of less than two hectares, are a crucial group that requires special attention (Khatri-Chhetri, Aryal, Sapkota, & Khurana, 2016). In addition to potential advantages and disadvantages of implementing the technologies, it was discovered that farmer preferences for CSA technologies and their effects varied depending on location, agro-climatic zones, and socio-economic characteristics of the farmer (size of landholding, total income, access to credit services, etc.). Khatri-Chhetri *et al.* (2017) conducted a primary survey in 16 villages from four dryland districts of Rajasthan and discovered that the most preferred technologies were crop insurance, weather-based crop advisories, rainwater harvesting, site-specific integrated nutrient management, contingent crop planning, and laser land levelling. This suggests that the schemes for promoting these technologies should be site-specific and most suitable for the local farmers.

4. Impacts on Food Safety and Human Health

Due to changes in temperature and precipitation patterns, humidity, an increase in the frequency and intensity of extreme weather events, and changes in contaminant transport pathways, climate change can have an impact on food safety by altering the population dynamics of contaminating organisms. As the climate changes, modifications to food and farming systems, such as intensification to sustain supply, may potentially raise vulnerabilities (Tirado *et al.* 2010). Mycotoxin-producing fungi, bacteria that cause disease in aquatic food chains (like dinoflagellates and Vibrio), and increasingly heavy rainfall and floods that contaminate pastures with enteric microbes (like Salmonella) that can enter the human food chain are just a few examples of contaminating organisms whose biology has changed as a result of climate change. Changes in humidity and temperature outside of cold chains can also affect how well-preserved goods deteriorate and spoil. This is especially true for microbial decay, but it can also happen with pests like mites, beetles, and moths (Moses *et al.*, 2015). Climate change will have an impact on the range of mycotoxin-producing fungus, raising dangers in some regions (such as mid-temperate latitudes) and decreasing them in others (such as the tropics) (Paterson and Lima, 2010). Aflatoxin contamination of maize in Southern Europe will significantly increase (Battilani *et al.*, 2016), and deoxynivalenol contamination of wheat in Northwestern Europe will increase by up to three times current levels, according to strong evidence from process-based models of specific species (Aspergillus/Aflatoxin B1, Fusarium/deoxynivalenol). These models also project future climatic conditions (van der Fels-Klerx *et al.* 2012a, b). Despite the fact that downscaled climate models make any specific projection for a given geography uncertain (Van der Fels-Klerx *et al.* 2013), experimental evidence on the small scale indicates that the combined effect of changing temperature and rising CO2 levels can be much greater than temperature alone (Medina *et al.*, 2014). Aflatoxin risks are anticipated to change, but precise forecasts are challenging since they depend on regional circumstances (Vaughan *et al.* 2016). Foodborne infections in the terrestrial environment typically result from enteric contamination (from people or animals) and can be dispersed by wind (blowing contaminated soil) or flooding, both of which are projected to become more common as a result of climate change (Hellberg and Chu 2016). Additionally, the amount of water kept in storage for irrigation, which some locations may do as a method for adaptation, can play a significant role in the spread of diseases (as well as other pollutants). Diarrheal illnesses and contaminated water pose serious concerns to food security (Bond *et al.* 2018). Although there isn't much concrete data (in the form of projected models), numerous reviews and expert groups' findings indicate that risks from foodborne infections are anticipated to rise through a variety of methods (Kirezieva *et al.* 2015; Hellberg and Chu 2016). Another way that climate change may affect human health is via modifying human exposure levels through the biology of plants. This could involve how crops adapt to shifting insect pressure or how they store heavy metals as a result of climate change (Rajkumar *et al.* 2013, for example) (e.g., cassava produces hydrogen cyanide as a defence against herbivore attack). Each of these elements will result in distinct geographical effects on food safety (Paterson and Lima 2011). For instance, in Europe, it is anticipated that mycotoxins formed on plant products in the field or during storage, pesticide residues in plant products affected by changes in pest pressure, trace elements and/or heavy metals in plant products depending on changes in soil availability and abundance, polycyclic aromatic hydrocarbons in foods as a result of changes in long-range atmospheric transport and deposition, and preservatives in foods will have the greatest impact on food safety (Miraglia *et al.* 2009). In conclusion, there is moderate evidence and strong consensus that changes in food safety (and possibly changes in access to food due to food loss) will be impacted by climate change, primarily by increasing risks, but there is low confidence in the specific ways in which they may change for any given place.

5. Climate Change Impacts on Food Stability

Food stability is related to people's capacity to get and consume food steadily, preventing intermittent hunger. Increasing extreme occurrences brought on by climate change can affect the stability of food supplies.

5.1. Impacts of Extreme Events

An investigation of the prevalence of undernourishment (PoU) by FAO *et al.* (2018) revealed that in 2017, the average PoU for all countries exposed to climate extremes was 15.4%. The PoU was 20% for nations whose agricultural yields and production are also very vulnerable to climate change, or 22.4 percent for nations with a high PoU sensitivity to severe drought. The PoU is 9.8 points higher when there is a high vulnerability of agricultural production/yields and a high PoU sensitivity to extreme drought (25.2 percent). When countries depended heavily on agriculture, as shown by the amount of people employed in the industry, these vulnerabilities were found to be greater. Due to the central organisation of the national food supply chains in the capital city, Bangkok endured significant flooding in 2011 and 2012. (Allen *et al.* 2017). The IPCC predicts that in the future decades, some severe occurrences will become more frequent, longer, and more intense (IPCC 2018, 2012). Tigchelaar *et al.* (2018) tested these effects on food security and demonstrated increased instability in global grain commerce and international grain prices, particularly harming the approximately 800 million people living in extreme poverty who are most susceptible to increases in food prices. To estimate how yield variability will alter in the world's largest producers and exporters of maize under 2°C and 4°C of global warming, researchers integrated worldwide databases on maize output and climate variability with forecasts of future temperatures. According to Tesfaye *et al.* (2017), compared to the baseline (1950–2000), the size of heat-stressed areas in South Asia might rise by up to 12 percent in 2030 and 21 percent in 2050. Another recent study discovered that populations in Sub-Saharan Africa are the most vulnerable, with drier regions anticipated to dry faster, more severely, and to a greater extent than humid regions (Lickley and Solomon 2018).

5.2. Food Aid

After a climatic disaster, food aid is crucial for preserving lives and ensuring food security. Activities aimed at reducing catastrophe risk, addressing climate change, or boosting resilience helped 14.5 million people in 2015. (WFP 2018). However, there is no consensus on how to employ emergency food assistance more effectively because it may have unforeseen repercussions for certain people, groups, regions, and nations (Barrett 2006). These can include price rises or the negative reliance of food receivers (Lentz *et al.* 2005). Some writers (Clay 2006) claim that tied food aid delivered as "in kind" by the donor country hinders local food production, however other authors (Ferrière and Suwa-Eisenmann 2015) found no evidence to support this. Untied cash aid can be used to purchase food locally or in neighbouring countries, which is less expensive and can help local farmers improve their standard of living (Clay 2006). Ahlgren *et al.* (2014) discovered that the Marshall Islands' reliance on food aid as a result of climate change impacts can have a negative influence on health because of the inadequate nutritional quality of food aid, which could lead to an increase in chronic diseases in the future. Mary *et al.* (2018) shown that nutrition-sensitive help can lessen the prevalence of undernourishment in this regard. In conclusion, based on the AR5 and SR15 assessments that the likelihood of extreme weather events will increase, decreases in food stability and consequent increases in food insecurity are likely to occur. Examples of these events include increases in heatwaves, droughts, inland flooding, and coastal flooding due to rising sea levels, depending on region (medium evidence, high agreement).

6. Conclusion

The strategy of reforming agricultural systems to assist global food security and poverty reduction is affected by climate change, which changes agricultural productivity and food systems. Climate change increases risk and uncertainty for farmers and policymakers, but it need not paralyse thought processes38. In order to achieve the scale and rate of change necessary, an integrated, evidence-based, and transformative approach to addressing food and climate security at all scales requires coordinated actions from the global to local levels, from research to policies and investments, and across the private, public, and civil society sectors. The farm sector may transition to CSA routes with the correct practises, policies, and investments, lowering food insecurity and poverty in the short term while helping to lessen climate change's threat to food security in the long run.

References

Ahlgren, I., Yamada, S., & Wong, A. (2014). Rising oceans, climate change, food aid, and human rights in the Marshall Islands. Health & Hum. Rts. J., 16, 69.

Allen, A., Griffin, L., & Johnson, C. (eds.), 2017: Environmental Justice and Urban Resilience in the Global South (pp. 1–307). Palgrave Macmillan, New York.

Amadu, F. O., McNamara, P. E., & Miller, D. C. (2020). Understanding the adoption of climate-smart agriculture: A farm-level typology with empirical evidence from southern Malawi. World Development, 126, 104692.

Arslan, A., McCarthy, N., Lipper, L., Asfaw, S., & Cattaneo, A. (2014). Adoption and intensity of adoption of conservation farming practices in Zambia. Agriculture, ecosystems & environment, 187, 72-86.

Barrett, C.B. (2006). Food Aid's Intended and Unintended Consequences. ESA Working Paper No. 06–05, Agricultural and Development Economics Division, The Food and Agriculture Organization of the United Nations. (pp. 27). Rome, Italy.

Battilani, P., Toscano, P., der Fels-Klerx, V., Moretti, A., Camardo Leggieri, M., Brera, C., & Robinson, T. (2016). Aflatoxin B1 contamination in maize in Europe increases due to climate change. Scientific reports, 6(1), 1-7.

Bond, H., Bjornlund, H., Clench, C., Nickum, J., & Stephan, R. (2018). Developing a global compendium on water quality guidelines. Report of the International Water Resources Association (IWRA), 1-79.

Challinor, A., Martre, P., Asseng, S., Thornton, P., & Ewert, F. (2014). Making the most of climate impacts ensembles. Nature Climate Change, 4(2), 77-80.

Clay, E. (2006). Is food aid effective? Id21 insights, 61, 6.

Duflo, E., Kremer, M., & Robinson, J. (2011). Nudging farmers to use fertilizer: Theory and experimental evidence from Kenya. American Economic Review, 101(6), 2350–2390.

FAO (2001). The state of food insecurity in the World 2001. Rome: Food and Agriculture Organisation (FAO) of the United Nations.

FAO (2010). Climate-smart agriculture: Policies, practices and financing for food security, adaptation and mitigation. Rome: Food and Agriculture Organization of the United Nations (FAO).

FAO (2013). Climate-smart agriculture sourcebook. Rome: Food and Agriculture Organization of the United Nations (FAO). Retrieved from http://www.fao.org/ publications

FAO (2014). FAO success stories on climate-smart agriculture. Rome: Food and Agriculture Organization of the United Nations (FAO).

FAO, IFAD, UNICEF, WFP and WHO (2018). The State of Food Security and Nutrition in the World 2018. Building climate resilience for food security and nutrition. Food and Agriculture Organization of the United Nations, Rome, Italy (pp. 202).

Ferrière, N., & Suwa-Eisenmann, A. (2015). Does food aid disrupt local food market? Evidence from rural Ethiopia. World Development, 76, 114-131.

Ghosh, M. (2019). Climate-smart Agriculture, Productivity and Food Security in India. Journal of Development Policy and Practice, 245513331986240.

Government of India (2018). Economic survey 2017–18. New Delhi: Ministry of Finance, Department of Economic Affairs, Economic Division.

Muench, J., Jarvis, K., Gray, M., Hayes, M., Vandersloot, D., Hardman, J., & Winkle, J. (2015). Implementing a team-based SBIRT model in primary care clinics. Journal of Substance Use, 20(2), 106-112.

Herrero, M., Thornton, P. K., Bernués, A., Baltenweck, I., Vervoort, J., van de Steeg, J., & Staal, S. J. (2014). Exploring future changes in smallholder farming systems by linking socio-economic scenarios with regional and household models. Global Environmental Change, 24, 165-182.

IPCC (2012). Managing the Risks of Extreme Events and Disasters to Advance Climate Change Adaptation. A Special Report of Working Groups I and II of the Intergovernmental Panel on Climate Change. [Field, C.B., V. Barros, T.F.

Stocker, D. Qin, D.J. Dokken, K.L. Ebi, M.D. Mastrandrea, K.J. Mach, G.-K. Plattner, S.K. Allen, M. Tignor, and P.M. Midgley (eds.)]. Cambridge University Press, Cambridge, UK, (pp. 582).

IPCC (2018). Global Warming of 1.5°C an IPCC Special Report on the Impacts of Global Warming of 1.5°C Above Pre-Industrial Levels and Related Global Greenhouse Gas Emission Pathways, in the Context of Strengthening the Global Response to the Threat of Climate Change [Masson-Delmotte, V., P. Zhai, H.-O. Pörtner, D. Roberts, J. Skea, P.R. Shukla, A. Pirani, W. Moufouma-Okia, C. Péan, R. Pidcock, S. Connors, J.B.R. Matthews, Y. Chen, X. Zhou, M.I. Gomis, E. Lonnoy, T. Maycock, M. Tignor, and T. Waterfield (eds.)].

Khatri-Chhetri, A., Aggarwal, P. K., Joshi, P. K., & Vyas, S. (2017). Farmers' prioritization of climate-smart agriculture (CSA) technologies. Agricultural systems, 151, 184-191.

Khatri-Chhetri, A., Aryal, J. P., Sapkota, T. B., & Khurana, R. (2016). Economic benefits of climate-smart agricultural practices to smallholder farmers in the Indo-Gangetic Plains of India. Current Science, 1251-1256.

Kirezieva, K., Jacxsens, L., van Boekel, M. A., & Luning, P. A. (2015). Towards strategies to adapt to pressures on safety of fresh produce due to climate change. Food Research International, 68, 94-107.

Kirsten, J. (Ed.). (2009). Institutional economics perspectives on African agricultural development. Intl Food Policy Res

Lentz, E., Barrett, C. B., & Hoddinott, J. (2005). Food aid and dependency: implications for emergency food security assessments. IFPRI Discussion Paper No. 12–2, SSRN, New York, US, (pp. 1–50).

Lickley, M., & Solomon, S. (2018). Drivers, timing and some impacts of global aridity change. Environmental Research Letters, 13(10), 104010.

Lipper, L., Thornton, P., Campbell, B. M., Baedeker, T., Braimoh, A., Bwalya, M., & Torquebiau, E. F. (2014). Climate-smart agriculture for food security. Nature climate change, 4(12), 1068-1072.

Marenya, P. P., Gebremariam, G., & Jaleta, M. (2020). Sustainable intensification among smallholder maize farmers in Ethiopia: adoption and impacts under rainfall and unobserved heterogeneity. Food Policy, 95, 101941.

Mary, S., Saravia-Matus, S., & y Paloma, S. G. (2018). Does nutrition-sensitive aid reduce the prevalence of undernourishment?. Food Policy, 74, 100-116.

McCarthy, N., Lipper, L., & Branca, G. (2011). Climate-smart agriculture: smallholder adoption and implications for climate change adaptation and mitigation. Mitigation of Climate Change in Agriculture Working Paper, 3(1), 1-37.

Medina-Vayá, Á., Rodriguez, A., & Magan, N. Effect of climate change on Aspergillus flavus and aflatoxin B-1 production. Frontiers in Microbiology, 5, 348.

Miraglia, M., Marvin, H. J. P., Kleter, G. A., Battilani, P., Brera, C., Coni, E., & Vespermann, A. (2009). Climate change and food safety: an emerging issue with special focus on Europe. Food and chemical toxicology, 47(5), 1009-1021.

Moses, J. A., Jayas, D. S., & Alagusundaram, K. (2015). Climate Change and its Implications on Stored Food Grains. Agricultural Research, 4(1), 21–30

Neate, P. J. H. (2013). Climate-smart agriculture: Success stories from farming communities around the world. Copenhagen: Consultative Group on International Agricultural Research (CGIAR) Research Program on Climate Change, Agriculture and Food Security (CCAFS), and Wageningen: Technical Centre for Agricultural and Rural Cooperation (CTA).

Paterson, R. R. M., & Lima, N. (2010). How will climate change affect mycotoxins in food?. Food research international, 43(7), 1902-1914.

Rajkumar, M., Prasad, M. N. V., Swaminathan, S., & Freitas, H. (2013). Climate change driven plant–metal–microbe interactions. Environment international, 53, 74-86.

Sayer, J., Sunderland, T., Ghazoul, J., Pfund, J. L., Sheil, D., Meijaard, E., & Buck, L. E. (2013). Ten principles for a landscape approach to reconciling agriculture, conservation, and other competing land uses. Proceedings of the national academy of sciences, 110(21), 8349-8356.

Taylor, M. (2018). Climate-smart agriculture: what is it good for?. The Journal of Peasant Studies, 45(1), 89-107.

Tesfaye, K., Zaidi, P. H., Gbegbelegbe, S., Boeber, C., Rahut, D. B., Getaneh, F., & Stirling, C. (2017). Climate change impacts and potential benefits of heat-tolerant maize in South Asia. Theoretical and Applied Climatology, 130(3), 959-970.

Thornton, P. K., Ericksen, P. J., Herrero, M., & Challinor, A. J. (2014). Climate variability and vulnerability to climate change: a review. Global change biology, 20(11), 3313-3328.

Tigchelaar, M., Battisti, D. S., Naylor, R. L., & Ray, D. K. (2018). Future warming increases probability of globally synchronized maize production shocks. Proceedings of the National Academy of Sciences, 115(26), 6644-6649.

Tirado, M. C., Clarke, R., Jaykus, L. A., McQuatters-Gollop, A., & Frank, J. M. (2010). Climate change and food safety: A review. Food Research International, 43(7), 1745-1765.

Van der Fels-Klerx, H. J., Olesen, J. E., Naustvoll, L. J., Friocourt, Y., Mengelers, M. J. B., & Christensen, J. H. (2012b). Climate change impacts on natural toxins in food production systems, exemplified by deoxynivalenol in wheat and diarrhetic shellfish toxins. Food Additives & Contaminants: Part A, 29(10), 1647-1659.

Van der Fels-Klerx, H. J., van Asselt, E. D., Madsen, M. S., & Olesen, J. E. (2013). Impact of climate change effects on contamination of cereal grains with deoxynivalenol. PloS one, 8(9), e73602.

Van der Fels-Klerx, H. J., Olesen, J. E., Madsen, M. S., & Goedhart, P. W. (2012a). Climate change increases deoxynivalenol contamination of wheat in north-western Europe. Food Additives & Contaminants: Part A, 29(10), 1593-1604.

Vaughan, M., Backhouse, D., & Del Ponte, E. M. (2016). Climate change impacts on the ecology of Fusarium graminearum species complex and susceptibility of wheat to Fusarium head blight: A review. World Mycotoxin Journal, 9(5), 685-700.

Vermeulen, S. J., Campbell, B. M., & Ingram, J. S. (2012). Climate change and food systems. Annual review of environment and resources, 37(1), 195-222.

Vermeulen, S. J., Challinor, A. J., Thornton, P. K., Campbell, B. M., Eriyagama, N., Vervoort, J. M., & Smith, D. R. (2013). Addressing uncertainty in adaptation planning for agriculture. Proceedings of the National Academy of Sciences, 110(21), 8357-8362.

Vervoort, J. M., Thornton, P. K., & Kristjanson, P. (2014). F örch W, Ericksen PJ, Kok K, *et al.* Challenges to scenario-guided adaptive action on food security under climatechange. Global Environmental Change, 28, 383-394.

WFP (2018). Climate action. World Food Programme. Rome, Italy, www1.wfp.org/climate-action.

World Bank (2013). Turn down the heat: Climate extremes, regional impacts, and the case for resilience. Washington DC, WA: World Bank.

World Bank (2015). Future of Food: Shaping a Climate-Smart Global Food System; World Bank: Washington, DC, USA.

Chapter 11

Recent Managemental Practices for Improvement of Dairy Production and Productivity

Rakhshan Jeelani, Biswajit Brahma, Dipanjali Konwar and Asma Khan

Division of Livestock Production and Management, Sher-e-Kashmir University of Agricultural Sciences and Technology of Jammu, Jammu and Kashmir, India 181104. email: jitt77@gmail.com

Dairy farming is a class of agriculture for long-term production of milk, which is processed for eventual sale of a dairy product. The rising demand for fresh and packaged dairy products and ethnic dairy specialties is broadening the base of India's modern dairy sector, which accounts for almost 17% of India's expenditure on food. The Indian dairy market was valued at USD5.4 billion, in the year 2010, which raised to USD11.8 billion in the year 2015. By the year 2020 the dairy industry market in India is expected to be valued at USD22.5 billion based on the projections and past trends. Affluent farmers are seeking to establish dairying units on a commercial level, with a big herd size of 30-50 animals and modern machinery. These large commercial farms are required to cope with a growing population, higher incomes, increasing health consciousness, and the consequent explosion in demand for pouch milk and value added products across the country. Thus, farmers will have to shift towards adopting modern dairy farming techniques in order to increase their production. India ranks first among the world's milk producing Nations since 1998 and has the largest bovine population in the World. Milk production in India during the period 1950-51 to 2017-18, has increased from 17 million tonnes to 176.4 million tonnes as compared to 165.4 million tonnes during 2016-17 recording a growth of 6.65%. FAO reported 1.46% increase in world milk Production from 800.2 million tonnes in 2016 to 811.9(Estim) million tonnes in 2017. The per capita availability of milk in the country which was 130 gram per day during 1950-51 has increased to 374 gram per day in 2017-18 as against the world estimated average consumption of 294 grams per day during 2017. Most of the milk is produced by animals reared by small, marginal farmers and landless labours. Of the total milk production in India, about 48% milk is either consumed at the producer level or sold to non-producers in the rural area. The balance 52% of the milk is marketable surplus available for sale to consumers in urban areas. Out of marketable surplus it is estimated that about 40% of the milk sold is handled by the organized sector (20% each by Cooperative & Private Dairies) and the remaining 60% by the unorganized sector. According to a UN report released **(27 May 2015, Rome)** The number of hungry people in the world has dropped to 795 million – 216 million fewer than in 1990-92, or around one person out of every nine. But because of increase in population world's 70% of the population will be in hunger in 50 years from now (2060) according to the estimation by Food and Agriculture Organisation of the United Nations (FAO). The world human consumption for the animal protein was 29 g per capita daily (or 10 kg per capita consumption). However, there is an increasing trend towards per capita demand for animal source foods and it is occurring primarily in developing countries (80% of world population).India is the world's largest producer and consumer of dairy. The dairy industry in India was worth INR 5,000 billion in 2016. India is also globally the largest milk producing country since 1997. In India, the co-operatives and private dairies have access to only 20% of the milk produced. Approximately, 34% of the milk is sold in the unorganized market while 46% is consumed locally. This is in comparison to most of the developed nations where almost 90% of the surplus milk is passes through the organized sector. The Indian ice cream industry is one of the fastest growing segments of the dairy or food processing industry. Currently the ice cream market in India is estimated to be over INR 4,000 crores, and is growing at a rate of 15-20% year-on-year. It is projected that by 2019, the market will reach a value of approximately INR 6,198 crores. Growing population, changing lifestyles, expanding urbanization and accelerated climate changes are creating new challenges in Bovine breeding systems. In the past, the challenge was to ample feed, but now it is to provide essential nutrients to promote health especially reproductive health; and in the future, the challenge would be to provide optimal nutrients based on an animal's genetic profile and productivity. Fortunately, along with challenges, the developments in science are creating new avenues for tackling the challenges.

Livestock Sector in J&K: A Brief Overview

Dairy sector has emerged as one of the key components of poverty lessening, rural growth and nutritional security policies. In J&K, the dependence over livestock is far more due to topographical challenges that make other activities difficult and unsatisfactory. Agriculture and allied sectors contribute about 38 per cent to the state gross domestic product of which 11 per cent is contributed by the livestock. The economy of Jammu and Kashmir is still agriculture dependent with main occupation of majority of the rural people who earn their livelihood from it. The contribution of Agriculture and livestock to total GSDP has increased from 5.04%% in 2007-08 to 18.52% in 2013- 14. Integrated Sample Survey (2011-12) revealed that total livestock population of Jammu region had demonstrated an increase of 8.40% over the year 2009-10 and reached to 78.908 lakhs from 72.79 lakhs. Kashmir region also witnessed an increase of 4.63% as livestock population had reached to 74.994 lakhs from 71.67 lakhs. However, in Ladakh region livestock population slipped by 1.44% from 6.60 lakhs to 6.505 lakhs during the same period. J&K state witnessed 30% increase in its contribution towards national milk production. Approximately 95.38 thousand metric tones (TMT) of liquid milk have been imported through Lakhanpur inter-state border terminal, up to February in 2014-15, from 76.61 TMT in 2013-14 and 72.16 TMT in 2012-13. The total milk production by the state in 2011-12 from total milch animal population of 1228 thousand animals (703 thousand crossbred cows of age above 2.5 yrs, 524 thousand indigenous cows of age above 3 yrs and 417 thousand buffalo females of over 3 yrs age) was 1614 thousand tones. The dairying marketing channels in Jammu and Kashmir is largely under-developed. Jammu and Kashmir Milk Producers Cooperative Limited (JKMPCL) is the oldest organization with two milk plants, one each at Srinagar and Jammu, and is the state level federation of 186 and 130 dairy cooperative societies, respectively. The pasteurization capacity of these two plants is 30000 litres / day (LPD) each and presently, 10000 -12000 LPD and 7000–8000 LPD are sold from Srinagar and Jammu, respectively. Khyber Agro farms Pvt. Ltd is the biggest milk plant of Jammu and Kashmir functioning in the Kashmir Valley with pasteurization capacity of 1 lakh LPD. Presently, it sells 50000–60000 LPD, of which 60 per cent is procured from the neighbouring state of Punjab and 40 per cent from the local network of vendors / wholesalers, etc. Milk distribution system in J&K has been quite exploitative and traditional with the organized sector handling only 1.70 and 5.0 per cent of the total liquid milk production and marketed surplus. The situation is alarming in the state and to reverse these trends, it is essential that an integrated dairy development project is planned and implemented to boost milk production, to provide a regular and remunerative market to the milk producers and capture a major share of urban milk markets with regular supply of quality milk.

As per the Integrated Sample Survey Report (ISS), 2011-12, total estimated meat production of State was worked out to be 322.781 Lakh Kg consisting of

80.79% of (260.762 Lakh Kg) Red Meat and 19.21% (62.019 Lakh Kg) of White Meat. As per the latest Integrated Sample Survey Report (ISS), 2010-11, total estimated meat production of State was worked out to be 308.986 Lakh Kg. As per estimated results of ISS survey 2011-12, the meat availability was worked out to be 2.701 Kg per person per year against 2.650 Kg for 2010-11.Meat production registered a growth from 308.986 Lakh Kg to 322.781 Lakh Kg in the year 2011-12 over the previous year resulting into 4.47% growth. Average wool yield per annum is 26.218 Kg for cross-breed sheep and 17.495 Kg for local-breed sheep. During the year 2011-12 the total wool production estimated for the State was 75.295 Lakh Kg which consisted of 61.586 Lakh Kg (81.79%) from Crossbreed Sheep and 13.709 Lakh Kg (18.20%) of Local-breed Sheep. During the year 2010-11, the total wool production estimated for the state was 73.819 Lakh Kg.

Animal diseases continue to be a major constraint in livestock productivity and agriculture development. Rinderpest has been eliminated in the state, whereas Anthrax, Black Quarter, Hemorrhagic septicemia, Foot and mouth disease and helminthic infections have been controlled substantially. However, ectoparasitic infestation and protozoan infections are still a major issue hampering livestock health status in animals. One of the major reason for low production in dairy sector is Jammu and Kashmir's diversified climate e.g. cold desert in Ladakh region and temperate climate in Kashmir region and humid sub-tropical climate in Jammu region. Small size of land holding is another hurdle for limited milk production in the state as state does not produce enough fodder because of lack of modern technologies in small and marginal farmers. Area specific dairy mineral mixture should be given to the dairy farmers to fulfill the diet of dairy animals. The dairying marketing channels in Jammu and Kashmir is largely under-developed. Lack of a regular outlet of milk has made the producers extremely dependent on the milk vendors/ shops, who exploit them by not paying remunerative prices, thereby discouraging the producers to increase their milk production. The situation is alarming in the state and to reverse these trends, it is essential that an integrated dairy development project is planned and implemented to boost milk production, to provide a regular and remunerative market to the milk producers and capture a major share of urban milk markets with regular supply of quality milk, which is otherwise being captured rapidly by the neighboring state of Punjab. The population of sheep has stagnated except a marginal increase between 2003 and 2007. The wool production has declined to around 40 million kg, of which fine wool is only 10%. The major bottlenecks in improving sheep and goat production are lack of pastures and fodder shrubs and trees in the area where the sheep and goat thrive The higher stocking rates, excessive grazing pressure, change in plant composition of grazing areas and reduced biomass availability have rendered migratory system of rearing difficult to sustain. Excessive parasitic load in migratory flock, higher energy spent while covering greater distance during migration along with other factors lead to body weight loss resulting into poor remunerative price realization by migratory Sheppard community.

Main Challenges and Problems of the Dairy Industry

Shortage of feed and fodder

There is an excessive number of unproductive animals which compete with productive dairy animals in the utilization of available feeds and fodder. The grazing area is being reduced markedly every year due to industrial development resulting in shortage of supply of feeds and fodder to the total requirement. Ever increasing gap between demand and supply in feeds and fodder limits performance of dairy animals. Moreover, provision of poor quality of forage to dairy cattle restricts animal production system. The low capability of purchasing feeds and fodder by the small and marginal farmers and agricultural labourers engaged in dairy development result in inadequate feeding. Non-supplementation of mineral mixture results in mineral deficiency diseases. High-cost Feeding reduces the profits of the dairy industry.

Breeding system

Late maturity, in most of the Indian cattle breeds, is a common problem. There is no effective detection of heat symptoms during oestrus cycle by the cattle owners. The calving interval is on the increase resulting in a reduction in efficiency of animal performance. Diseases causing abortion leads to economic loss to the industry. Mineral, hormone and vitamin deficiencies lead to fertility problems.

Education and Training

A vigorous education and training programmes on good dairy practices could result in the production of safe dairy products, but to succeed they have to be participative in nature. In this regard, education and training of all the employees is essential so that they understand what they are doing and develop a sense of ownership. However developing and implementing such programs in the dairy sector requires a strong commitment from the management, which at times, is a stumbling block.

Health

Veterinary health care centres are located in far off places. The ratio between cattle population and veterinary institution is wider, resulting in inadequate health services to animals. No regular and periodical vaccination schedule is followed, regular deworming programme is not done as per schedule, resulting in heavy mortality in calves, especially in buffalo. No adequate immunity is established against various cattle diseases.

Animal and Product Hygiene

Many cattle owners do not provide proper shelter to their cattles leaving them exposed to extreme climatic conditions. Unsanitary conditions of cattle shed and milking yards, leads to mastitis. Unhygienic milk production leads to a reduction in storing quality and spoilage of milk and other products.

Less emphasis on indigenous breeds

The native cattle is more resistant to diseases and environmental conditions, but very less attention is given to indigenous cattle and their breeding programmes. More emphasis is given to cross breds which are not adapted to the local climatic conditions.

Marketing and Pricing

Dairy farmers are not getting remunerative price for milk supply. Due to the adoption of extensive crossbreeding programme with Holstein Friesian breed, the fat content of crossbreed cow's milk is on the declining condition and low price is offered as the milk price is estimated on the basis of fat and solid nonfat milk content. There is also a poor perception of the farmers, due to lack of marketing facilities and extension services, towards commercial dairy enterprise as an alternative to other occupation.

Welfare problems

Despite the relaxing image of outdoor farming, several industry practices negatively affect dairy cows. To meet production demands, dairy cows are subject to a continuous cycle of impregnation, induced calving and milking.

Tail-docking and horn removal are routinely performed without pain relief. Lameness is another major animal welfare problem, often the result of environmental pressures, such as tracks, herd size and handling. The average lifespan of a dairy cow is six to seven years, whereas generally cows can live for 20 to 25 years.

One of the most controversial issues is young "bobby" calves. A bobby calf is a newborn calf, less than 30 days old, who has been purposely separated from their mother. Immediately after separation, cow and calf call out and search for each other. Most bobby calves are slaughtered within the first week of their life. Handling and transport pose added problems for young calves who have not developed herding behaviours, are vulnerable to stress, and are forced to go without their mother's milk. Each year, 450,000 bobby calves are slaughtered.

The environmental impact

The effect of thermal stress on livestock is multidimensional, affecting metabolism, production, product quality, health and reproduction negatively (Babinszky *et al.*, 2011). The feed intake of animal is reduced and as a result the energy is used for maintenance purpose of the animal and nothing is left for the production purpose. In India, an estimated loss of 1.8 million tonnes of milk annually, is attributed to heat stress amounting to a fiscal loss of approximately Rs. 2661 crore (Upadhayay, 2010).

High perishability of dairy products

In India, ice cream industry is mostly regional and there is a multitude of brands focusing on only one or two districts or in some case only one state. There are very few national brands and the major reason behind slow growth of the smaller players is the high perishability of ice cream products.

Unavailability of records

As per 19th Livestock census, there are 88 million In-Milk animals whose records are unavailable on an annual basis even. Records of those in breeding stage, their productivity, treatment and vaccination are also not properly maintained by State Animal Husbandry Departments. This is because the system for maintaining records on the above aspects has not yet evolved in complete shape due to lack of prioritization. Impediments like lack of animal identification and traceability, inability to meet sanitary and phyto sanitary conditions also need to be addressed in this connection.

A brief history of Dairy Development Programmes in India:

Government of India has made efforts for strengthening infrastructure for production of quality milk, procurement, processing and marketing of milk and milk products through following Dairy Development Schemes in the recent pasts.

Military Dairy Farms

The earliest attempts at dairy development can be traced back to British rule, when the Defense Department established military dairy farms to ensure the supply of milk and butter to the colonial army. The first of these farms was set up in Allahabad in 1913 and subsequent facilities were established at Bangalore, Ootacamund and Karnal. These farms were well maintained and even in early stages, high quality milch animals were raised. But it failed to supply milk to urban consumers and it catered to only the needs of the military personnel.

The Royal Commission on Agriculture

The first attempt to conceive a set of policies for livestock development in India was the Royal Commission on Agriculture (1928). Following the report of the Royal Commission on Agriculture,interest in promoting development of animal husbandry gradually came to focus on ways to increase milk production and to improve the quality of milch cattle. However, in the pre-independence period attempts to improve the quality of the bovine population in the country were limited in scale and geographical coverage.

Key Village Scheme

Key Village Scheme (KVS) was the most important component of he animal husbandry development programmes during the first three Five Year Plans.

Initially, its main focus was on increasing the supply of breeding bulls in the country by setting up bull breeding farms in the major cattle tracts. Gradually the KVS was transformed into a more comprehensive programme for general cattle development intended to improve the productivity of cattle by giving simultaneous attention to better feeding, improved breeding, effective disease control measures, scientific management practices and organized marketing facilities. Towards the end of the second plan nearly 600 KVS centres were functioning in the country covering an estimated 6 million cows and she-buffaloes, that is, about 10 per cent of the total stock. However in the absence of stable and remunerative market for milk, production remained more or less stagnant. During the two decades between 1951 and 1970 milk production grew by barely 1 per cent annually while per capita milk availability declined by an equal ratio.

Intensive Cattle Development Project

The perceived failure of the KVS to make significant impact and the shortage of milk in the rapidly growing urban areas led to the formulation of the Intensive Cattle Development Project (ICDP). Its primary purpose was to increase the production of milk to feed public sector dairy plants in the Hinterlands of the main urban centres. Consequently they placed great emphasis on cross-breeding in indigenous cows with exotic dairy breeds and tended tobe concentrated in milk shed areas of large cities and towns. Animal health and breed improvement remained common elements in the project. The scheme envisaged provision of all necessary inputs and services simultaneously to milk producers. The ICDP is, thus, distinguished from KVS by the shift of attention to the cross-breeding component and closer link up with dairying and urban milk supply programmes. By 1960, 62 centers were functioning in the country under the ICDP. In areas where ICDP existed, the KVS was merged into the ICDP. In areas where ICDP did not exist, KVS was continued in theoriginal form.

Establishment of Cattle Colonies and Milk Schemes

During 1960s various state governments tried out different strategies to develop dairying, including establishing dairies run by their own departments,setting up cattle colonies in Bombay, Calcutta and Madras. These government projects had extreme difficulties in organizing rural milk procurement and running milk schemes economically, yet none concentrated on creating an organized system for procurement of milk, which was left to contractors and middlemen. Milk's perishable nature and relative scarcity gave the milk vendors leverage, which they used to considerable advantage. This left government -run dairy plants to use large quantities of relatively cheap, commercially imported milk powder, which resulted in a decline in domestic milk production.

AMUL and the Evolution of the Anand Model

AMUL formed the basis for the Anand pattern of dairying, referring to its origin in Anand District, in the state of Gujarat. Under the Anand pattern structure individual farmers are joined in village level dairy co-operative societies which are joined to form district level unions which, in turn, are joined in state level marketing federations. In each state the Anand Pattern has the following features:

a. Decentralized milk production by the small milk producers
b. Milk procurement by the village level dairy co-operative societies.
c. Centralized milk processing by the district-level unions.
d. Marketing of milk and milk products by the state level federation.

The primary milk producers democratically govern this entire federal co-operative structure to ensure that higher-tier organizations servethe purpose of the lower levels and that the gains at all levels go back to themilk producers in significant measure. The core feature of the Anand pattern model is farmer control of the three stages following production, that is,procurement, processing and marketing of milk and milk products.

Emphasis on Cross-Breeding Strategy

The Board of Agriculture and Animal Husbandry Wing of the Government of India, in 1958, and the expert committee appointed to evaluate the KVS, in 1959, have recommended a shift of policy to the cross-breeding of indigenous non-descript cattle with exotic stock for rapid increase in milk production. Though the Third Five Year Plan referred to the need to evolve better dual-purpose breed for increasing the work capacity and milk production potential of the Indian cattle, its programmes gave high priority to cross-breeding for increased milk production. Such a shift of policy was the need of the time to meet the increasing demand of milk in the urban and sub-urban centres of the country.

National Dairy Development Board (NDDB)

It was constituted by the Ministry of Agriculture and Irrigation, under the Societies Registration Act, in 1965. Its headquarters were established at Anand. During its initial stages NDDB was assisted financially by the Government of India, the Danish Government and by AMUL. In 1969, NDDB formulated an integrated dairy development programme under Dr. Varghese Kurien, its founding Chairman. It was accepted by Government of India in 1970 and it functioned as a corner stone for the Operation Flood I.

Indian Dairy Corporation (IDC)

In 1970 the Government of India established a public sector company, the Indian Dairy Corporation (IDC). It functioned as a finance and promotion link for the Phase I programme of NDDB. The IDC was given responsibility for receiving the project's donated commodities, testing their quality, their storage and transfer to user dairies and receiving the dairie's payments.

Dairying in Period of Operation Flood (1970-1996)

Government of India launched a massive dairy development programme popularly known as Operation Flood (OF) from 1970 to 1996. The programme was initially started with the help of the World Food Programme(WFP) and later continued with diary commodity assistance from the European Economic Community (EEC) and a soft loan / credit from the World Bank. The OF programme established milk producers' cooperatives in villages and made modern technology available to them. The broad objectives were to increase milk production ("a flood of milk")2, augment rural incomes and transfer to milk producers the profits of milk marketing that were hitherto enjoyed by well-to-do middlemen. This kind of innovative effort has greatly increased milk production and ushered in a "White Revolution", making India the world's largest milk producer.

Implementation of Operation Flood

The programme was implemented in three phases:

- ❒ Operation Flood I [1970-1981]
- ❒ Operation Flood II [1981-1985]
- ❒ Operation Flood III [1987-1996]

Operation Flood I [1970-1981]

e. Started in 1970, it envisaged certain specific targets:

f. The organization of one crore farmers into 30,000 village cooperatives in virtually all the states of India.

g. Establishing a national bufferstock of skim milk powder and butter oil.

h. The evolution of a national milk grid covering all parts of India and connecting all the major consumption and production centres.

i. To help the State Dairy Co-operative Federations to set upprocessing facilities and to develop the National Milk Herd of one crore improved buffaloes and dairy cattle.

j. To increase the daily per capita consumption of milk from107 grams in 1970 to 144 grams in 1985.

During its first phase, the project aimed at linking India's 18 best milk sheds with the milk markets of the four metropolitan cities of Delhi,Mumbai,

Calcutta and Madras. The programme visualized organizing dairy cooperatives at the village level, creating the physical and institutional infrastructure for milk procurement, processing and marketing services at the union level and establishing dairies in India's major metropolitan centre's.

Operation Flood II [1981-1985]:

The second phase of the programme was implemented between 1981and 1985. It was an extension and intensification of the first phase to cover more cities and districts in India. Both in terms of the financial outlay involved[Rs. 4800 million] and of the geographical coverage (160 districts) it was one of the biggest dairy development projects ever undertaken in India by the National Dairy Development Board. The main objectives of the project are:

a. to build up the infrastructure for the development of a growing and self-reliant dairy industry consisting of 10 million rural families of milk producers and a national milch herd of about14 million cross bred cows and she-buffaloes by middle 1980s;

b. to link up the rural supply sources and urban demand centre's (with a population of 150 million) through the establishment of approximate marketing arrangements; and

c. to increase per capita consumption of milk in the national diet.

With Operation Flood II, a self-sustaining system of 43,000 village co-operatives covering 4.25 million milk producers had become a reality. Phase II mainly emphasized to build infrastructure for technical input services and management services. Phase II built on the foundations established by Phase I.

Operation Flood III (1987-1996):

The third phase of the Operation Flood (1987-1996) enabled dairy co-operatives to expand and strengthen the infrastructure required to procure and market increasing volumes of milk. Phase III consolidated India's dairy co-operative movement by adding 69,600 new dairy co-operative societies and thereby covering 90 lakhs milk producer members. These co-operatives form part of the National Milk Grid which today links the milk producers with consumers in more than 799 towns and cities,bridging the gap between the seasonal and regional variation in the availability of milk while at the same time ensuring a remunerative price to the producers and supplying quality milk and milk products to the consumers. For the five years ending March, 2003, the average milk procurement by dairy co-operatives grew at 7.3 per cent whereas the marketing of milk by co-operatives grew at3.2 per cent. Phase III gave increased emphasis to research and development in animal health and animal nutrition. Innovations like vaccine for The ileriosis, bypass protein feed and urea-molasses mineral blocks, all contributed to the enhanced productivity of milch animals.

Introduction of Milk and Milk Products Order

The Government introduced the Milk and Milk Products Order(MMPO) in 1992 under the essential commodities act of 1955 to regulate the production of milk and dairy products. The order required permission from State/Central registration authorities to set up units handling more than 10,000litres of milk per day or milk solids upto 500 tons per annum, depending on the capacity of the plant. The order included sanitary and hygienic regulations to ensure product quality. However, concerns were raised about these government controls and licensing requirements for restricting large Indian and multinational firms from making significant investments in this sector. The government, therefore,amended the MMPO in March, 2002, and restrictions on setting up milk processing and milk product manufacturing plants were removed and the concept of milk shed was abolished. This amendment is expected to facilitate the entry of large companies, which would definitely increase competition in the domestic markets.

Uruguay Round Agreement on Agriculture

The second major development in Indian dairy sector policy came when India signed the Uruguay Round Agreement on Agriculture (URAA) in1994 and became a member of the World Trade Organization (WTO), which made India open up its dairy sector to world markets. The import and export of dairy products were delicenced and decanalized and trade in dairy products was allowed freely with certain inspection requirements. The first major step was taken in 1994-95, when the import of skim milk powder and butter oil were decanlized. Restrictions on the remaining products were removed in April 2002. Now India has bound its import tariffs for dairy products at low levels according to the Uruguay Round decisions.

Strengthening Co-operative Business

Perspective 2010 aims to recruit, train and motivate increasing number of women to work for co-operatives, to achieve significant improvements in dairy husbandry, as they primarily shoulder animal husbandry related responsibilities in rural India. It visualizes the consolidation and grow thin milk and milk product marketing, promoting better equity for regional co-operative brands and developing qualified and skilled manpower. It also aims to persuade the State and Central Governments to remove the shackles on cooperative laws so that co-operatives can compete on equal terms with other forms of enterprise. Expanding the market is a major target of perspective 2010. It offers financial and technical help to milk unions and federations in areas such as sales promotion, consumer education, infrastructure development etc. As part of sales promotion it recommends standardization of artwork, colour, logo and retail outlet design across regional co-operative brands with a view to promote better recall by consumers under a common mnemonic umbrella. Another target is increasing women membership in dairy co-operatives to 50 per cent and improving women participation in the governance of dairy co-operatives at all levels.

Production Enhancement

Perspective 2010 stresses to improve the production potential of indigenous breeds of cattle such as Sahiwal, Gir, Rathi and Kankrej and breeds of buffalo such as Murrah, Mehsana and Jaffarbadi through appropriate selection programme. It gives proper direction to cross breeding technology to increase production in such a way that crossing of non-descript cattle with Holestein Friesian in areas with adequate feed and fodder and with Jersey in resource poor areas. As a step to increase production and availability of fodder, it appeals to unions, NGOs and co-operatives to put common property area under improved pasture and fodder tree. It promotes first aid coverage through village level societies and disease Free Zones in the country.

Quality Assurance Programmes

As part of increasing quality it facilitates improvement of hygiene,sanitation, food safety and operating efficiency in the dairy plants and sensitize dairy personnel to product quality aspects as per international standards. It promotes encouragement of quality incentives supported by educational programmes for dairy co-operative society staff, transporters and farmer producers. Quality is assured through facilitating dairy co-operatives in ISO9000-2000 (Quality Management Systems), ISO HACCP (Safety Management Systems) certification and maintain the required plant conditions under the accreditation on a sustainable basis.

Information and Development Research

'Perspective 2010' plans to link large Co-operatives, Unions,Federations and NDDB in a national network that collects and disseminates information to all. It ensures the availability of analytical information for policy planning and implementation. The integrated dairy industry information service facilitates decision making at various levels in co-operative institutions with the help of an extensive on-line computer network that analyses relevant data obtained from Dairy Co-operative Societies, District Milk Producers' Union,State Milk Marketing Federations, NDDB and research institutions.

Perspective 2010 also proposes the need of a National Database that generates data on milk supply (producer, animal and village data) data on milk and milk product demand (consumer and urban data) performance data (societies, unions and federations) and secondary data (domestic and international).

Role of Modern Management

Management of Diary:

The scheme for diary, farming should include information on land, livestock markets, availability of water, feeds, fodders, veterinary aid, breeding facilities, marketing aspects, training facilities, experience of the farmer and the type of assistance available from State Government, dairy society/union/federation.

(A) Technical Feasibility – this would briefly include:

1. Nearness of the selected area to veterinary, breeding and milk collection centre and the financing bank's branch.
2. Availability of good quality animals in nearby livestock market.
3. Availability of training facilities.
4. Availability of good grazing ground/lands.
5. Green/dry fodder, concentrate feed, medicines etc.
6. Availability of veterinary aid/breeding centres and milk marketing facilities near the scheme area.

(B) Economic Viability – this would briefly include:

1. Cost of for feeds and fodders, veterinary aid, breeding of animals, insurance, labour and other overheads.
2. Output costs i.e. sale price of milk, manure, gunny hags, male/female calves, other miscellaneous items etc.

Modern and well established scientific principles, practices and skills should be used to obtain maximum economic benefits from dairy farming.

Reducing Pre-weaning Mortality

A nutritional intervention for decreasing mortality in young livestock would include the use of colostrum for neonates. Colostrum contains high levels of energy and nutritionally important proteins, and neonates depend on the early infusion of nutrients to maintain body temperature because their energy stores are low at birth.

Improving Grass Forage

Many forage-fed animals in the tropics grow slowly and produce small amounts of milk because their diets are inadequate in protein, energy, and micronutrients. The types of forage available to the animals are mainly the C4 grasses (so named for the metabolic pathway used to fix carbon dioxide). The C4 grasses that predominate in the tropics are less digestible than temperate C3 grasses and have low energy and protein content. Until recently, there was little work on characterization of forage traits that need improvement, and temperate forage still receives far more attention than that grown in the tropics. In general, tropical forage plants have received little attention from plant breeders with a few notable exceptions: alfalfa, which is grown in some tropical highlands; *Brachiaria*spp.; *Pennisetumpurpureum* (elephant or Napier grass); and *Panicum maximum* (Guinea, colonial, or Tanganyikagrass. The collections of germplasm of tropical forage are poorly funded, and loss of current accessions is threatened. The collections include diverse accessions that may be important sources of disease resistance, increase in digestibility, or increase in biomass production. For example, the most recent outbreak of a smut (*Ustilago kamerunensis*) is affecting Napier grass. In much of eastern and southern Africa, farmers rely heavily on Napier grass because it produces copious amounts of reasonably high-quality forage. The smut has the potential to affect the small-holder dairy industry seriously and has already reduced forage yields in much of the Kenyan highlands (Farrell *et al.*, 2002). Research to understand smut biology and to develop resistant strains of Napier grass is important for the rapidly growing dairy industry in the Kenyan highlands. A better understanding of plant chemistry and lignin synthesis could help plant breeding programs to improve the nutritional value of forage because the lignin cross-linkages affect whether plants are easily digested. There may be advantages in attaching work on this problem to the burgeoning international interest in biofuels, such as switch grass. There is a common interest in understanding how lignin cross-linkages can be broken down, whether in the context of biofuels or with respect to the processes occurring in forage digestion by ruminants.

Improving Legume Forage

In temperate areas, legumes, especially alfalfa and clover, are high protein, highly digestible forage that permit cows to sustain milk production as high as 20 kg/day on forage alone. In the tropics, however, many promising legume species contain high concentrations of anti-nutritionalfactors (such as proanthocyanidins, hydrolyzable tannins, alkaloids, and terpenoids) that confer disease resistance on the plants. Condensed tannins have both beneficial and deleterious effects on domestic animals. The adverse effects of consuming high-tannin forage include lower feed intake, lower protein and dry matter digestibility, inhibition of microbial and mammalian enzymes, reduced live weight gain and milk yield, and systemic effects that are due to absorption of phenolics and are sometimes offset by lower urinary nitrogenloss, greater parasite resistance, and improved efficiency of nutrient use. The apparently contradictory research results are due largely to the heterogeneity of tannin structures and to variation in the quantities ingested. Achieving the goal of developing disease resistant legumes that provide animals with needed nutrients requires research on tannin chemistry linked to legume breeding programs. Progress has been made in understanding some aspects of tannin synthesis, but the polymerization process that affects tannin chemistry and anti-nutritive effects remains poorly understood.

Evolving Technologies

Improving Animal Germplasm

Since the beginning of domestication of animals, substantial progress has been made in improving their characteristics as food and fiber producers by selectively mating individual animals that had advantageous traits (phenotypes). The importance of phenotypic information is sometimes lost in this age of genomics, and it is astonishing to recognize that animal breeders could triple average milk yield of dairy cattle in 50 years without knowing a single gene involved or having any genome sequence information to guide them. They simply needed to know the milk production traits of members of the dairy cattle family and select the right mates to breed.

Although it is possible to practice breeding of that type on a farm or village scale, small-herd owners in SSA and SA are likely to have difficulty in systematically improving the genetic potential of their livestock by using only locally available germplasm. Nor can small-holders apply modern quantitative breeding practices on the basis of the knowledge of genotypic associations with specific traits; information systems to collect phenotypic and genotypic data from populations of the desired species or breeds systematically have not been put into place.

The use of quantitative phenotypic methods to improve breeding requires collecting data on a large number of animals in a family that exhibit wide variation in the traits of interest. That kind of effort typically takes place in breeding centers, where resource populations of animals can be developed over a decade or two and individual phenotypes can be collected and recorded to make it possible to identify genetically superior animals. Infrastructure is needed to distribute the germplasm to farmers through artificial insemination and embryo transfer techniques. In industrialized countries, that approach has been used over the last 50 years to develop animals with superior genetics, and it is the model used in developing countries, often successfully. However, the scientific community is now in a position to bypass many of the heavily resource-dependent approaches used in the industrialized world. Emerging technologies offer potentially practical approaches for more rapidly discovering superior livestock genetics and delivering them to subsistence farmers in SSA and SA.

Leapfrogging Selective Breeding with Molecular Sampling : DNA-Derived Pedigrees

The buffalo is a primary source of milk protein and is used for draught and as a supplementary source of meat in parts of Asia.Ten major breeds exist in India, some having been selected and maintained for each of the three functions, which are essential to the farm economy. National programs to improve milk and meat production have been initiated and are being termed the "white" and "red" revolutions, respectively, in keeping with the name of the Green Revolution.

Genetic improvement for production traits and disease resistance in buffalo does not benefit from the availability of the powerful genomic tools recently generated for domestic cattle in the United States, for two mainreasons: the areas of the world where buffalo are economically important lack the financial resources for genomic research, and the application of genomic research to identify genetically meritorious individual animals can be applied only within families of animals. There is no such information on *Bubalusbubalis*, the Asian water buffalo, or on any farm animals (such as goats and hair sheep) raised by subsistence farmers in SSA and SA, so the use of well-established quantitative genetic tools is precluded. However, it may be possible to construct an equivalent dataset from the bottom up with the aid of molecular genetic tools. To implement that approach, a reference genome of the breed of interest would need to be generated with DNA sequencing, and DNA samples and phenotypic data would have to be collected from several thousand animals in geographic regions that have common environmental stresses. Single nucleotide polymorphisms (SNPs) would be generated from the DNA samples by sequencing regions of the genome that have proved to be informative in related species. The database of tag SNPs generated from the sequencing data would be aligned with the reference sequence to build family pedigrees. With pedigrees in hand, traditional quantitative tools could be applied to identify animals of superior genetic merit. The approach requires several lines of research. An inexpensive field kit for preserving DNA in tissue samples (ear snips, buccal swabs, or the equivalent) that does not require refrigeration would have to be developed, as would an effective questionnaire for gathering trait phenotypes. Whole genome sequencing (6X coverage) of *BubalisBubalis*, the African buffalo (*Synceruscaffer*), *Bosindicus*cattle, sheep, and at leasta few representative milk- and meat-producing breeds of goats and sheep should be included in the sequencing project. Finally, substantial investment in developing informatics algorithms for what is essentially reverse engineering of pedigrees from SNP data would be needed. With today's sequencing capability, it might take 2 years to generate a reference genome sequence for a species. It might take a year each to develop a DNA tissue-sample preservation kit and a phenotype questionnaire, 3 to 5 years to collect DNA samples and phenotypic data, and a year to build pedigrees and test the hypothesis that animals of high genetic merit can be identified with this approach. All the steps except the last can be conducted in parallel, the overall time frame of the project to reach proof of concept would be 6 to 10 years. If the project were successful, its impact would be large. DNA-enabled approaches to building a pedigree would leapfrog existing approaches by eliminating the decades of breeding needed to create resource populations, the usual starting point of contemporary genetic-genomic analyses. The substantial costs of housing and feeding such a population would be eliminated. Most important, it would provide a tool to identify genetically superior animals without having to develop the enormous infrastructure currently used in the developed world.

Genetic Engineering

For the first 8 to 10 millennia since animal domestication began, selective breeding has been the method by which desirable phenotypes were enriched in a population. The dramatic diversity generated in dog breeds and improvement in the efficiency of producing dairy cattle are just two examples of the power of selective breeding. However, the approach has limitations, of which the largest is the inability to introduce a trait if genetic information on the trait does not exist in the species of interest. For example, endowing swine with the ability to synthesize lysine *de novo*, which would eliminate the need to supplement feed with what is now an essential amino acid, is impossible because the biochemical pathway does not exist in any breed of swine. The pathway does exist in bacteria and yeast, but that is of no use to the animal breeder. Furthermore, selective breeding lacks precision. There are many examples of selecting for one important economic trait at the expense of another, such as sacrificing the reproductive performance of dairy cattle for increased milk production. But, as previously described, the most serious constraint in applying modern tools of genetic selection is the time needed to build resource populations and harvest the phenotypic information needed to populate the genetic algorithms for each breed of interest. In 1981, a new method for altering the genetic makeup of mammalian offspring led to the ability to place genes from one species (transgenes) into the genome of another (Gordon and Ruddle, 1981). Transgenes can encode completely novel natural or synthetic information, modulate the level of gene expression, or switch transcription on or off conditionally or permanently. In the last 2 decades, genetically engineered cattle, chickens, goats, pigs, rabbits, and sheep have been produced. Transgenic livestock applications have been diverse and range from projects focused on animal well to drug manufacturing. Transgenic animal technology has now advanced to the point where specific genetic information can be introduced precisely into any desired location of the genome. In theory, the technology provides a vast array of new ways to address challenges in animal agriculture.

Engineering Animals for Disease Resistance

Transgenic technology can be used in many ways to reduce susceptibility to disease in animals. It can be directed at a specific pathogen or a wide variety of pathogens, depending on the protein encoded by the transgene. There are dozens of examples of successful application of genetic engineering to protect mice, and many are expected to be predictive of outcomes of transgenic livestock experiments. However, to date there is only one example of genetic engineering that has protected a livestock species from disease. In general, a transgenic strategy for disease resistance involves identifying an anti-pathogenic protein to be expressed in the animal and determining in which tissue and at what developmental stage expression should occur. The transgenic protein should cause no harm to the animal itself or to the consumer that eats it. If possible, the transgene product should avoid interfering with endogenous homeostatic feedback loops. The recombinant protein produced should be benign to the environment. Finally, a plan should be devised to prevent the target pathogen from developing resistance to the transgene product.

RNA Interference

RNA interference (RNAi), is an evolutionarily conserved mechanism of plants and animals that processes microRNA (miRNA) and destroys double-stranded RNA, targeting, in a sequence-specific manner, both messenger RNA and retroviral genomes. Theoretically, it should be possible to target any virus with this mechanism. Retroviruses, with their RNA genomes, are an obvious potential target but DNA viruses could be targeted if their provirus encodes a unique mRNA that could serve as a target. The RNAi approach to targeting HIV-1 infection has been demonstrated in cell-culture studies. And RNAi approaches have been devised to inhibit viral evolution of resistance. The idea of using RNAi as a viral therapeutic is not new but has not been fully explored, possibly because of a number of potential hurdles (Silva *et al.*, 2003) and because the focus has been on using the technology in basic research. But it is clear that targeting gene expression can be achieved with this approach in mammals and RNAi has recently been shown to work against influenza virus in mice.

Germ Cell Distribution

No matter how the genetics of indigenous farm animals are improved, there needs to be a means of distributing the improved genetics to farmers. Whether by quantitative genetics (selective breeding) or transgenic technology, the germplasm of the lineage progenitors will be produced at a hightechnology center that resembles a modern-day artificial insemination (AI) stud farm. A distribution system must be in place to allow farmers access to the superior genetic material; the lack of a distribution system seriously constrains the improvement of the genetic potential of subsistence farmers' livestock. Since the 1950s, genetic improvement of the livestock herds in industrialized nations has been achieved primarily by distributing gametes (spermatozoa)from outstanding sires and more recently by distributing embryos from meritorious females (Foote, 1998; Thibier, 2005). As currently practiced, AI and embryo transfer (ET) require a ready supply of inexpensive liquid nitrogen. Liquid nitrogen must be available during the initial gamete- or embryo-freezing process and thereafter as a storage medium. On-farm storage dewars require replenishment (commonly every 4 to 6 weeks) that depends on use and environmental temperature. Furthermore, the use of preserved spermatozoa or embryos requires a specific knowledge of female reproductive physiology and highly skilled and practiced peopleto transfer the gametes or embryos into the recipient females. Even in developed countries, AI and ET are practiced mainly in intensively managed livestock operations, such as medium to large dairy, swine, and poultry farms. Less intensively managed enterprises, such as beef-cow and -calf operations or small farms, do not use these reproductive strategies, primarily because of the cost of estrous cycle management, the lack of availability of skilled AI technicians, and the low return on investment. Attempts have been made to eliminate the dependence on liquid nitrogen for preservation of gametes. At first, freeze-dried spermatozoa were not very successful (Foote *et al.*, 1962). Recent refinements in freeze-drying protocols have resulted in live-born mice from spermatozoa stored for over a year at 4°C (McGinnis *et al.*, 2005). Such freeze-dried spermatozoa are not viable in the traditional sense, but viable offspring can be produced by injection of their nuclei into an oocyte (intracytoplasmic sperm injection, ICSI). Storage at –20°C or –80°C improves the success rate of ICSI over storage at 4°C (Li *et al.*, 2007). Most recent attempts to adapt this technology to a non-rodent species have been unsuccessful (Nakai *et al.*, 2007). Modern cryobiology may eventually develop room-temperature storage methods that will preserve the genome of gametes. But it is also likely that any such technique will require some highly sophisticated method, such as ICSI, to introduce the stored genetics into a living organism. Such an approach would be impracticalfor production of breeding stock.

Spermatogonial Stem Cell Transplantation

The use of an emerging technology called spermatogonial stem cell (SSC) transplantation may be able to overcome the infrastructure and technical skill deficiencies that will inhibit subsistence farmers from taking of meritorious germplasm when it is available. SSC transplantation (also known as male germ-cell transplantation or germline stem cell transplantation) involves transplanting self-renewing male germ-cell stem cells from one male to another. The recipient male becomes the mechanism for spreading the genetics through a herd. In the early 1990s, it was demonstrated that the stem cells that give rise to spermatozoa in a male mouse could colonize the testes of another mouse, and the recipient mouse could sire offspring with the donor-derived spermatozoa. The SSCs can also be frozen and stored at –196°C and, on thawing, be used to colonize a recipient's testes after transplantation. The technique also works in rats and it has been shown that it can be used to restore fertility in two mouse models of infertility. Most relevant here are reports that demonstrate the potential of applying the technology to livestock. SSCs isolated from immature pig testes have been transplanted and the transplanted spermatogonia remained in the seminiferous tubules of the recipients for at least a month; somewhat surprisingly, not only did SSCs colonize the seminiferous tubules and generate spermatozoa capable of siring offspring, but this was possible in recipients unrelated to the host. Several studies also demonstrated the feasibility of the approach in goats . One attempt has been made to demonstrate its efficacy in cattle: testicular cells were isolated from *Bostaurus*bull calves and transferred, after fluorescent staining, into *Bosindicus* prepubertal recipient calves; *Bostaurus* fluorescently labeled cells were found in the testes of recipients up to 6 months after transfer. It is envisioned that once this technology has been refined and adapted to local breeds, SSCs harvested from males with superior genetic merit will be distributed to males of average genetic merit but good libido. The recipient males, harboring the "good genetics," could then be distributed to farmers. Alternatively, because the transplantation procedure can be performed in the field, the SSCs from genetically superior males could be frozen, transported to farms (or villages), and transplanted into farmers' (villages') own breeding males by someone skilled in the surgical procedure. The main constraint limiting the technology is the acquisition of enough SSCs. It is possible on rare occasions to harvest enough SSCs from a donor to distribute to four recipients (Ina Dobrinski, University of Pennsylvania, presentation to the committee, October 15, 2007). Protocols are needed for multiplying SSCs in culture so that dozens, if not hundreds, of males can be serviced by a single genetically superior donor. In addition to a propagation system, there is a need for more efficient SSC-enrichment methods. Finally, methods to improve the ability of the newly acquired SSCs to dominate the testes of the recipient will need to be devised. The current literature on this new procedure is not vast, but there are enough examples to suggest that SSC transplantation should be applicable in a wide array of species. It will be necessary to confirm that SSC transplantation can be implemented in the breeds and species of interest in SSA and SA and, if so, that it can be optimized for each species.

Improving Animal health

Improving the health of animals can have a substantial impact on the livelihood of farmers, especially subsistence farmers that rely on animals for labor, food, and additional income. Some of the ways to improve animal health discussed below include fortification of neonatal passive immunity, development of animal vaccines for diseases affecting SSA and SA, and use of animal disease surveillance. Not explored by the committee are the development of novel drugs and drug delivery strategies for animal disease infections in SSA and SA, two areas in which innovations have been lacking.

Neonatal Passive Immunity

In Kenya, calf mortality after weaning ranges from 6 to 70 percent, depending on health and nutritional management (Lanyasunya *et al.*, 2006). Almost all the primary causes of high preweaning mortality—including failure to ensure that the young receive colostrums within 6 hours of birth, respiratory diseases, diarrhea, and inadequacy of maternal milk production can be substantially reduced with currently available, low-cost management interventions. These include giving young animals the same salt- and sugar-based rehydration solutions made with clean water as are given to children who have diarrhea. The application of existing knowledge to raising calves, lambs, and kids could reduce mortality to below 9 percent, a commonly accepted target, and improve animal productivity and profitability. Technologies are used to enhance colostrum quality by vaccinating pregnant dams. Also, the preparation and preservation (freeze-drying) of serum antibody extracts are used as artificial colostrums substitutes. There are some truly nutritional interventions to prevent preweaning mortality,such as enhancing the nutrition of the lactating dam and providing nutritional supplements to the diets of lactating animals. The issue of access to veterinary services by the poor in SSA and SA is critical. The delivery of appropriate medicines and information on livestock health has been compromised by privatization of veterinary services in many countries. Now only those able to pay have access to veterinary services, so livestock productivity is low and the risk of outbreaks of zoonotic and other diseases is high. Very simple interventions in providing education, medicines, and vaccines could have a major impact in protecting the health and productivity of animal populations in SSA and SA. Reversing the situation would make meeting the projected increased demand for meat products more feasible without large increases in the number of breeding females.

Animal Vaccine Development

Disease is a major constraint on livestock productivity in developing Countries. Infact, when the committee asked several experts what could have a major effect on improving the life of poor farmers, they noted that effective vaccines already exist to prevent globally endemic disease, such as brucellosis, leptospirosis, and bovine virus diarrhea (Hans Draayer and Raja Krishnan, Pfizer Animal Health, presentation to committee, September 24, 2007). However, some factors that affect the use of current vaccines in SSA and SA include strain variations, costs of vaccines, and the need to have an effective cold-chain for transportation, marketing, distribution, and delivery to the animals in the field. Technologies to develop thermo-stable vaccines, such as the development of a thermo-stable attenuated vaccine for Newcastle disease in chickens in Australia and Malaysia, can compensate for the lack of a cold-chain. A focus on infections for which vaccines exist and on others that cause respiratory and intestinal diseases in young, pre weaned animals could reduce mortality and improve productivity. The other two categories of opportunity identified by experts the committee consulted were zoonotic diseases, particularly those associated with foodborne illness, and endemic infections peculiar to SSA and SA (Guy Palmer, Washington State University, and Roy Curtiss, Arizona State University, presentation to committee, September 24, 2007). SSA and SA are home to the most severe vector-borne diseases, including trypanosomiasis, babesiosis, and theileriosis (the last two of which are parasitic diseases spread by ticks). Diseases, such as trypanosomiasis, have been the subject of vaccine investigations for many years and have thwarted vaccine effectiveness because of the great variability of surface proteins of the parasite, which the organism is able to "switch" under the pressure of the host immune system. The greatest challenge for these vaccines is thediscovery of antigens that will result in a protective immune response in the host. Such a discovery will be assisted by the complete genome sequences that have been completed for all six major vector-borne pathogens in the last 2 years, including *Anaplasmamarginale, Babesiabovis, Ehrlichiaruminantium, Theileriaparva, T. annulata,* and *Trypanosomabruci*. For example, it is known that immunity to East Coast fever, caused by the tick-borne parasite *T. parva*, can be created by inoculating a host with sporozoites of the parasite in conjunction with long-acting oxytetracycline. However, until the *T. parva*genome was available, the antigens involved in the immune response were unknown. Using gene prediction methods, investigators were able to identify candidate genes in the genome that were associated with a secretion signal on the basis of the idea that secreted proteins would be the first to become associated with the host major his to compatibility complex apparatus. Screening of the protein products of those genes narrowed the search to the ones involved in establishing immunity and paved the way to future vaccine development (Graham *et al.*, 2006). Zoonotic diseases were suggested as targets for disease control because of their implications for limiting the spread of diseases to humans and back to animals. Effective vaccines exist for some of the diseases, including brucellosis, salmonellosis, and listeriosis; but problems related to supply, cost, and delivery mechanisms slow their widespread use.

DNA Vaccines

DNA vaccination stimulates the immune response by introducing into the host naked DNA that codes for antigens of a pathogen. The protein synthesis machinery of the host cell expresses the antigen and stimulates a response by the host's immune system. In theory, DNA vaccines can be manufactured far more easily and less expensively than vaccines composed of inactivated pathogens, protein subunits, or recombinant proteins. Other potential advantages include stability, resistance to extreme temperatures, efficacy as an oral vaccine, and the ability to introduce multiple antigens. However, substantial development is needed before DNA vaccines become an alternative to conventional methods. Most of the experimental DNA vaccines have not shown as great protective immunity as conventional vaccines, but new technologies, such as the coating of colloidal gold with DNA, that are in development could improve effectiveness. If future research can deliver a DNA vaccine that offers protective immunization, this approach would add flexibility to the custom designing of vaccines for regional needs. For instance, it is easier to change the sequence of an antigenic protein or to add heterologous epitopes. The protective immunity of the expressed protein can be easily evaluated after the DNA is injected into a model animal, such as the mouse. This simple,elegant method could quickly allow researchers to learn about the effectiveness of candidate antigens. The final goal of effective DNA vaccines is considered to be far in the future because of the many unresolved problems,but the potential high payoff will continue to draw investment.

Animal Disease Surveillance

It is pointless to develop and deliver drugs and vaccines without knowing which syndromes are present in a region, because protecting an animal against one pathogen only to have it succumb to another will not reduce the burden of disease on a small-holder farmer. Developing a database of such information will require field research, trained technicians, and diagnostics. The relatively new World Animal Health Information Database managed by the World Organization for Animal Health (OIE) is a significant database that tracks disease prevalence in all regions of the world. In cooperation with the Food and Agriculture Organization of the United Nations (FAO), the OIE is investigating disease rumors that surface on ProMED or other non-scientific sources of information; these early warning systems serve as good alert systems for emerging disease outbreaks. The use of satellite-based remote sensing technologies could be useful as early warning systems for the emergence of serious infectious diseases, particularly those that are transmitted by arthropods. The FAO's Emergency Prevention System (EMPRES) for Transboundary Animal and Plant Pests and Diseases program currently uses remote sensing technologies to determine the Normalized Difference Vegetation Index (NDVI), and theuse of such data has led to the successful advanced prediction of Rift Valley fever outbreaks. Similar technologies have been used for the advanced notification of blooms of desert locus and of outbreaks of Venezuelan Equine Encephalomyelitis. Inexpensive diagnostic tests, like that developed for rinderpest are needed for disease detection and vaccination campaigns. Other similar rapid pen-side tests for the recognition of infectious diseases have been developed and are in use, such as the field diagnosis of human and avian influenza outbreaks. Increasing in greater numbers are the development, validation, and deployment of rapid RT-PCR technologies for accurate diagnosis of a variety of diseases affecting SSA and SA. These tests require only a nasal swab as a sample and are not sensitive to the effect of higher temperatures in the transportation to diagnostic laboratories. Furthermore, emerging technologies, such as biosensors are promising because of their sensitivity, speed, portability, and ease of use and could be developed for a variety of surveillance efforts and especially useful in resource-constrained countries in SSA and SA. Moreover, if farmers have tools to detect the presence of disease, they are more likely to seek out a drug or vaccine. Farmers' confidence in medical treatment and vaccination depends on their seeing a benefit, which they will not if a problem is not solved by a drug or vaccine that targets a single pathogen.

Conclusions

The sustainability prospects for livestock farming remain depressing unless the main stream perceptions about the problems are not changed. The development strategies are essential for formulating farmer responsive plans, giving due consideration to the nature of marginality, fragility, diversity and niches of each area. It will help to ameliorate the impact of the marginalization of livestock farmers and in achieving social equity by building on the comparative advantages of key land resources.

References

Anderson, J. and. Akkina R. 2005. CXCR4 and CCR5 shRNA transgenic CD34+ cell derived macrophages are functionally normal and resist HIV-1 infection. *Retrovirol.*, 2 : 53-64

Anderson, J. and. Akkina R.2007. Complete knockdown of CCR5 by lentiviral vector-expressed siRNAs and protection of transgenic macrophages against HIV-1 infection*Gene.Ther.*, 14:1287-1297

Babinszky, L., Dunkel, Z., Tóthi, R., Kazinczi, G. and Nagy, J. 2011. The impacts of climate change on agricultural production. *Hungarian Agricultural Res.*, ISSN: 1216-4526.

Farrell, G., S. A. Simons, and R. J. Hillocks. 2002.Ustilagokamerunensis on napier grass in Kenya. *Int. J. Pest Manage.* 48:25-28.

Foote, R.H. 1998. Artificial insemination to cloning: Tracing 50 years of research. Ithca: New york State College of Agriculture & life.

Foote, R.H., R. W. Bratton, C. R. Handerson, F.M. Shantz, and J. K, Pollard. 1962. Survival of Bovine spermatozoa at room temperature in citrate and Cornell University and tris extenders containing whole and fractionated cocconut milk. *J. Dairy Sci.*45:1383-1391.

Gordon, J.W. and Ruddle, F.H. (1981). Integration and stable germ line transformation of genes injected into mouse pronuclei. *Science* 214:1244-1246

Graham, W. J., Cairns, J., Bhattacharya, S., Bullough, CHW.,Quayyum, Z., Rogo, K., In: Jamison, D.T., Breman, J.G., Measham, A.R., Alleyne, G., Claeson, M., Evans, D. B,. Jha, P., Mills, A., Musgrove, P., editors. 2006. Maternal and Perinatal Conditions. Disease Control Priorities in Developing Countries. 2nd ed.. Washington (DC): World Bank.

Lanyasunya, T. P., W. H. Rong, S. A. Abdulrazak and E. A. Mukisira. (2006). Effect of supplementation on performance of calves on small holder dairy farms in Bahati Division of Nakuru District, Kenya Pakistan. *J Nutrition*5:141-146.

Li, H.G., Ding, X.F., Liao, A.H., Kong, X.B., Xiong, C. L. 2007 Expression of CatSper family transcripts in the mouse testis during post-natal development and human ejaculated spermatozoa: relationship to sperm motility. *Mol Hum Reprod.*13:299–306.

McGinnis, L.K., Zhu, L., Lawitts, J. A., Bhowmick, S., Toner, M., Biggers, J.D., 2005. Mouse sperm desiccated and stored in trehalose medium without freezing. *BiolReprod.* 73(4):627-33.

Nakai, M., Kashiwazaki, N., Takizawa, A., Maedomari, N., Ozawa, M., Noguchi, J., Kaneko, H., Shino, M., and Kikuchi, K. (2007). Effects of chelating agents during freeze-drying of boar spermatozoa on DNA fragmentation and on developmental ability *in vitro* and *in vivo* after intracytoplasmic sperm head injection. *Zygote* 15, 15–24.

Niemann, H. Kues W. A.2007. Transgenic farm animals: An update. *Reprod. Fertil. Dev.*19:762–770.

Silva, J. M., Sachidanandam, R. & Hannon, G. J. 2003. *Free energy lights the path toward more effective RNAi. Nature Genet.*35, 303–305.

Silva, J.M. Li, M.Z. Chang, K. Ge, W. Golding, M.C. Rickles, R.J. Siolas, D. Hu, G. Paddison, P.J. Schlabach, M.R., *et al.* 2005. Second-generation shRNA libraries covering the mouse and human genomes.*Nat. Genet.*, 37:1281-1288

Thibier M. 2005. The zootechnical applications of biotechnology in animal reproduction: current methodsand perspectives. *ReprodNutr Dev.* 45:235-42.

Upadhayay, R.C. 2010. Annual milk production loss due to global warming, Animal Physiology, National Dairy Research Institute (NDRI) Press Trust of India / New Delhi, September 26.

Chapter 12

Entrepreneurship Opportunities from Dairy Farm Waste Management

Nazam Khan[1], Vikas Mahajan[1], Suraj Amrutkar[1], Manpreet Kour[2] and Gagandeep Kour[3]*

[1]Division of LFC, [2]Division of Agronomy, [3]Division of Animal Nutrition SKUAST-J, R.S. Pura, Jammu, India
Corresponding author: dr.vikasmahajan@gmail.com

Introduction

Cow is regarded as blessed/sacred animal by 1.3 billion inhabitants of India and other Hindu dominated nations. Despite that, increase in number of unproductive animals, is an economical burden. Not only its slaughter is banned, but its intentional killing and/or slaughter is a cognizable, bailable offence under section 429 of the Indian Penal Code (IPC) and is punishable by either a term of imprisonment of any description, up to five years, or by a fine, or by both. To overcome the financial burdens, farmers especially marginal one is consequently increasingly releasing unproductive cows into the wild because they cannot afford to feed them, leading to marauding herds that destroy farms, obstruct traffic, pollutes environment, etc. The efficacy of using male animals for farming has also been questioned as a result of the mechanization that has limited it to just hilly terrains. Urbanization and artificial insemination both have decreased the male animal use for draught work and breeding purposes. Therefore, it's imperative to look for alternative benefits from these animals.

As per the 20th livestock census, a total of 50.21 lakh cattle *i.e.*, 2.61% of the 192.50 million cattle in the country were reported as stray. The numbers of stray cattle have declined by around 5% in 2019 in comparison to 2012 census. Despite the fact that the number of stray cattle decreased at the national level, it gets increased in some key states. Of the total stray cattle population in India, 25.40% were in Rajasthan (12.7 lakh), followed by 23.60% in Uttar Pradesh (11.84 lakh). Along with Madhya Pradesh, these three states alone account for almost 66% of the stray cattle in the country. Another problem that may arise by unproductive animals is the shortage of feed and fodder. Although, India has a large number of both recognized and unrecognized cow shelters, but they are all overrun by animals.

Thus, it's the need of hour to develop revenue/income from numerous dairy farm wastes generated in the farm, which will generate employment opportunities also besides doubling farmers income. It is clearly mentioned in Ayurveda (an ancient Indian science) that cow dung/Gomeya and cow urine/Gomutra are not wastes, rather they are designated as purifier of all the natural wastes. Since the times of Vedic era, Gomeya have often been utilized as fuel and organic fertilizer in most of India. Its insect repellent and thermal insulation property is exploited to line the floor and wall of earthen houses since ancient past. It is also used in Indian tradition for different religious ceremonies as the purifying agent and also used in fire sacrifices (Yajnas) as dung cakes, to pure atmosphere when burnt. It is explained in the light of fact that reaction of cow dung and ghee when mixed releases oxygen into the air (Hawan). In Ayurveda, its use is also advocated in the form of fuel to purify metals to be used as medicines. Now-a-days vermicompost and vermiwash prepared from cow dung are gaining popularity particularly for organic produce and is looked upon as an alternative for synthetic pesticides/fertilizers. These products have dual benefit of increasing fertility and humus to the soil. In Veda, cow's urine was compared to the nectar. In Sushrut, several medicinal properties of cow's urine have been mentioned, particularly for weight loss and to cure leprosy, cardiac and kidney problems, indigestion, stomach ache, edema, *etc*. Recently, two US patents has been granted for the bio-enhancer properties of cow urine, especially for early cure of tuberculosis and cancer in human beings. United States Patent and Trade mark Office had granted Patents No 6410059 and No. 6896907 to an "Indian innovation which has proved that cow urine can make antibiotics, anti-fungal agents and also anti-cancer drugs more effective". Other studies also revealed the immunomodulatory properties of cow urine. Besides, cow urine is said to be a very effective insect repellent when mixed with certain herbs (Nair, 2002). Gomutra is useful for numerous human ailments like, cancer, osteoarthritis, allergies, kidney failures, skin diseases, healing of wounds, *etc*. The above facts signify that cows are not meant only for milk and draught purposes, but they also protect our health and sustain the agriculture. An ancient shlok "Matradev bhutnaam gava sarv such pradha" implies that cow being mother of all living entities gives pleasure to everyone.

As per India times 21st January 2018, Gujarat state's "Gauseva Ayog" has started a cow tourism project to deliver knowledge on cow rearing and making products with cow urine and dung. They disseminate information on some basic products such as bio gas and medicines using cow dung and urine. Many products containing cow dung and urine *i.e.* soap (23 % cow dung content and 7% urine), toothpaste (10 % cow dung content and 10% urine, 2% Ghee), floor cleaners (Gaunyle: 11% cow urine), hair oil, incense shaving cream, face wash, talcum powders and dark circle removers are available in the market. Many people use cow urine in rituals like yagna and panchgavyam during 'janaau' ceremony.

As mentioned earlier, cow dung and urine have high demand in organic farming also. There is one report that government run Maharana Pratap University of agriculture and technology in Udaipur has roped dairy farmers across the state for supplying cow urine for its use in organic farming project (Times of India, July 24, 2018).

Thus, need of hour is to bring revolution by extensive dissemination of information regarding scientific and judicious use of animal waste management for doubling the income of farmers in an eco-friendly manner besides popularizing it for self-employment generation.

Cow Urine Composition and its Benefits

Cow urine, synonymously called Gomutra is known as "water of life" and/or "Amrita". It contains all the essential elements which aids in maintaining appropriate balance between Vaata, Pitta and Kapha. Cow urine consists of 95% water, 2.5%- urea and mineral salts and 2.5%- hormones and enzymes. It contains Vitamin A, B, C, D, E, sodium, nitrogen, sulphur, manganese, iron, silicon, chlorine, magnesium, citric acid, succinic acid, calcium salts, phosphate, lactose, carbolic acid, enzymes, creatinine and hormones in right composition. Therefore, consumption of cow urine reinstates the equilibrium of above-mentioned vital constituents. But urine should be collected from healthy and indigenous cows only. Although, any age group can be designated for the said purpose, but its collection should be preferably done in pots which may be earthen, glass and/or china clay made.

Benefits of Cow Urine

Its different benefits are discussed in different subheadings and are presented underneath:

1. Maintenance of soil fertility: Cow urine along with Neem leaves is not only a pleasing biopesticide, but it is safe to use, has no harmful effects and also doesn't accumulate in the food chain/food web. Preparation of liquid manure from cow urine is very easy and is good for plants when compared with artificial fertilizers. A new scheme has been launched in Chhattisgarh (2022) in which the purchased urine will be used in making pest control products and 'jeevamrut' (liquid organic fertilizer). It in turn will minimize the use of expensive chemical pesticides and simultaneously promote organic farming by ensuring the availability of natural pesticides to farmers at nominal prices. 'Gauashish Phenyl' a natural disinfectant from cow urine was developed by Allahabad-based gaushala in 2015. It has been found to have anti-microbial and anti-fungal properties. Another preparation of biopesticide is the anaerobic digestion of cow urine with fresh chopped Neem leaves in earthen clay pot for 21 days beneath the ground. It is then boiled to prepare concentrated extract which can be stored up to 6 months.

2. Therapeutic and Prophylactic cure of diseases: Gomutra has proven useful for curing different ailments such as colic, abdominal tumor, abdomen enlargement and flatulence. Besides it is used for numerous therapies also *viz.*, purgation, decoction, leprosy and hemorrhoid therapy. It has immune-stimulant and anti-convulsant characteristics too. Cow urine has shown promising results for different skin diseases *viz.*, itching, sunburns, eczema, psoriasis, acne etc. Also, it is designated as best cures for stomach diseases, heart diseases, kidney ailments and tuberculosis. There is indigenous traditional knowledge saying that a glass of fresh cow urine consumed for 21 consecutive days as empty stomach can dissolve stones to a manageable size. Cow urine distillate/Goark also has anti-diabetic property and reduces blood glucose, serum cholesterol and serum triglycerides. For treating liver cirrhosis, warm cow urine can be used whereas malignant jaundice can be cured by using antimony sulphide supplemented with cow urine.

3. Bioenhancer/Adjuvant property: Bioenhancers are the substances, which do not have their own drug activity, but promote and augment the bioactivity or bioavailability or the uptake of drugs in combination therapy. Cow urine has this bioenhancer property and it can promote the activity of various drugs (antibiotics, anti-fungal and anti-cancer), thus has supra-additive action. Gomutra also has anti-septic, antioxidant and antimicrobial properties, and thus promotes early wound healing.

Keeping in view the above-mentioned facts, it is inferred that collection of cow urine in an organized manner may help the farmers to generate additional income to support their livelihood/social security in the form of employment generation and helps to motivate youth for scientific and sustainable disposal of dairy farm waste.

Merits of Cow Dung

Cow dung/Gomeya plays important role for maintaining the soil fertility, environmental safety, energy source, etc. and are enlisted below:

1. Role in soil health: Depletion of soil fertility is an important challenge for food security in recent times due to modernization and production load. The use of synthetic pesticides/fertilizers has increased production, but also has deleterious effect on soil productivity, besides reducing the per capita income of farmers for a given land area. The use of cow dung compost/manure/ vermicompost may substitute these chemical substances as it provides three major plant nutrients, nitrogen, phosphorus and potassium (NPK), along with different essential nutrients such as Ca, Mg, S, Zn, B, Cu, Mn etc. It also improves soil tilth, aeration, organic matter content, and water holding capacity of the soil and promotes growth of beneficial soil organisms, apart from increasing cation exchange capacity. Cow dung can be processed by following methods before to be used for improving soil health by:

2. Composting: During composting, solid waste of dairy farm is collected in a 1.5 m deep pit with dimensions of 3×4 m or larger depending upon the herd strength of farm and/or need. Generally, 3 m3 per animal unit is used. Two pits with walls on three sides and a temporary roof to prevent desiccation are used. It is suggested to fill and empty pits on alternate basis. A gutter containing cresol and water to control fly breeding should be present on the front side besides a vertical sliding shutter so that debris don't fall into the gutter. Alternatively, the manure is dumped and then packed tightly in both the pits such that when one is being filled and packed, fermentation and decomposition occurs in the other which was filled earlier. Turning the manure periodically is a must to ensure uniform mixing and decomposition. It also aids in the destruction of parasitic larvae which are otherwise present normally in the dung. The process of piling increases the temperature to 50°C within 24 hrs and 70°C within 3-8 days, thereafter falling back to 50°C. The maintenance of proper C: N ratio and moisture is important for the process to occur efficiently.

3. Vermicompost: Farm organic waste includes animal waste, fodder waste (green/dry fodder), plant leaves, weeds etc are the common solid waste which is abundantly available in India and constitutes 40-70% of solid waste with Carbon/Nitrogen ratio ranging between 800 and 1000 kcal/kg. Availability of the farm wastes in India is 500 million tons per year, whereas, dung production is estimated to be 3 million tons per year. Thus, efficient disposal of these wastes into environmentally compatible product is most desirable issue today. Vermi-composting of dairy farm waste/cow manure using earthworm species *E. andrei* and *E. foetida* favored nitrification, resulting in the rapid conversion of ammonium-nitrogen to nitrate-nitrogen and improves the available nitrogen for plants. It is an organic manure made from animal and agricultural waste by the cast of earthworms, particularly *Eisenia, Eudrillus, Perionyx* and *Dolvin* species. It is high in N, P and K as well as several micronutrients, enzymes (proteases, amylases, lipases, cellulose, and chitinases), plant growth hormones (auxins, cytokinins, and gibberellins) and some nitrogen-fixing bacteria (*Pseudomonas, Actinomycetes* etc.) that enhance the nutritional value of vermicompost. Though vermicompost is not comparable to chemical fertilizer, but its application to soil improves soil structure and texture, water holding capacity, facilitates aeration and prevents soil erosion. As a result, healthy and non-toxic food can be produced from the soil without having a negative impact on the health of people or animals. The preparation process is quite simple and economical such that rural farmers can adapt easily to gain benefits. It enables the farmers to process and utilize large amounts of agricultural and livestock waste to useful material which aids in the prosperity of rural life by improving health, wealth and the environment. It can be prepared in either proper sheds (Above and below ground model) or in open heap method under the tree shade/or covering by gunny bags (Technology given by Division of Livestock Farm Complex, Faculty of Veterinary Sciences and Animal Husbandry, SKUAST-Jammu).

Heap method used for vermicompost production under tree shade

Salient Features/Description of technology

a. A heap of farm organic waste (10 days old gobar) is made under the tree shade and/or covered by gunny bags followed by bunds making all around is one method. Alternatively, pit is dug around 10 inches deep and bunds are made all around. Raw culture of earth worms is needed to be added in simple heap. It is low-cost sustainable composting technology and can decompose waste rapidly in 84 daysby making heaps of dumping waste over Katchha floor and using tree shade or covering the heap with gunny bags.

b. It has potential for increasing the farm income by transforming waste into wealth and can reduce the dependence on chemical fertilizers besides enhancing nutrient use efficiency.

c. The compost prepared by heap method contains approximately 1.37% N, 0.4% P and 0.92% K and was prepared in 84 days. It is comparable to conventional compost that contains NPK as 1.92% N, 0.58% P and 1.57% K and requires 72 days for preparation.

Benefits

a. The compost so produced is a store house of major and micro plant nutrients and also contains growth promoting hormones "auxins", "cytokinins" and flowering hormone "gibberellins" which are secreted by earth-worms.

b. Minimize environmental pollution and cost involved in the disposal of waste.

2. Environment safety: Global warming is a burning issue now-a-days, as it leads to climate change which in turn has negative impact on our ecosystem. One of the major causes is the combustion of fossil fuels for energy generation, but it can be replaced by deployment of renewable fuels such as biogas. It is one of the alternative sustainable resources that may significantly reduce greenhouse gas emissions. Biogas is harvested by the anaerobic digestion or fermentation of biodegradable materials such as biomass, manure, sewage, municipal waste and organic matter. But it is mainly generated by converting cow manure via anaerobic digestion into methane biogas. A study states that one cow in a day can generate three kilowatt hours of electricity from its dung whereas only 2.4 kilowatt hours of electricity is needed to power one-hundred-watt light bulb per day (State Energy Conservation Office, 2009). Thus, it may substitute fossil fuel to certain extent and may solve the arising environmental issues.

3. Energy resource: Due to shortage of fuel wood, cow dung/upalas are used mainly in rural areas as a source of energy. Around 72% of Indian population relies on traditional biomass for cooking. A new approach of making cow dung logs can be popularized among the rural masses to overcome the issue of energy deficiency. It has a uniform central hole all along the length of the log, which facilitates its drying in 5-12 days depending upon the season and also helps in fine burning. Division of Livestock Farm Complex, Faculty of Veterinary Sciences and Animal Husbandry, SKUAST-Jammu conducted a study on making of cow dung logs as well as its enrichment with different materials (saw dust, straw, coal dust and their combinations) under university funded project by using cow dung log making machine. The value addition of these logs can be done by using saw dust, straw, coal dust and/or their combinations. It has been realized that burning efficiency is enhanced by using different ingredients and can be sold at premier price.

Salient Features/Description of Technology

a. A single cow dung log of approx. 3.0 kg is prepared by taking 2.8 kg gobar (5-7 days old; 50-60% moisture) and mixed with either 0.2 kg straw or saw dust or coal dust. Whereas in combination gobar is mixed with 0.08 kg straw, 0.08 kg saw dust and 0.04 kg coal dust. After thorough mixing, the material so prepared is passed through cow dung log machine. The length/shape of wood log can be managed by using different dye/ collecting pipe. The wet cow dung log so obtained are dried under sun light and can be sold either loose or after proper packaging by using thin wrapper film.

Cow dung logs

Cow dung logs enriched with 20% straw

Cow dung logs enriched with 20% saw dust

Cow dung logs enriched with 20% coal dust

Cow dung logs enriched with (8 % straw + 8% saw dust + 4% coal dust)

b. It is low-cost sustainable technology which can ensure income generation in small time span.

c. It has potential for increasing the farm income by transforming waste into wealth.

Benefits

a. Raw gobar usually goes waste, and may be sold @ Rs 1/kg. But cow dung logs prepared by using different ingredients resulted in value addition and can be sold at premier prices.

b. Cow dung logs enriched with different ingredients increases its demand in cremation areas, brick kiln, road side dhabas, etc.

c. It minimizes nuisance besides lowering the dumping/storage space.

4. Disease prevention: Cow dung is very effective for reducing the bacterial and fungal pathogenic diseases in agriculture sector. It showed positive response in suppression of mycelial growth of plant pathogenic fungi like *Fusarium solani, F. oxysporum* and *Sclerotinia sclerotiorum.* Also, cow dung extract spray was reported to be effective for the control of bacterial disease of rice.

Its use is extended for humans also, as dung is now-a-days used for making soap which is claimed to be good for dandruff, skin diseases, to counteract bad odors of body and to make skin soft. Cow-dung medicine is in liquid or powder form, and now also available in capsule or tablet form. Along with neem leaves it is smeared on the skin to cure boils and heat rash. Antibacterial properties of cow-dung extend its use as disinfectant. It is also used as a body pack to detoxify the body. The ash produced from cow-dung was used to clean the teeth as it is known to strengthen the gums. Also, this ash when mixed with dried powder of used citrus, acts as an excellent dish cleanser.

5. Other uses: Manure consists of a great deal of fiber, whose source is either from undigested animal feed and some is from straw, sawdust, or other bedding that gets mixed in with the manure. Manure fiber has been used to produce a number of specialty consumer products like plant growth medium (similar to peat moss), seed starter pots, fertilizer garden sculptures, paper, and building materials.

a. **Paints**: Certified by the Bureau of Indian Standards after passing prerequisite test parameters like thinning properties, drying time, finish etc, the 'Khadi Prakritik Paint' is a first-of-its-kind. It is available as both distemper and emulsion in white base and different colors can be developed by mixing suitable colorants. Made from actual cow dung the paint is cost effective, non-toxic, free from heavy metals, odorless and eco-friendly. It has natural thermal insulator properties and is quick to dry (less than four hours) leaving a uniform and smooth finish. The paint has anti-bacterial and anti-fungal properties.

b. **Cosmetics**: A number of cooperatives, entrepreneurs, and non-government organizations have set up units for the manufacture of various beauty products made from cow dung and urine. Cow dung has been found to have certain anti-bacterial properties and has been used in dry powdered form to treat skin disease since ancient times. It is even said that it has the tendency to improve circulation and reduce infection. Cow dung and urine have been included in a number of products like cleansers, masks, scrubs, hair and massage oils etc. These products are gaining popularity not only in India but in other countries like the US, the UAE, the UK, Singapore and Germany. More than 50 products are exported to different countries. Besides a large number of face and body care products, the ash obtained from cow dung is said to strengthen gums and is therefore used to clean teeth. Govardhan Goutpadan, a cottage industry set up to manufacture various toiletries like soaps, shampoos, tooth powders and pain relief balms and medicinal products besides carrying out research regarding the benefits of cow therapy.

c. **For growing mushrooms and algae**: Cow dung contains nearly 24 different minerals and harbours rich microbial diversity that produces various enzymes and other nutritive and non-nutritive substances. For many years cow manure has been used to grow mushrooms and microalgae being rich in nutrients and moisture. Cow dung has the tendency to retain nutrients and moisture for long which is regarded as essential for the suitable growth of healthy mycelia.

d. **Bioremediation of pollutants**: *Acinetobacter, Bacillus, Pseudomonas, Serratia* and *Alcaligenes* species present in cow dung makes it suitable for microbial degradation of pollutants.

e. **Production of enzymes**: cow dung harbours a rich microbial diversity thatis capable of producing a number of enzymes like proteases, lipases, esterases, etc.

f. **Different panchgavya variants** are available with different composition namely swalpa panchgavya ghrita, maha panchgavya ghrita and panchgavya ghrita. The therapy has also demonstrated its therapeutic potential against severe pathological conditions like cancer, AIDS, and diabetes. It also plays a crucial role in promoting organic farm practices and enhancing soil fertility by zero or minimal usage of harmful synthetic fertilizers, pesticides, insecticides, and *antibiotics.*

Conclusion

Dairy farm wastes have a lot of potential for employment generation among the entrepreneurs and also curbing stray animal menace. There is a need to generate awareness among the farmers/society regarding the benefits of harvesting cow dung and urine in a scientific way. But there is need to promote research in cowpathy for its scientific validation and benefits. Many institutes and NGOs are working on different products of cowpathy and revealed very encouraging results for the betterment of the society. The effective use of dung and urine may contribute in the form of energy security and reduction in environmental degradation. Further, it will also help in empowering the youth community to exploit it for social and economic strengthening.

References

19th Livestock Census- All India Report, (2012). Ministry of Agriculture, Department of Animal Husbandry, Dairying and Fisheries. Krishi Bhawan, New Delhi.

Bhadauria, H. 2002. Gomutra-Ek Chamatkari Aushdhi (Cow urine – A Magical Therapy). *Vishwa Ayurveda Patrika*, 5:71-74.

Department of Animal Husbandry, Dairying and Fisheries, Ministry of Agriculture, Govt. of India (Basic Animal Husbandry Statistics, 2006, 2010, 2012, 2013, 2014). 2015. State at a Glance: Jammu and Kashmir, Vol. 1(3), http://dahd.nic.in/dahd/WriteReadData /Final%20BAHS% 202014%20 11.03.2015.pdf (accessed on 16.11. 2015).

Gulhane, H., Nakanekar, A., Nilesh, M., Sunanda, B. and Amrut, S. 2017. Gomutra (cow urine): A multidimensional drug review article. *Int. J. Res. Ayurveda Pharm. 8 (5)*, DOI: 10.7897/2277-4343.085231.

Harnessing the energy potential of cattle dung in India: A policy memorandum to the ministry for new and renewable energy 2018. *Journal of Science Policy & Governance*, 12(1).

Kumar, A., Kumar, S., Rastogi, A. and Bhat, Z.F. 2015. Organic Dry Farming: an Upcoming Trend. A manual on "Organic Livestock Farming". Pp. 58-65.

Raj, A., Jhariya, M.K. and Toppo, P. 2014. Cow dung for ecofriendly and sustainable productive farming. *International Journal of Scientific Research*, 3(10): 2277 – 8179.

Sahu, R., Lalchand, R., Gupta, R. and Rout, O.P. 2017. Benefits of cow urine – A review. *International Journal of Recent Advances in Multidisciplinary Research*, *04*(09):2833-2835.

Sahu, R.K., Krishnaiah, N., Ramya, P. and Anusha, P. (2016) Cow Urine - Therapeutic Value. *International Journal of Livestock Research*, 6(11): 93- 99. doi:10.5455/ijlr.20160 930083917.

State Energy Conservation Office (Texas). 2009. "Biomass Energy: Manure for Fuel." 23 April 2009. Web. 3, October 2009.

Times of India, July 24, 2018. https://www.google.com/amp/m.times of india.com/city/jaipur /cow-urine/new/revenue/stream for dairy farmers/amp article show/ 65111454.cms.

Chapter 13

Poultry Farming: A Means for Entrepreneurship Development

Dr. Suraj Amrutkar[1], Dr. Nazam Khan[1], Vikas Mahajana[1], Bharti Deshmukh[2] and Dr. M.A. Malik[1]

[1]Division of LFC, FVSc & AH, SKUAST J, Jammu, [2]Assistant Professor, Division of AGB, GADVASU, Ludhiana, Punjab

Introduction

Poultry farming is the form of animal husbandry which raises domesticated birds such as chickens, ducks, turkeys and geese to produce meat or eggs for food. Poultry mostly chickens are farmed in great numbers. Poultry production is the main part of the farming in India. People depend on poultry for food and it serves as an additional occupation to supplement the income of small and marginal farm families. The poultry sector has developed such that large scale production is being practiced and a lot of people derive their means of livelihood from poultry and its associated industries. Egg and poultry meat are inexpensive sources of protein, especially across different socio-economic groups in India. In India, rural poultry farming contributes to about 40% of total egg production in India. It receives Government support through the State-run co-operatives and largely depends on rearing native breeds like Chabro, Aseel and Kadaknath. Rural poultry farming provides livelihood and food security for local rural populations while promoting the agrarian economy of the country. Private enterprises and businesses in India have also been proactive in the research & development of parent stock (both indigenous and international breeds), meat breeds and egg layers. Venkateshwara Hatcheries, Suguna Foods and Skylark Hatcheries have developed international poultry breeds to suit Indian environmental conditions. A number of bird hybrids including hyline are reared for rapid growth, feed efficiency and higher profits. Kegg Farms has developed a hybrid, high-yielding variety bird called the Kurolier.

Present Status of Jammu and Kashmir

Poultry farming is not new in not new in J & K and it has been practiced for the farmers since the time immemorial. Poultry farming has immense potential in Jammu and Kashmir. Despite the strenuous efforts, poultry farming is not picking up pace in Jammu and Kashmir. In J & K, there is 69 percent shortfall in the poultry meat and 97 percent shortfall in the poultry egg production. The gap between demand and supply of poultry meat and eggs fulfill by other states of the country specially Punjab state. In J & K UT, every year around Rs. 2000 crores of meat is imported from the outside of the JK. A little progress has been made on the broiler production in the state but poor progress has been made in commercial egg laying industry in the J & K. Some outstanding dual purpose poultry strains such as Chabro and Vanraja have been introduced in the J & K which have raised hope for mitigating the eggs and meat need in the J & K.

Most of the backyard poultry production in the J & K comprises the indigenous birds. The indigenous poultry birds produce only 60-80 eggs per year and meat production is also very less. Commercial poultry production can be easily boosted in the J & K with introduction of improved varieties of poultry which can lay 160-200 eggs per year and also produce more meat. In J & K, Poultry farming is emerging an important enterprise and is assisting to create income and employment for the rural youths of the state. There is immense potential of poultry farming in the J & K. Government is offering the various incentives to motivate the aspiring poultry farmers of the J & K. All Krishi Vigyan Kendra (KVKs) of the J.&K. are imparting training programmes on advance poultry farming to the farmers and creating interest on poultry farming. Veterinary College, RS Pura has introduced black meat poultry variety namely Kadaknath in the J & K which is very low in cholesterol (0.73-1.05%), rich in protein (25%), vitamins and helpful for the asthmatic and nephritis patient. The meat and eggs of this poultry variety are costly and can help to raise the socio-economic status of the farmers. Now, many farmers of the area become the poultry entrepreneurs.

The cost of quality proteins is increasing rapidly. Due to poor purchasing power of the farmers, their family's members suffer malnutrition especially of animal protein. The poultry sector can ensure the proper nourishment of the farmer's families besides income and employment. Poultry birds can increase the income of the farmers as poultry farming requires low inputs, have short gestation period and give better and quick return. In the domestic market, the consumption of poultry meat has been low due to many reasons such as low purchasing power of people. Only 25 percent of the population living in the urban areas consume about 75-80 percent of the eggs and poultry meat. *Per capita consumption of egg is 100 and poultry meat is 1.2 kg per person per annum in urban area. But in rural area, it is only 15 eggs and 0.15 kg poultry meat.*

Poultry Housing

Select well raised land for poultry sheds. Land with hard rock or "murram" is more suitable. Avoid water logging and flooding near the sheds. Ensure adequate facility for water, electricity, approach road, supply of chicks, feed, veterinary aid and nearness to market for sale of live birds and dressed chicken. Obtain training/experience in broiler farming before starting a farm. You should be prepared to stay on the farm and have constant supervision. Provide adequate floor space, feeding space and watering space per bird. BIS specifications for construction of poultry sheds are available. Construct sheds in such a way that the end walls face East-West direction and the side walls face North-South direction, so that direct sun rays will not enter the sheds. Provide strong roof and hard flooring. Raise plinth of the shed at least one feet above the outside ground level. Provide 3 to 4 feet overhang of the roof to avoid entry of rainwater inside the shed. Provide at least 50 feet distance between two sheds. Provide adequate light and ventilation and comfortable housing conditions during all seasons (cool in summer and warm in winter). Construct sheds in such a way that predators (cats/dogs/snakes) will not enter the shed. Avoid entry of rats by constructing rat proof civil structures. Keep the shed clean and free from flies/mosquitoes etc. After disposal of every batch of birds the dirty litter material and manure should be removed, walls and floors should be cleaned, white washed with lime and disinfected with 0.5% malathion or DDT insecticide spray. If deep litter system is followed, always use dry and clean litter material (sawdust, paddy husk, rice husk etc.). Spread 4" layer of litter on the floor, keep clean/disinfect brooding, feeding and watering equipment and then introduce chicks in the house. The litter material should be always kept loose and dry. Stir the litter twice a week. Any wet litter/droppings etc. should be removed and replaced with fresh/clean dry litter.

Poultry Equipment

Use scientifically designed cages and equipment for brooding, feeding and watering purposes. BIS specifications for equipment are available. A good design can be shown and manufactured locally, so that cost can be reduced.

Chicks

Purchase of improved strain of one day old healthy broiler type chicks from a reputed hatchery. Usually 2-5% extra chicks are supplied. Clean, wash and disinfect all equipments with 0.5% malathion spray after every batch of birds is disposed off.

Brooding Management

Brooding is an art and science of rearing large number of baby chicks in the absence of a broody hen. A newly hatched chick has not developed the thermoregulatory mechanism fully and takes about two weeks to develop this mechanism and homeostasis. Therefore, they can not maintain the body temperature properly for the first few weeks of life and may be subjected to chilling, if not properly taken care of. Hence, artificial brooding is mainly aimed at, providing the right temperature to the chicks. In addition to the temperature, adequate floor, feeder space, water space, relative humidity, ventilation and light should be provided for optimum comfort and growth of the chicks. Maintaining the correct temperature, light, ventilation, space, feeding, watering etc. Of these, the temperature is the most important criterion. Room temperature of about 35°C has to be maintained during the first week of age. Gradually reduced by about 3°C every week until the room temperature or 18-20°C is reached

Types of brooders:

- Incandescent bulbs and other types of bulbs
- Heater coil with thermostat
- Gas brooder
- Kerosene stove
- Charcoal stove
- Centralized heating system
- Battery brooder
- Biogas brooder

Behaviors of Chicks

If they crowd under or near the source of heat, then the warmth given is not sufficient. If the chicks have moved to the periphery and are reluctant to come to the center under heat source, then temperature in the environment is higher than required. If the chicks feel comfortable at a given temperature, they walk actively throughout the area unmindful of heat provided and some take rest setting their head down on the side, the posture being given the name as "Chick comfort".

Feeding

Use high quality balanced feeds. With proper knowledge/experience, the feed can be prepared at the farm. BIS feed formulae and specifications are available. Store the feed in clean, dry, well ventilated room. A wet feed may bring fungus infection. Use properly designed feeders and control the rats to avoid feed wastage. Keep proper records on feed consumption per bird for each batch. Compare with the standard feed consumption pattern. Too low feed consumption may be due to disease condition, low quality/unpalatability of feed, high temperature in poultry shed.

Watering of Birds

Always give fresh and clean drinking water. Water should be always available at birds. Use properly designed watering equipment. Provide adequate watering space per bird. Always keep water-pots clean. Avoid birds entering inside pots. Provide cool water during summer. Store the water in tanks that are not exposed to hot sun in summer.

Disease Prevention/Control

Clean sanitary conditions of poultry sheds and equipment, balanced feed, fresh clean water, healthy chicks are essential to prevent diseases. Avoid entry of visitors to farm, especially inside the sheds. If visitors come, ask them to dip their feet in a disinfectant solution, wash and clean hands and to wear apron/ boots provided by the farm. Use proper vaccination schedule. Use high quality vaccines purchased from reputed manufacturers. Keep vaccines in cool, dry conditions away from sunlight. Any left-over vaccine should be properly disposed off. Vaccines should not be used after their expiry date is over. Any dead bird should be immediately removed from the shed and sent to laboratory for post-mortem or buried/burnt suitably away from the poultry sheds. The waste of farm should be suitably disposed off. Any bird showing advanced signs of a disease, should be removed from the shed and culled. It can be sent to laboratory for diagnosis. Birds showing advanced signs of a disease should be shown to a qualified veterinarian and suitable medication/treatment be given as per his/drug manufacturers recommendations. Poultry manure, if infected, can spread disease, from one batch to another. Keep the litter dry, remove it after flock is sold and dispose the manure properly and quickly. Keep proper records on mortality and its causes and the treatment given to birds. Dates of vaccination for each flock should be properly recorded. Rats are important carriers of poultry disease. Avoid rats. Use suitable rat poisons/rat traps. Many poultry medicines can be given in drinking water. When medication is to be given, remove the waterers in poultry sheds on the previous evening. Next morning give medicine in measured quantity of water, so that entire medicine will be quickly consumed and there will be no wastage of medicines.Mild infection of disease may not cause mortality but it will reduce growth. Keep sample record of body weight and mortality rate. Study the possible causes, if weight is low take steps to improve the management of the subsequent batches. A Constant vigil and analysis of records/results is necessary to keep up the efficiency in farming.

Processing/Marketing

Ensure the constant and steady demand for broiler meat is available and the market is nearer to the farm. Study the market demand for particular live weight of the birds. Birds should not be kept on the farm beyond 5-6 weeks of age, as their feed efficiency will go down considerably. If birds are sold after dressing (processing) use clean dressing hall and processing equipment. Dressed birds should be chilled in the ice-cold water for 3-4 hours and excess water removed. Birds should then be packed in clean plastic bags and the mouth of bag sealed. Processed birds should be marketed as early as possible. If they have to be preserved, deep freezing equipment (-10 to -20°C) be used. Refrigerated vans may be required for long distance transportation.

Chapter 14

Pearl Millet-A Climate Resilient Millet for Livelihood Security

Vikas Gupta[1], A. P. Singh[1], Neeraj Gupta[2] and Meenakshi Gupta[3]*

[1]Advanced Centre for Rainfed Agriculture (ACRA), Skuast-J, Rakh Dhiansar, Samba, J&K UT-181133
[2]Division of FST, SKUAST-J, Chatha, J&K UT-180009
[3]Division of Agronomy, Skuast-J, Chatha, J&K UT-180009
**email: vikasadr@gmail.com*

Abstract

Pearl millet (Pennisetum glaucum L.) is an important crop for food security in Asia and Africa's arid and semi-arid regions. It is widely grown as a staple cereal grain for human consumption and livestock fodder. Pearl millet is the sixth most important cereal in the world. Today, pearl millet is grown on more than 30 million ha mainly in West and Central Africa and the Indian sub-continent as a staple food for more than 90 million people in agriculturally marginal areas. It is rich in proteins and minerals and has numerous health benefits such as being gluten-free and having slow-digesting starch. It is drought and heat tolerant, and a climate-smart crop that can withstand unpredictable variability in climate. However, research on pearl millet improvement is lagging behind other major cereals mainly due to limited investment in terms of man and money power. So far breeding achievements include the development of cytoplasmic male sterility (CMS) and nuclear fertility restoration genes (Rf) for hybrid breeding, dwarfing genes for reduced height, improved input responsiveness, photoperiod neutrality for short growing season, and resistance to important diseases. Further improvement of pearl millet for genetic yield potential, stress tolerance, and nutritional quality traits would enhance food and nutrition security for people living in agriculturally dissolute environments. The recent tools, technologies and adoption of pearl millet crop will not only improve the resilience of global food system through on-farm diversification but also dietary intake which depends on diminishingly fewer crops.

Keywords: Biofortification, Climate resilient, Pearl millet, rainfed

Introduction

Climate-smart agriculture is a pathway towards development and food security built on three pillars:

- Increasing productivity and incomes,
- Enhancing resilience of livelihoods and ecosystems
- Reducing and removing greenhouse gas emissions from the atmosphere

What are the objectives of smart agriculture?

Climate-smart agriculture (CSA) is an integrated approach to managing landscapes-cropland, livestock, forests and fisheries-that addresses the interlinked challenges of food security and accelerating climate change

CSA aims to tackle three main objectives

- Sustainably increasing agricultural productivity and incomes;
- Adapting and building resilience to climate change and
- Reducing and/or removing greenhouse gas emissions, where possible.

What CSA means

- It contributes to achievement of sustainable development goals;
- It integrates social, economical and environmental development to meet challenge of providing sustainable (a) livelihood to farmers (b) food security to hungry millions, and (c) eradication of poverty.

CSA is composed of four pillars

- Sustainably increasing agriculture productivity and income
- Adapting and building resilience to climate change
- Reducing and/or removing green house gas emission wherever possible
- It uses agriculture as a major tool for mitigation of GHG–CO2 by laying emphasis on its unique capacity to absorb CO2 and release oxygen through photosynthesis process.

- It envisages to achieve this through increased cropping, by reducing rain fed areas through integrated water and river basin management and expansion of agriculture on wasteland, wetland, degraded fallow areas and introducing urban agriculture

Climate Smart Technologies are a host of clean low green house gas emission technologies to facilitate adaptation and mitigation to achieve resilience directly and indirectly against climate change impacts globally in short, medium and long term in across all sectors.

Climate-smart crop production practices and technologies

- Use of quality seeds and planting materials of well-adapted crops and varieties.
- Biodiversity management.
- Integrated Pest Management.
- Improved water use and management.
- Sustainable soil and land management for increased crop productivity.
- Sustainable mechanization.

CSA can benefit smallholder farmers directly by increasing efficiency of precious inputs such as labour, seeds and fertilizers, increasing food security, and opportunities for income generation. By protecting ecosystems and landscapes, CSA helps protect natural resources for future generations

Impact of climate change on agriculture

Climate change is very likely to affect food security at the global, regional and local level. Climate change can disrupt food availability, reduce access to food and affect food quality (Brown *et al.*, 2015). Many advance technologies, such as improved crop varieties, pest control methods, genetically modified organisms and irrigation systems, are widely adopted for crop improvement. Due to rainfall variability, changes in the frequency and severity of droughts and floods could pose challenges for farmers and ranchers and will ultimately threaten global food security (Ziska *et al.*, 2014). Along with effects of climate change on agriculture, other evolving factors that affect agricultural production, such as changes in farming practices and technology, need to be addressed.

Why is Climate Smart Agriculture (CSA) Needed?

As per FAO estimate, by the year 2050 world population will increase by one-third and food required for food security by 60 per cent. Already cumulative impact of climate change since last decade has a great effect on productivity. Agriculture has become a high risk profession-farmers increasingly prefer to migrate. In India as per National Sample Survey Organization 2005 estimate 60% farmers like to leave farming. But with available knowledge and experience, it is possible to make agriculture a sustainable livelihood means-but this will require intensive efforts at ground level-local level where agriculture exists and it has to be made climate smart (Shelat, 2014).

To deal with the impact of climate change, the potential adaptation strategies are as follows:

- Developing climate-ready crops
- Crop diversification
- Alteration in land-use pattern.
- Changing cropping season
- Conservation agriculture
- Efficient utilization of resources
- Integrated nutrient management

In India, nearly 60 per cent net sown area is under rain-fed farming, which is completely dependent on monsoon rains. About 55 per cent of India's farmers rely on agriculture as a source of food, income and employment. Over the years, farmers in rain-fed areas have been facing several adversities such as climate variability, crop failure, non-remunerative prices, etc. Their adverse impact on livelihood of farmers is remarkable, as India's economy is more dependent on agriculture. To make agriculture a viable and adaptable proposition in the country, there is a need to do more research and development in rain-fed agriculture. Presently, climate change and hidden hunger are the major bottlenecks that hamper the productivity, healthiness and economic potential of people living on the margins. Crop and dietary diversification of the rain-fed region is limiting due to seasonality and socio economic status. Under this perspective, rain-fed agriculture is risk prone activity and creating panic situation in the backdrop of observed erratic weather and threatening food production and rural livelihoods. In this sense, it is very crucial to elect right crops, which have the ability to buffer crop production against these multiple stresses linked with climate change and produce sustainable yield with minimum inputs to grab the maximum profitability. (Narasimhulu *et al.*, 2020).

In the present-changing climatic scenario, abiotic stresses entail a huge risk for plant growth and development leading to an over 50 per cent decrease in the yield among the popular cereal crops. Almost 90 per cent of the cultivable land is affected by various abiotic stresses globally, while only 10 per cent of the agricultural land is free from these abiotic stresses (Dita *et al.*, 2006). Drought and heat are the two most significant production constraints existing among the different environmental stresses. In this context, a crop species like pearl millet, which is resilient to higher temperatures and lower rainfall, can play a crucial role in fulfilling the increasing food demands of the growing population of the world. Pearl millet is mainly cultivated on marginal lands facing untimely and irregular rainfall patterns and environmental stresses due to its natural inbuilt capacity to survive in such areas and withstand abiotic stresses such as drought, salinity, heat, etc. (Serba and Yadav, 2016). Genetic improvement of pearl millet through increased production was realized using hybrid technology and conventional breeding methods of selection (Yadav and Rai, 2013; Yadav *et al.*, 2021) but later, various biotechnological and genomic approaches were used for further improvement (Varshney *et al.*, 2017; Bollam *et al.*, 2018; Ambawat *et al.*, 2020).

Developing climate-ready crops

Pearl millet (Pennisetum glaucum L.) is cultivated in dry regions of arid and semi-arid tropics where no other cereal can be successfully grown. India is the largest producer of millets in the world, harvesting about 11 million tonnes per year, nearly 36% of the world's output. Pearl millet, which accounts for about two-thirds of millet production in India, is grown in the drier areas of the country, mainly in the states of Rajasthan, Maharashtra, Gujarat, Uttar Pradesh and Haryana. In India, pearl millet is the fourth most widely cultivated food crop after rice, wheat and maize. It occupies an area of 6.93 million hectares with a production of 8.61 million tonnes and a productivity of 1243 kg/ha (Directorate of Millets development, 2020). Pearl millet is a nutritious dryland cereal, rich in protein, fibre and essential micronutrients like iron, zinc and folate. Studies have shown that this cereal has the potential to fight iron deficiency. In India, Rajasthan and Uttar Pradesh are the highest producers of pearl millet.

One of the goals of the research was to bring pearl millet to the mainstream as this requires less input cost. It is also a good solution for food security and climate change.

Status of Pearl millet Production

Pearl millet is a descendent of the wild West African grass and was domesticated over 4,000 years ago in the West African Sahel, spreading later to East Africa and India (Sharma *et al.*, 2021). Now it is being cultivated over 30 million hectares worldwide, with the majority of the crop grown in Africa (>18 million hectares) and Asia (>10 million hectares) (Raheem *et al.*, 2021). Around 90 million people in the Sahelian region of Africa and north-western India consume pearl millet grain as a staple food (Srivastava *et al.*, 2020). In terms of annual production, pearl millet is the sixth most important cereal crop in the world (Brunken *et al.*, 1977; Chandana *et al.*, 2018). Pearl millets, therefore, play a critical role in improving food security for resource-poor farmers. Pennisetum glaucum, a C4 grass, has excellent photosynthetic efficiency and biomass production potential. Millet grain is highly nutritious, with 8–19% protein, low starch, high fibre (1.2%) (Nambiar *et al.*, 2011) and higher micronutrient concentrations (iron and zinc) than rice, wheat, maize and sorghum (Tako *et al.*, 2015). It is grown on 7.5 million ha with an average production of 9.73 million ton and productivity of 1305 kg/ha during 2016-17 (Directorate of Millet Development, 2018).

Global Warming and Climate Change

Climate change (CC) is continuing as a major threat to sustainable agricultural production by embracing unpredictable extreme climatic events such as fluctuation in temperature and uneven rainfall patterns. This climate variability is expected to increase crop vulnerability in different agro-ecologies with drastic consequences on food security and economic growth. Pearl millet is a climate-smart crop with nutritious grains (Anuradha *et al.* 2017) and ideal for environments prone to drought and extreme heat, thereby reinforcing the fight against food insecurity in the arid and semiarid environments (Jukanti *et al.* 2016).

Withstanding higher temperatures

"Most cereals like rice and maize cannot support temperatures over 30 to maximum 35 degrees Celsius when they start forming their grains whereas pearl millet will fill its grain in air temperatures of up to 42 degrees Celsius," says Rajeev Varshney of ICRISAT.

The analysis of reference genome and 1,000 lines of the pearl millet genome have provided the clue as to how this dry land cereal survives in temperatures over 42 degrees Celsius. Results highlighted that wax bio-synthesis genes present in the crop is the reason for this high level of heat resistance.

Pearl millet can survive and produce a large quantity of grain, whereas other cereals such as rice, wheat, maize, sorghum and barley may fail to provide economic benefits under adverse conditions and poor soil. It can provide multiple securities in form of food, fodder, livelihood, nutrition health and ecological benefits, whereas wheat and rice provide only food security, thus making it a crop of agricultural security. Its ability to withstand higher temperature and survival in drought-prone areas and cultivation in parts of Gujarat and eastern Uttar Pradesh of India during hot summers makes it a climate resilient crop for overcoming the adverse effects of the changing climate (Gupta *et al.*, 2015).

Photoperiodism

Days to flowering is one of the most important plant characters that influence the cropping pattern in relation to the natural environment (Takei and Sakamoto, 1987). Besides being an adaptive trait, flowering time and photoperiodic sensitivity are influencing yield and yield stability of pearl millet in arid and semi-arid regions of West and Central Africa (WCA). Pearl millet is generally a photoperiod sensitive crop and almost all landraces flower in short-day condition (Dave, 1987). As Photoperiodism affects the growth and development, it allows for flexible sowing dates by maintaining flowering and grain maturity at the end of the growing season. For instance, pearl millet varieties grown in the Sahel can mature in less than 90 days. In relation with the short duration of the rainy season in the Sahel, Photoperiodism has an advantage in such hot and dry condition where the onset of rain varies from year to year.

In temperate regions, the minimum temperature during the light period of 12 h and less are mostly below optimum for pearl millet growth and development. This low suboptimal temperature for the crop can potentially extend the number of days from floral initiation to anthesis and slow down grain maturity (Burton, 1965) which exposes the crop to likely frost in the autumn. Therefore, pearl millet breeders are intended to develop early maturing day-neutral varieties. So, the day-neutral cultivars can flower in the long-day condition of the summer by accumulating growing degree days and mature earlier before the onset of frost (Serba *et al.*, 2020).

Root System

The root system of vascular plants plays the key roles in acquiring water and mineral nutrients required for the survival as well as accumulation of metabolites for yield and nutritional quality. Moreover, the roots play a significant role in holding soil in place, carbon sequestration, reducing emissions of greenhouse gasses, and prevent the eutrophication of water bodies associated with the application of mineral fertilizers (White *et al.*, 2013)

Sustainable crop production requires root systems optimized for growing conditions in the field. Many of the traits related to abiotic stress tolerance such as water and nutrient use efficiency, and yield performance are linked to the root properties. Deeper and profuse root systems could withstand drought effects by tapping extra water from the lower soil profile. Thus, root architecture has long been suggested mainly to improve crop adaptation and performance in water limited environments. Drought tolerance in pearl millet is in large part related to the root system (McIntyre *et al.*, 1995). Sustained water uptake ability by increasing total root length and maintenance of high leaf water status under soil drying conditions are the main factors for drought tolerance in pearl millet (Matsuura *et al.*, 1996). Therefore, breeding pearl millet varieties with improved root traits promises to deliver benefits in water and nutrient acquisition in dry environments. The hidden part of the plant, root system makes phenotyping very challenging. Recently, imaging technologies have become available that allow to illuminate the dynamics of root structure and function in the soil. Rhizotrons and micro-computed tomography (micro CT) phenotyping approaches characterized early stage pearl millet root system development as a fast growing primary root and three distinct types of lateral roots formation on both primary roots and crown roots (Passot *et al.*, 2016). Pearl millet inbred lines studied for the root system architecture exhibited significant variation in primary root length and lateral root density. It is presumed that primary root growth is associated with early stage drought stress tolerance. Search for genetic markers associated with primary root growth in a large panel of genetically fixed pearl millet inbred lines is a priority research area for drought tolerance.

Drought and Heat Tolerance

High temperature stress can affect pearl millet plant at two stages; either at germination when soil temperatures can be very high or at flowering/ reproductive stage. Field studies in the Sahel indicated that pearl millet seedlings are most vulnerable to high temperatures during the first 10 days of sowing; this was confirmed by field studies in the Indian Thar Desert (Yadav *et al.*, 2012). Genetic differences in seedling survival under high soil surface temperatures were identified (Peacock *et al.*, 1993). Pearl millet has good degree of tolerance to high temperatures of up to 42°C during flowering. Hence, it has occupied considerable areas (about 1 m ha) in the hot and dry post-rainy season (locally referred to as summer) in the northern and western parts of India. Heat tolerance is required in summer season when air temperatures of >42°C coincides with flowering period which leads to reproductive sterility and finally drastic reductions in grain yield. Large genetic variability for seed set at daily maximum air-temperature of 42°C during flowering in pearl millet was reported (Gupta *et al.*, 2015). They indicated that seed set in pearl millet started declining when maximum air temperatures reaches 42°C and decreased in curvilinear fashion to 20 per cent till 46°C. High temperature stress imposed for season long, or during reproductive stages of development for different durations (18 or 45 days) caused significant decrease in number of seeds per panicle, individual seed weight and seed yield per panicle (Djanaguiraman *et al.* 2017). Two periods (10–12 and 2–0 days before anthesis) were identified as most sensitive to short episodes (2 d) of high temperature stress, causing maximum decreases in pollen germination percentages and seed numbers per panicle. The various genotypes available for abiotic stress management are summarized here as under table 1.

Table 1: Available genotypes for abiotic stress resistance in pearl millet

Abiotic stress	Genotypes	References
Drought	CZP 9802; 863B, PRLT 2/89-33ICMP 83720,	Yadav et al, 2004; Dwivedi *et al.* 2012
	ICMV 9413, and ICMV 94472	
Heat	H77/833-2, H77/29-2, CVJ 2-5-3-1-3, 77/371XBSECT	Sankar *et al.*, 2013,
	CP1, 96AC-93, 1305, 77/371, Togo II, 99HS-18,	2014, Arya *et al.* 2014
	G73107, 77/371	
Salinity	ICMB 02111, ICMB 94555, ICMB 95333, ICMB 00888, ICMB 01222, ICMP 451, IP 3732, IP 3757, IP8210, and PRLT 2/89-33, 10876 and 10878 (Sudan), 18406 and 18570(Namibia), and ICMV93753 and ICMV 94474 (India); 863-B, CZI 98-11, CZI 9621, HTP 94/54, DPR-18	Ali *et al.* 2004; Dwivedi *et al.*2012, Raipuria *et al.*, 2012

Salinity Tolerance

In arid and semiarid regions, the extent of salt affected soils is steadily increasing mainly because of flooding irrigation system. Today, more than one billion hectares of land area are affected by soil salinity around the globe (Saade *et al.*, 2016), and has become one of the major constraints on agricultural production. Pearl millet is well-suited to grow in harsh conditions including saline soils. It is generally considered as an alternative crop for salt affected areas because of its fair tolerance to salinity (Ali *et al.*, 2006). Research conducted on salt tolerance in semi-arid areas found a significant interaction of NaCl concentration and pearl millet varieties (Yakubu *et al.*, 2010). Though both plant growth parametres and nutrient contents significantly decreased with increasing soil salinization, they found a variety named Maiwa that was relatively tolerant to soil salinization as it expressed superior nutrient content and root and shoot growth when compared to the control. But the usefulness of these growth parametres for salinity response with respect to grain and forage yield needs to be further investigated.

Cuticular Wax

Abiotic stress arises from exposure to climatic extremes such as drought, heat, cold and frost. Plants have evolved to exist in conditions which are not ideal for maintenance of normal physiology and may be at the limit for survival. The above ground surfaces of terrestrial plants are covered with cuticular wax (Jetter *et al.*, 2007). Epicuticular waxes are complex mixtures of hydrophobic molecules form the outermost layer of the hydrophobic cuticle, which is composed of cutin polyester membrane and provides the last barrier to prevent uncontrolled water loss (Kosma and Jenks 2007). Cuticular wax plays a significant role in plant abiotic and biotic stress tolerance and has been implicated in defense mechanisms against excessive ultraviolet radiation, high temperature, bacterial and fungal pathogens, insects, high salinity, and low temperature tolerance (Xue *et al.*, 2017). Cuticular waxes consist of homologous series of very-long-chain fatty acids, alcohols, aldehydes, alkanes, esters, and cyclic organic compounds (Cameron *et al.*, 2006). Primary plant surfaces are coated with hydrophobic cuticular waxes to minimize non-stomatal water loss. Wax compositions differ greatly between plant species, organs, tissues, and developmental stages (Guo *et al.*, 2017).

A comparison study between sorghum, maize and pearl millet leaves showed no glossiness even on the pearl millet twelfth leaf (Traore, 1985). The glossy character was correlated with a reduction or absence of observable wax deposits on the leaf surfaces and higher cuticular water loss than non-glossy leaves. A recent genomic study unfolded a substantial enrichment for wax biosynthesis genes (Varshney *et al.*, 2017). Therefore, it is believed that cuticular wax deposition in the leaves that minimizes water loss and reflect radiation probably contributed to heat and drought tolerance in pearl millet.

It was also demonstrated that changes in cuticle gene expression in drought exposed plants (Suh *et al.*, 2005) provide evidence for alteration of cuticle traits for drought tolerance.

Climate Resilient varieties/hybrids of pearl millet

To accelerate progress towards ending hunger and safeguard food security and improved nutrition, there is a dire need to promote nutrition sensitive agriculture, where in cultivation of biofortified varieties require no more water and fertilizers than the regular varieties. Consumption of biofortified pearl millet can minimize the deleterious effects of nutrient deficiencies by providing significant amount of the iron and zinc needed daily by young children and women, further resulting in profound positive impacts on livelihoods. (Narasimhulu *et al.*, 2020).

Table 2. Important climate resilient biofortified varieties/hybrids of pearl millet

S. No.	Name of variety/ hybrid	Bred at	Areas of adaptation	Salient features	Grain yield (kg/ha)	Fe (ppm)	Zn (ppm)
Varieties							
1	Dhanshakti	MPKV, Dhule, Maharashtra	Maharashtra, KN, AP, TN, Raj, Haryana, Guj, Punjab, UP, Jammu	Early maturing variety, bold, globular, shining slate grey coloured seed, cylindrical resistant to downy mildew	2199	81	43
2	ABV 04 (MP552)	ARS, ANGRAU, Ananthapura mu, Andhra Pradesh	Maharashtra, KN, AP, Telangana, TN	Medium maturing, resistant to downy mildew, smut and blast diseases; tolerant to drought	2500	70	63
Hybrids							
3	HHB 299	CCSHAU, Hisar	Haryana, Raj, Guj, Punjab, Delhi, Maharashtra, TN	Medium maturing, resistant to major diseases and insect pests	3274	73	41
4	AHB 1200 (AHB1200Fe) (MH 2072)	VNMKV, Parbani, Maharashtra	Raj, Guj, Haryana, Punjab, Delhi, Maharashtra, Telangana, AP, TN	Medium maturing, high Fe content, resistant to downy mildew, resistant to stem borer, highly responsive to fertilizers	3170	77	39

S. No.	Name of variety/ hybrid	Bred at	Areas of adaptation	Salient features	Grain yield (kg/ha)	Fe (ppm)	Zn (ppm)
5	AHB 1269 (AHB1269Fe) (MH 2185)	VNMKV, Parbani, Maharashtra	Raj, Guj, Haryana, Punjab, Delhi, Maharashtra, Telangana, AP, TN	Medium maturing, high Fe content, resistant to downy mildew	3200	91	45
6	Phule Mahash akti (DHBH 1211/MH 2078)	MPKV, Dhule, Maharashtra	Maharashtra	Compact earhead, bold globular grains with gray colour, resistant to downy mildew	2900	87	43

Table 3. Major reasons for gap in productivity of pearl millet and action required to be taken

S. No.	Reasons for gap	Action to be required
1	Imbalance use of chemical fertilizer *i.e.* higher use of N fertilizer as compared to P & K fertilizers	Follow INM to improve soil fertility and productivity
2	Low water and fertilizer use efficiency	Improve water and fertilizer use efficiency through the efficient farm management and micro irrigation system
3	Plant density cannot be maintained properly as per recommendation	Optimum plant density should be maintained with appropriate seed rate
4	Poor adoption of plant protection measures and weed management	Adopt integrated IPM and IWM schedule
5	Problem of lodging at harvesting stage in *kharif* season	Carry out earthing up operation at 35 DAS

Nutritional Importance

Pearl millet as such is already a high-iron crop along with high Zn content; however, commercially available cultivars have lower Fe and Zn contents. Biofortification priority index (BPI) indicates that pearl millet is a major target crop for iron and zinc Biofortification (Kanatti *et al.*, 2014). The first high-iron pearl millet variety "Dhanashakti" was released for the state of Maharashtra in 2013, and subsequently, it was notified and released by Central Variety Releasing Committee in April 2014, for growing in all pearl millet-growing regions of India. Inclusion of nutritional quality traits for Fe (42 ppm) and Zn (32 ppm) in the varietal promotion criteria is first of its kind in any of the food crop and the world too (Satyavathi, 2017). Since then, many hybrids/varieties were developed as micronutrient-rich pearl millet cultivars, including the seven biofortified cultivars identified from the special biofortification trial.

Pearl millet crop envisages less yield losses which might be occur due to the effects of climate change, changes in temperature, environmental stresses, poor crop management and adverse or erratic rainfall in comparison to other crops (Ausiku *et al.*, 2022).

Health Benefits/ nutritional security

Food grain production has increased but still, to feed the growing population and meeting out good health of the people in the present situation, nutritional security is very important. Intake of all essential macronutrients and micronutrients (vitamins and minerals) through a balanced and diversified diet in sufficient quantities is very vital for an active and healthy life. Nutritional insecurity is a major challenge for the world's growing population, which mainly depends on a micronutrient deficient cereal-based diet. Pearl millet is the primary source of energy for the semi-arid tropics and drought-prone regions of Asia and Africa after the major cereals such as wheat, rice, maize and sorghum due to its nutritionally superior grain enriched with high amounts of essential amino acids, proteins, better fat digestibility, vitamins and minerals (Table 4). Pearl millet is rightly termed as "nutricereal" as it is a good source of energy, carbohydrate, protein, fat, ash, dietary fiber, iron and zinc. It is a rich source of energy (361 Kcal/100 g) comparable to sorghum (349 Kcal/100 g), wheat (346 Kcal/100 g), rice (345 Kcal/100 g) and maize (325 Kcal/100 g). The carbohydrate content of pearl millet is 67.5 g/100 g with 56 to 65 per cent starch comprising 20 to 22 per cent amylase and 2.6 to 2.8 per cent free sugars-mainly sucrose. It is high in fibre (1.2 g/100 g) and (▯) amylase activity. The protein content of pearl millet is (11.6/100 g), comparable to wheat but higher than rice. It is rich in methionine but poor in sulphur containing amino acids. With low prolamine fraction, pearl millet is gluten free grain and is the only grain that retains its alkaline properties after being cooked which is ideal for people with gluten allergy. Pearl millet is rich in fat content (5 mg/100 g) with better fat digestibility and rich in unsaturated fatty acids (75%) with higher content of nutritionally important N-3 fatty acids. It is a rich source of vitamins and minerals with overall mineral content of 2.3 mg/100 g constituting potassium, phosphorous, magnesium, iron, zinc, copper and manganese. It is rich in B-vitamins (thiamine, riboflavin and niacin). Pearl millet is used in different forms at a global level: unleavened bread (roti or chapatti), porridge, gruel, dessert etc. and it is generally defined as a poor man's bread. Its flour can substitute (10–20%) for wheat flour in "whole-grain" breads, pretzels, crackers, tortillas and dry and creamed cereals (Dahlberg *et al.*, 2003). In India and Africa, it is primarily grown for food and forage, while in the American continents; it is a main component of poultry and livestock sector (Serba *et al.*, 2020).

Value addition is also very useful to promote its consumption, and several value-added products are gaining popularity among people. It can be used for making various traditional food products such as khichri, roti, sakarpare, gulgule etc., while industries are also using it for making products such as noodles, pasta, vermicelli, biscuits, bread, cookies, cakes, puffs etc in India (Satyavathi *et al.*, 2021).

In Nigeria and Niger, the thin porridge "Fourra" and Masa, a fried cake are very popular, while "Soungouf," "Sankhal" and "Araw" are mainly popular in Senegal (Kaur *et al.*, 2014; Ajeigbe *et al.*, 2020). Its grains are also locally brewed to produce non-alcoholic or alcoholic beverages in Asia and Africa (Dwivedi *et al.*, 2012).

Table 4. Nutritional comparison of pearl millet with sorghum, rice, and wheat (in 100 g grain)

Contents	Crops			
	Pearl millet	sorghum	Rice	Wheat
Carbohydrates	61.8	67.7	78.2	64.7
Protein (g)	10.9	9.9	7.9	10.6
Fat (g)	5.43	1.73	0.52	1.47
Energy (kcal)	347	334	356	321
Dietary fibre (g)	11.5	10.2	2.8	11.2
Calcium (mg)	27.4	27.6	7.5	39.4
Phosphorus (mg)	289	274	96	315
Magnesium (mg)	124	133	19	125
Zinc (mg)	2.7	1.9	1.2	2.8
Fe (mg)	6.4	3.9	0.6	3.9
Thiamine (mg)	0.25	0.35	0.05	0.46
Riboflavin (mg)	0.20	0.14	0.05	0.15
Niacin (mg)	0.9	2.1	1.7	2.7
Folic acid (μ g)	36.1	39.4	9.32	30.1
Source: Satyavathi *et al.*, 2021				

Medicinal use of pearl millet

Pearl millet has been attributed to have several health promoting abilities due to its chemical composition, which are listed below

Table 5: Medicinal Use of Pearl Millet

Disease	Benefits	Positive Factors in Perl Millet
Anemia	May help in increasing Hb	High iron content (8 mg/100 g), high zinc content (3.1 mg/100 g)
Constipation	May help in dealing with constipation	High fibre (1.2 g/100 g)
Cancer	Anti cancer property, inhibit tumor development	
Diabetes	Help in dealing with diabetes	Has low glycemic index
Celiac	Anti Allergic	Gluten free
Diarrhea	Probiotic treatment	Lactic acid bacteria
NCDs	Inhibits DNA scission, LDL cholesterol, liposome oxidation and proliferation of HT-29 adenocarcinoma cells	Flavonoids, phenolics Omega 3 fatty acids

Disease	Benefits	Positive Factors in Perl Millet
Helps in bone growth development and repair	Pearl millet has a large amount of phosphorus. Phosphorus is very essential for bone growth and development as well as for development of ATP which is the energy currency of our body	Due to large amount of phosphorus
Stomach ulcers	Pearl millet is recommended for curing stomach ulcers. The most common cause for stomach ulcers is excess acidity in the stomach after food intake. Pearl millet is one of the very few foods that turns the stomach alkaline and prevents formation of stomach ulcers or reduces the effect of ulcers	Prevents formation of excess acidity
Respiratory problems for asthma patients	Pearl millet contains high concentration of magnesium which helps reduce severity of respiratory problems for asthma patients and is also effective in reducing migraine attacks	Due to high amount of magnesium
Weight loss (Obesity)	Pearl millet can aid the process of weight loss as it is high in fibre content. Owing to its fibre content it takes longer for the grain to move from the stomach to the intestines. This way, pearl millet satiates hunger for a long period of time and thus helps in lowering the overall consumption of food	Due to high fibre content
Preventing Gall stones	The high fibre content in pearl millet is also known to reduce the risk of gall stone occurrence. The insoluble fibre content in pearl millet reduces the production of excessive bile in our system. Excessive amount of bile secretion in our intestine often leads to aggravate the condition of gall stones	Due to high fibre content
Anti allergic properties	Pearl millet is a treasure trove of beneficial properties. The grain is very digestible as such and has a very low probability of causing allergic reactions. Due to its hypo allergic property, it can be safely included in the diets of infants, lactating mothers, elderly and convalescents.	Due to its hypo allergic property
Source: Malik, 2015		

Challenges for pearl millet

Despite the breeding efforts, most of breeding programs fail to deliver hybrids due to a vast variation in microclimate (day and night temperature and humidity) and soil apart from rainfall, which requires proper quantification. Further, narrow cultivar diversity in drought-prone ecology also is another factor for this. Thus, there is a high need to give higher priority to the below mentioned areas to promote its production and utilization:

- Development of hybrids/varieties of pearl millet with better regenerative capacity on reversal of dry spell for harsh environment/ drought-prone areas (for A1 zone in India).
- Development of hybrids/varieties resistant/tolerant to salt/high temperature.
- Shift in focus of breeding from productivity improvement to the identification of end product-specific traits.
- Mainstreaming of biofortification in pearl millet for iron and zinc.
- Enhancement of shelf life of pearl millet flour and overcome rancidity to promote its products.
- Development of screening protocols and control measures against different diseases such as downy mildew, blast, rust, ergot, smut.
- Generating authentic data on nutritional benefits of pearl millet and bioavailability studies.
- A study on demand survey for pearl millet.

Future Perspectives

Today, fewer crop species are feeding the world than several decades ago. This scenario raises a serious concern about the resilience of the global food system in the wake of global climate change. With lack of on-farm species diversity and dietary intake, more and more people are now exposed to harvest and health failures than ever. Many people are affected by gluten allergy. Gluten free, protein and micronutrient rich crops that can withstand the climatic variability would have a great chance of coming to the main stream production in certain parts of the world where the consequence of climate change would be impactful (Serba *et al.*, 2020)

Potential for Expansion of Productivity

Among crops with high potential of production expansion, pearl millet is expected to come to the forefront. This is attributed to its exceptional heat and drought tolerance and high nutritional value. Productivity has been low; however, some improvements have been made in India (Serba *et al.*, 2020)

Improved hybrids have been widely adopted by Indian farmers with the result that the crop productivity has gone up from 305 kg/ha during 1951–1955 to 998 kg/ha during 2008–2012, registering a 227% improvement which assumes greater significance is being given despite more than 90% of pearl millet is grown as rainfed and often on marginal lands (Yadav and Rai 2013). Again productivity f pearl millet gone up to 1237 kg/ha in 2017 (Anonymous, 2018) and 1243 kg/ha in 2020 (Directorate of Millets, 2020). Launching hybrid breeding in major pearl millet growing countries in west Africa have a glimmer of hope to bring about a paradigm shift in productivity in the region.

Conclusions and Way forward

In the face of leading water scarcity, pearl millet is most preferred crop due to its adaptive climate-resilience and nutritional features that can really help the policymakers and farmers to combat drought threat situation. Promotional efforts for millets have made huge demand and significant differences in production. The productivity may be further enhanced by adoption of improved production technologies and may increase economic returns from pearl millet-based crop-livestock production systems. In an overview, pearl millet "a power house of nutrients" is easily accessible and affordable to everyone to address the issues of food security, malnutrition and climate change, while protecting the livelihoods of farmers.

References

Anonymous, 2018. Agricultural Statistics at aGlance. Directorate of Economics and Statistics, Government of India.

Ajeigbe, H.A., Angarawai, I.I., Inuwa, A.H., Akinseye, F.M. and Abdul Azeez, T. 2020. Hand Book on Improved Pearl Millet Production Practices in North Eastern Nigeria, 9885-9888.

Ali, G.M., Murtaza, N., Collins, J.C. and McNeilly, T. 2006. Study of salt tolerance parameters in pearl millet (Pennisetum americanum). Journal of Century European Agriculture, 7: 365-376.

Ambawat, S., Singh, S., Shobhit, Meena R.C. and Satyavathi, C.T. 2020. Biotechnological applications for improvement of the pearl millet crop. In: Pearl Millet: Properties, Functionality and its Applications, eds S.K. Gahlawat, S. Punia, A.K. Siroha, K.S. Sandhu, and M. Kaur (Boca Raton, FL: Taylorand Francis, CRC Press), pp: 115-138.

Anuradha, N., Satyavathi, C.T., Bharadwaj, C., Nepolean, T., Sankar, S.M., Singh, S.P., Meena, M.C., Singhal, T. and Srivastava, R.K. 2017. Deciphering genomic regions for high grain iron and zinc content using association mapping in pearl millet. Frontiers in Plant Sciences, 8: 412.

Ausiku, P.A., Annandale, J.G., Steyn, J.M. and Sanewe, A.J. 2022. Crop model parameterization of three important pearl millet varieties for improved water use and yield estimation. Plants, 11: 806. https://doi.org/10.3390/plants11060806.

Bollam, S., Pujarula, V., Srivastava, R.K. and Gupta, R. 2018. Genomic approaches to enhance stress tolerance for productivity improvements in pearl millet: genomic approaches. In: Biotechnologies of Crop Improvement, Vol. 3, eds. S. S. Gosal and S. H. Wani. (Basel: Springer International Publishing AG, part of Springer Nature), pp: 239-264.

Brown, M.E., Antle, J.M., Backlund, P., Carr, E.R., Easterling, W.E., Walsh, M.K., Ammann, C., Attavanich, W., Barett, C.B., Bellemare, M.F., Dancheck, V., Funk, C., Grace, K., Ingram, J.S.I., Jiang, H., Maletta, H., Mata, T., Murray, A., Ngugi, M., Ojama, D., Neill, B. and Tebaldi, C. 2015. Climate change, Global food security and the US Food System. USDA, pp: 146.

Brunken, J., De Wet, J. and Harlan, J. 1977. The morphology and domestication of pearl millet. Economic Botany, 31: 163-174.

Burton, G.W. 1965. Photoperiodism in pearl millet, Pennisetum typhoides. Crop Science, 5: 333-335.

Cameron, K.D., Teece, M.A. and Smart, L.B. 2006. Increased accumulation of cuticular wax and expression of lipid transfer protein in response to periodic drying events in leaves of tree tobacco. Plant Physiology,140: 176-183.

Chandana, P., Lata, A.M., Khan, M.A. and Krishna, A. 2018. Influence of nutrient management practices on growth and yield of pearl millet in Melia dubia based agri-silvi system. International Journal of Current Microbiology and Applied Sciences, 7: 443-454.

Dahlberg, J.A., Wilson, J.P. and Snyder, T. 2003. Sorghum and pearlmillet: health foods and industrial products in developed countries. In: Alternative Uses of Sorghum and Pearl Millet in Asia. Proceedings of Expert Meeting (Patancheru: ICRISAT) pp: 42-59.

Dave, H.R. 1987. Pearl millet hybrids. In: Witcombe JR, Beckerman SR (eds) Proceedings of the Iinternational pearl millet workshop. ICRISAT, Patancheru, AP 502324, India, pp: 121-126.

Serba, D.D., Yadav, R.S., Varshney, R.K., Gupta, S.K,. Mahalingam, G., Srivastava, R.K., Gupta, R., Perumal, R. and Tess, T.T. 2020. Genomic Designing of Pearl Millet: A Resilient Crop for Arid and Semi-arid Environments. In: Pearl millet: A Resilient Crop for Arid and Semi-arid Environments (eds.Ratan S. Yadav, Rajeev K Varshney, S. K. Gupta, Mahalingam Govindraj) https://www.researchgate.net/publication/339124201

Directorate of Millets Development. 2018. Available online at: http://millets.dacfw.nic.in

Directorate of Millets Development. 2020. Available online at: http://millets.dacfw.nic.in

Dita, M.A., Rispail, N., Prats, E., Rubiales, D. and Singh, K.B. 2006. Biotechnology approaches to overcome biotic and abiotic stress constraints in legumes. Euphytica, 147: 1-24. doi: 10.1007/s10681-006-6156-9

Djanaguiraman, M., Perumal, R., Ciampitti, I.A., Gupta, S.K. and Prasad, P.V.V. 2017. Quantifying pearl millet response to high temperature stress: thresholds, sensitive stages, genetic variability and relative sensitivity of pollen and pistil. Plant, Cell Environment, 41: 993-1007.

Dwivedi, S., Upadhyaya, H., Senthilvel, S. and Hash, C. 2012. Millets: genetic, and genomic resources. In: Janick J (ed) Plant Breeding Reviews. John Wiley and Sons, Hoboken, NJ, 247-375.

Guo, Y., Busta, L. and Jetter, R. 2017. Cuticular wax coverage and composition differ among organs of Taraxacum officinale. Plant Physiology and Biochemistry, 115: 372-379.

Gupta, S.K., Rai, K.N., Singh, P., Ameta, V.L., Gupta, S.K., Jayalekha, A.K., Mahala, R.S., Pareek, S., Swami, M.L. and Verma, Y.S. 2015. Seed set variability under high temperatures during flowering period in pearl millet (Pennisetum glaucum L.). Field Crops Research, 171: 41-53.

Jetter, R., Kunst, L. and Samuels, A.L. 2007. Composition of plant cuticular waxes. Annals of Plant Reviews, 23: 145-181.

Jukanti, A., Gowda, C.L., Rai, K., Manga, V. and Bhatt, R. 2016. Crops that feed the world: Pearl Millet (Pennisetum glaucum L.): An important source of food security, nutrition and health in the arid and semi-arid tropics. Food Security, 8: 307-329.

Kanatti, A., Rai, K.N., Radhika, K., Govindaraj, M., Sahrawat, K.L. and Srinivasu, K. 2014. Relationship of grain iron and zinc content with grain yield in pearl millet hybrids. Crop Improvement, 41: 91-96.

Kaur, K.D., Jha, A., Sabikhi, L. and Singh, A.K. 2014. Significance of coarse cereals in health and nutrition: A Review. Journal of Food Science and Technology, 51: 1429-1441. doi: 10.1007/s13197-011-0612-9.

Kosma, D.K. and Jenks, M.A. 2007. Eco-physiological and molecular-genetic determinants of plant cuticle function in drought and salt stress tolerance. In: Jenks MA, Hasegawa PM, Jain SM (eds) Advances in molecular breeding toward drought and salt tolerant crops. Springer, Netherlands, Dordrecht, pp 91-120.

Malik, S. 2015. Pearl millet-nutritional value and medicinal uses. International Journal of Advance Research and Innovative Ideas in Education, 1 (3): 414-418.

Matsuura, A., Inanaga, S. and Sugimoto, Y. 1996. Mechanism of interspecific differences among four gramineous crops in growth response to soil drying. Japan Journal of Crop Sciences, 65: 352-360.

McIntyre, B.D., Riha, S.J. and Flower, D.J. 1995. Water uptake by pearl millet in a semiarid environment. Field Crops Research, 43: 67-76.

Nambiar, V.S., Dhaduk, J., Sareen, N., Shahu, T. and Desai, R. 2011. Potential functional implications of pearl millet (Pennisetum glaucum) in health and disease. Journal of Applied and Pharmaceutical Sciences, 1: 62.

Narasimhulu, R., Tara Satyavathi, C., Govindaraj, M, and Veeraraghavaiah, R. 2020. Pearl millet-potential to boosts rainfed farmers livelihoods and nutritional security. Agriallias, 2 (1): 42-48.

Passot, S., Gnacko, F., Moukouanga, D., Lucas, M., Guyomarc, S., Ortega, B.M., Atkinson, J.A., Belko, M.N., Bennett, M.J., Gantet, P., Wells, D.M., Guédon, Y., Vigouroux, Y., Verdeil, J.L., Muller, B. and Laplaze, L. 2016. Characterization of pearl millet root architecture and anatomy reveals three types of lateral roots. Frontiers in Plant Sciences, 7: 1-11.

Peacock, J.M., Soman, P., Jayachandran, R., Rani, A.U., Howarth, C.J. and Thomas, A. 1993. Effects of high soil surface temperature on seedling survival in pearl millet. Experimental Agriculture, 29: 215-225.

Raheem, D., Dayoub, M., Birech, R. and Nakiyemba, A. 2021. The contribution of cereal grains to food security and sustainability in Africa: potential application of UAV in Ghana, Nigeria, Uganda, and Namibia. Urban Science, 5: 8. doi: 10.3390/urbansci5010008.

Saade, S., Maurer, A., Shahid, M., Oakey, H., Schmöckel, S.M., Negrão, S., Pillen, K. and Tester, M. 2016. Yield-related salinity tolerance traits identified in a nested association mapping (NAM) population of wild barley. Science Rep, 6: 325-386.

Satyavathi, C.T. 2017. Proceedings of 52nd Annual Group Meet, ICAR-AICRP on Pearl millet, Jodhpur held at PAU, Ludhiana (Punjab) w.e.f April 28-30, 2017.

Satyavathi, C.T., Ambawat, S., Khandelwal, V. and Srivastava, R.K. 2021. Pearl Millet: A Climate-Resilient Nutricereal for Mitigating Hidden Hunger and Provide Nutritional Security, A Review. Frontiers in Plant Sciences, 12: 659938. doi: 10.3389/fpls.2021.659938.

Serba, D.D. and Yadav, R.S. 2016. Genomic tools in pearl millet breeding for drought tolerance: status and prospects. Frontiers in Plant Sciences, 7: 1724 doi: 10.3389/fpls.2016.01724

Serba, D.D., Yadav, R.S., Varshney, R.K., Gupta, S.K., Mahalingam, G. and Srivastava, R.K. 2020. Genomic designing of pearl millet: a resilient crop for arid and semi-arid environments", in Genomic Designing of Climate-Smart Cereal Crops, ed C. Kole (Springer Nature), 221-286. doi: 10.1007/978-3-319-93381-8_6.

Sharma, S., Sharma, R., Govindaraj, M., Mahala, R.S., Satyavathi, C. T. and Srivastava, R. K. 2021. Harnessing wild relatives of pearl millet for germplasm enhancement: challenges and opportunities. Crop Sciences, (TSI) 61: 177-200. doi: 10.1002/csc2.20343

Shelet, K.N. 2014. Climate smart agriculture-the way forward The Indian perspective. Editor Dr. Kirit N Shelat. Ist Edition: 2014. Published by: NCCC, (NCCSD), Ahmadabad and CRIDA, Hyderabad.

Srivastava, R.K., Bollam, S., Pujarula, V., Pusuluri, M., Singh, R.B. and Potupureddi, G. 2020. Exploitation of heterosis in pearl millet: A Review. Plants, 9: 807. doi: 10.3390/plants9070807.

Suh, M.C., Samuels, A.L., Jetter, R., Kunst, L., Pollard, M., Ohlrogge, J. and Beisson, F. 2005. Cuticular lipid composition, surface structure, and gene expression in Arabidopsis stem epidermis. Plant Physiology, 139: 1649 LP-1665.

Takei, E. and Sakamoto, S. 1987. Geographical variation of heading response to day length in foxtail millet (Setaria italica P. BEAUV.). Japan Journal of Breeding, 37: 150-158.

Tako, E., Reed, S.M., Budiman, J., Hart, J.J. and Glahn, R.P. 2015. Higher iron pearl millet (Pennisetum glaucum L.) provides more absorbable iron that is limited by increased polyphenolic content. Journal of. Nutrition, 14: 11.

Traore, M. 1985. Physiological and morphological mechanisms of drought resistance in sorghum and pearl millet. I. Effects of leaf treatment with abscisic acid. II. Seed pre-treatment with abscisic acid. III. Comparative shoot and root development. IV. Leaf surface morphol. University of Nebraska, Lincoln.

Varshney, R.K., Shi, C., Thudi, M., Mariac, C., Wallace, J., Qi, P., Zhang, H., Zhao, Y., Wang, X., Rathore, A., Srivastava, R.K., Chitikineni, A., Fan, G., Bajaj, P., Punnuri, S., Gupta, S.K., Wang, H., Jiang, Y., Couderc, M., Katta, M.A.V.S.K., Paudel, D.R., Mungra, K.D., Chen, W., Harris-Shultz, K.R., Garg, V., Desai, N., Doddamani, D., Kane, N.A., Conner, J.A., Ghatak, A., Chaturvedi, P., Subramaniam, S., Yadav, O.P., Berthouly-Salazar, C., Hamidou, F., Wang, J., Liang, X., Clotaul, J., Upadhyaya, H.D., Cubry, P., Rhoné, B., Gueye, M.C., Sunkar, R., Dupuy, C., Sparvoli, F., Cheng, S., Mahala, R.S., Singh, B., Yadav, R.S., Lyons, E., Datta, S.K., Tom Hash, C., Devos, K.M., Buckler, E., Bennetzen, J.L., Paterson, A.H., Ozias-Akins, P., Grando, S., Wang, J., Mohapatra, T., Weckwerth, W., Reif, J.C., Liu, X., Vigouroux, Y. and Xu, X. 2017. Pearl millet genome sequence provides a resource to improve agronomic traits in arid environments. Natural Biotechnology, 35: 969-976.

Varshney, R.K., Shi, C., Thudi, M., Mariac, C., Wallace, J. and Qi, P. 2018. Pearl millet genome sequence provides a resource to improve agronomic traits in arid environments. Natural Biotechnology, 35: 969. doi: 10.1038/nbt.3943

White, P.J., George, T.S., Gregory, P.J., Bengough, A.G., Hallett, P.D. and McKenzie, B.M. 2013. Matching roots to their environment. Annals of Botany, 112: 207-222.

Xue, D., Zhang, X., Lu, X., Chen, G. and Chen, Z.H. 2017. Molecular and evolutionary mechanisms of cuticular wax for plant drought tolerance. Frontiers in Plant Sciences, 8: 621.

Yadav, O.P. and Rai, K.N. 2013, Genetic improvement of pearl millet in India. Agricultural Research, 2: 275-292.

Yadav, O.P., Rai, K.N. and Gupta, S.K. 2012. Pearl millet: genetic improvement in tolerance to abiotic stresses. In: Improving crop productivity in sustainable agriculture, pp 261-288.

Yadav, O.P. and Rai, K.N. 2013. Genetic improvement of pearl millet in India. Agricultural Research, 2: 275-292. doi: 10.1007/s40003-013-0089-z

Yadav, O.P., Gupta, S.K., Govindara, J.M., Sharma, R., Varshney, R.K. and Srivastava, R.K. 2021. Genetic gains in pearl millet in India: insights into historic breeding strategies and future perspective. Frontiers in Plant Sciences, 12: 645038. doi: 10.3389/fpls.2021.645038

Yakubu, H., Ngala, A.L. and Dugje, I.Y. 2010. Screening of millet (Pennisetum glaucum L.) varieties for salt tolerance in semi-arid soil of Northern Nigeria. World Journal of Agricultural Sciences, 6: 374-380.

Ziska, L., Crimmins, A., Aucliar, S., DeGrasse, J.F., Garofalo, A.S., Khan, I., Loladze, A.A., Perez de Leon, A., Sholer, J. and Thurston Wall, I. 2016. Food safety, nutrition and distribution. The impacts of climate change on human health in the United States: A scientific assessment, Washington. DC: US Global change research program, pp: 189-216.

Chapter 15

Resource Conservation Technology for Enhancing Productivity

A.K. Sinha[1], Neerja Sharma[2], Saurav Gupta[3], Vijay Kumar Sharma[4], Sunil Mishra[5], Vijay Kumar[6] and Shalini Khajuria[7]

[1,2,3,4,7] KVK Samba, SKUAST-Jammu
[5]Division of Agronomy, SKUAST-Jammu
[6]Rainfed Research Sub Station for Sub Tropical Fruits Raya, SKUAST-Jammu
Email: abhaysinha.1995@gmail.com

Resource conservation technology refers to the use of various techniques, practices, and technologies to reduce the consumption and waste of natural resources while achieving the desired economic and environmental goals. This approach is critical for sustainable development and the protection of our planet's ecosystems. Here are some key areas and examples of resource conservation technology:

1. Energy Conservation Technology

- Energy-efficient appliances and lighting systems.
- Smart thermostats and HVAC systems to optimize heating and cooling.
- Renewable energy sources like solar panels and wind turbines.
- Energy management systems for industrial processes.

2. Water Conservation Technology

- Low-flow toilets and water-saving faucets.
- Drip irrigation systems for agriculture.
- Rainwater harvesting systems.
- Water recycling and reclamation technologies for industrial use.

3. Waste Reduction and Recycling Technology

- Recycling facilities and sorting systems.
- Waste-to-energy conversion technologies.
- Composting systems for organic waste.
- 3D printing for on-demand production, reducing excess material waste.

4. Sustainable Agriculture and Farming Technology

- Precision farming techniques using sensors and data analytics.
- Controlled environment agriculture (CEA) systems like hydroponics and vertical farming.
- GMOs (genetically modified organisms) for drought-resistant and pest-resistant crops.
- Conservation tillage and no-till farming practices.

5. Sustainable Forestry and Land Management Technology

- GPS and GIS (Geographic Information System) for sustainable forest management.
- Sustainable logging practices, such as selective harvesting.
- Reforestation and afforestation techniques.
- Soil conservation methods, like terracing and contour farming.

6. Transportation and Mobility Technology

- Electric vehicles (EVs) and hybrid cars.
- Public transportation systems and ridesharing platforms.
- Intelligent transportation systems (ITS) for traffic management.
- Bicycle-sharing programs and pedestrian-friendly infrastructure.

7. Industrial Resource Efficiency Technology

- Lean manufacturing principles to minimize waste.
- Industrial symbiosis, where waste from one industry becomes a resource for another.
- Green chemistry and eco-friendly materials in production processes.
- Circular economy practices to extend product lifecycles.

8. Building and Construction Technology

- Green building standards like LEED (Leadership in Energy and Environmental Design).
- Passive solar design for energy-efficient homes.
- Sustainable building materials like recycled and locally sourced materials.
- Building automation systems for energy and resource optimization.

9. Information and Communication Technology (ICT)

- Smart grids for efficient energy distribution.
- IoT (Internet of Things) devices for monitoring and controlling resource use.
- Data analytics and AI for optimizing resource-intensive processes.
- Digital platforms for sharing and reusing resources, such as car-sharing apps.

Resource conservation technology plays a crucial role in mitigating the environmental impact of human activities and ensuring a more sustainable future for generations to come. These technologies not only reduce resource consumption but also often lead to cost savings and improved quality of life.

Super Seeder for Wheat Sowing

A super seeder, also known as a zero-till seed drill or a no-tillage drill, is a specialized agricultural implement used for sowing wheat and other crops while minimizing soil disturbance. This modern farming technology has gained popularity due to its ability to conserve soil moisture, reduce erosion, and enhance crop yields. Here is a detailed description of a super seeder for wheat sowing:

1. Purpose

Super seeders are primarily designed for sowing wheat and other small-grain crops in a no-till or minimum-till farming system. The main objectives are to plant seeds directly into untilled soil with minimal disruption to the existing crop residue and soil structure.

2. Components: A super seeder typically consists of the following components

- **Seed Hopper:** This is the container that holds the wheat seeds. It can vary in capacity depending on the size of the machine.
- **Metering Mechanism:** It precisely measures and dispenses the seeds at a desired rate. This ensures uniform seed spacing and optimal plant density.

- **Seed Tubes:** These tubes transport the seeds from the seed hopper to the ground level, where they are placed in the soil.
- **Disc Openers:** These are cutting discs or blades that make a narrow slit in the untilled soil for seed placement. They can be adjusted for depth and spacing.
- **Press Wheels or Closing Wheels:** These wheels follow the disc openers and help to close the seed furrow and ensure good seed-to-soil contact.
- **Depth Control System:** It allows the farmer to adjust the planting depth of the seeds to match the soil conditions and the crop's requirements.

3. Operation: The super seeder is attached to a tractor and pulled across the field. Here's how it operates

- As the tractor moves forward, the disc openers cut a narrow slit or furrow in the untilled soil.
- The metering mechanism releases seeds into the seed tubes, which drop the seeds into the furrows at a controlled rate.
- The press wheels or closing wheels follow immediately behind, pressing the soil around the seeds to ensure good soil-seed contact.
- The depth control system allows the operator to adjust the planting depth to match the desired depth for wheat cultivation.

4. Advantages: Using a super seeder for wheat sowing offers several advantages

- **Conservation of Soil:** Minimal soil disturbance helps retain moisture and organic matter in the soil, reducing erosion and preserving soil health.
- **Increased Yields:** Better seed placement and soil-seed contact lead to higher germination rates and improved crop yields.
- **Time and Fuel Savings:** Super seeders are efficient and reduce the need for multiple passes through the field, saving time and fuel costs.
- **Environmental Benefits:** Reduced soil erosion and the use of fewer inputs contribute to more sustainable farming practices.

5. Challenges: While super seeders offer numerous benefits, they also present some challenges

- Initial Investment: The cost of purchasing and maintaining a super seeder can be significant.
- Learning Curve: Operators need training and experience to use these machines effectively.
- Crop Residue Management: Managing crop residue from previous crops can be a challenge in no-till systems.

Super seeders have revolutionized wheat and small-grain crop cultivation by promoting sustainable and efficient farming practices. They help farmers maximize yields while minimizing the environmental impact of agriculture.

Zero Tillage for Wheat Sowing

Zero tillage, also known as no-till farming, is a modern agricultural practice that involves minimal or no disturbance of the soil during the planting and cultivation of crops. This method stands in contrast to conventional tillage, which involves plowing, disking, or other mechanical soil disturbance techniques. Zero tillage has gained popularity in recent years due to its numerous benefits for soil health, crop yield, and environmental sustainability. Here's a detailed description of zero tillage:

1. Soil Conservation

- Zero tillage helps preserve and improve soil quality. By not disturbing the soil, it reduces erosion, prevents soil compaction, and promotes better soil structure.
- The undisturbed soil acts as a natural barrier against wind and water erosion, which is particularly crucial in areas prone to soil erosion.

2. Improved Soil Health

- Zero tillage promotes the growth of beneficial microorganisms and earthworms in the soil, which enhance its nutrient cycling and overall health.
- The continuous presence of crop residues on the surface helps to increase organic matter content in the soil, leading to better water retention and nutrient availability.

3. Water Conservation

- Zero tillage reduces water runoff and evaporation by maintaining a protective cover of crop residues on the soil surface. This helps in conserving soil moisture and reducing the need for irrigation.

4. Reduced Fuel and Labor Costs

- Traditional tillage practices require the use of heavy machinery like plows and disks, which consume fuel and demand significant labor. Zero tillage reduces these costs as it requires fewer passes over the field.

5. Increased Crop Yields

- While the initial transition to zero tillage may result in slightly lower yields due to the adjustment period, over time, yields tend to equal or even surpass those achieved through conventional tillage.
- The improved soil health and water retention capabilities of no-till systems contribute to better crop growth and resilience to extreme weather conditions.

6. Weed Control

- Zero tillage can help control weed growth. By leaving crop residues on the soil surface, it creates a natural mulch that can smother weeds and reduce their competitiveness.

7. Environmental Benefits

- Zero tillage reduces the release of carbon dioxide (CO2) into the atmosphere by decreasing soil disturbance, which prevents the oxidation of organic matter.
- It also helps mitigate greenhouse gas emissions associated with fossil fuel consumption in conventional tillage practices.

8. Pest Management

- Some pests, such as certain soil-borne nematodes, can be reduced through zero tillage, as the undisturbed soil can disrupt their life cycles.

9. Time and Energy Efficiency

- Zero tillage can save farmers time and energy by eliminating the need for multiple passes with heavy equipment. This can be especially beneficial during busy planting and harvesting seasons.

10. Sustainable Agriculture

- Zero tillage is a key component of sustainable agriculture, as it aligns with the principles of conservation and environmental stewardship while maintaining or improving crop productivity.

In summary, zero tillage is a sustainable and environmentally friendly agricultural practice that offers numerous benefits, including improved soil health, water conservation, reduced costs, increased crop yields, and reduced environmental impact. However, successful implementation requires careful planning and adaptation to specific soil and climate conditions. Farmers often need to make adjustments to their crop rotation, weed management, and pest control strategies when transitioning to zero tillage to maximize its benefits.

Paddy Seeder

A paddy seeder is a specialized agricultural machine designed for planting rice seeds in paddy fields. Rice is one of the most important staple crops in the world, and efficient planting methods are essential to ensure a successful and productive rice harvest. Paddy seeders are widely used in rice-growing regions to automate the labor-intensive process of rice seed planting.

Here is a detailed description of a typical paddy seeder:

1. Structure and Components

- **Frame:** The main structure of the paddy seeder is usually made of sturdy metal or steel to provide stability and durability.
- **Seed Hopper:** This is a container located on top of the seeder where rice seeds are loaded. It has a capacity that varies depending on the model and can hold a significant amount of seeds.
- **Seed Metering Mechanism:** Within the seed hopper, there is a mechanism that regulates the flow of seeds to ensure uniform spacing between seeds during planting.
- **Seed Tubes or Chutes:** These tubes or chutes extend from the seed hopper to the ground and guide the seeds from the hopper to the planting furrows.
- **Planting Units:** These are devices responsible for placing the rice seeds into the soil at the desired depth and spacing. They may consist of rotating drums, seed plates, or other mechanisms.

2. Power Source

- Paddy seeders are typically powered by small gasoline engines or, in some cases, by electric motors. The power source drives the various components of the machine, such as the seed metering mechanism and the planting units.

3. Operation

- The operator fills the seed hopper with rice seeds before starting the machine.
- The paddy seeder is maneuvered across the paddy field by the operator or attached to a tractor, depending on the model.
- As the machine moves forward, the seed metering mechanism dispenses seeds at a predetermined rate into the seed tubes.
- The planting units then place the seeds into the prepared furrows in the soil at the desired depth and spacing. The depth and spacing can often be adjusted to accommodate different rice varieties and field conditions.
- The machine's wheels or tracks help maintain the correct row spacing as it moves through the field.

4. Advantages

- **Increased Efficiency:** Paddy seeders significantly reduce the labor required for planting rice, making the process more efficient and less physically demanding.
- **Uniform Planting:** The machine ensures uniform seed placement, which leads to consistent plant spacing and better crop management.

- **Faster Planting:** Paddy seeders can cover large areas quickly, reducing the time required for planting compared to manual methods.
- **Improved Yield:** Properly spaced and planted rice seeds result in better crop development and yield.

5. Maintenance

- Regular maintenance of the paddy seeder is essential to ensure its proper functioning and longevity. This includes cleaning, lubricating moving parts, and inspecting for wear and tear.

Conclusion

In summary, a paddy seeder is a valuable agricultural machine that automates the process of planting rice seeds in paddy fields. It improves efficiency, reduces labor, and contributes to better crop management, ultimately leading to increased rice yields. Farmers in rice-producing regions rely on paddy seeders to streamline the planting process and enhance their agricultural productivity.

Chapter 16

Smart Water Technologies for Sustainable Food Production Under Climate Scenario

Vijay Kumar[1], Shahid Ahamad[2], Manoj Kumar[3], Rakesh Kumar[1] and Neerja Sharma[4]*

[1]Rainfed Research Sub-Station for Sub-Tropical Fruits (Rrssf), Raya
[2]Directorate of Research, Foa, Chatha
[3]Regional Horticulture Research Sub-Station, Bhaderwah
[4]Krishi Vigyan Kendra (Kvk), Samba
Sher-E- Kashmir University of Agricultural Sciences and Technology, Jammu-180 009, Jammu & Kashmir, India
**Corresponding author E-mail: vijaykumar.1144@yahoo.com*

1.1 Abstract

Water is considered as the most critical resource for sustainable farming progress in the worldwide. The farming community has a face of rising food production by more than 70% to meet demand from the global population enlarge by the middle-21st century. Sustainable food production involves the continuous accessibility of resources, such as water and energy, to agriculture. The input challenges to sustainable food production are population increase, increasing demands for food, climate change, and climate variability, decreasing per capita land and water resources. Irrigated areas will increase in forthcoming years, while fresh water supplies will be diverted from agriculture to meet the increasing demand of domestic use and industry. Additionally, the efficiency of irrigation is very short, since less than 65 % of the useful water is really used by the crops. The sustainable use of irrigation water is a priority for agriculture in rainfed areas. Consequently, under scarcity conditions and climate change significant shot has been dedicated over time to introduce policies aiming to increase water efficiency based on the assertion that more can be achieved with less water through better management. Improved management frequently refers to improvement of water allocation and is closely related to adequate pricing, while the latter depends on the type of irrigation technology, environmental conditions and the scheduling of water purpose. Farming practices, such as soil management, irrigation and fertilizer application and disease and pest control are related with the sustainable water management in agriculture/ horticulture and security of the environment. Socio-economic pressures and climate change impose limitations to water allocated to agriculture/horticulture. Sustainable water management in agriculture/horticulture, which has a multi-functional role in India, can be achieved by adopting improvements in irrigation application, soil and plant practices, water pricing, reuse of treated waste water, farmers' contribution in water management and capacity structure.

Keywords: Irrigation schedule, Water use efficiency, Water reprocess, advance, Capacity building

1.2 Introduction

Water is considered as the most critical resource for sustainable development in most of India and worldwide countries. It is essential not only for agriculture, industry and economic growth, but also it is the most important component of the environment, with significant impact on soil and human health and nature conservation. Presently, the quick growth of population along with the extension of irrigation agriculture, industrial development and climate change, are stressing the quantity and quality aspects of the natural system. Because of the increasing problems, man has begun to realize that he can no longer follow a "use and discard" methodology either with water resources or any other natural resource. As a result, the need for a consistent policy of rational management of water resources has become evident. Soil quality is the needle of soil structure and also the entire agricultural management practices may be due to associated with several properties and processes dependable for agriculture production and ecological reliability. The most favorable soil structure supports plants with adequate water, aeration and release of available nutrient content (Kumar *et al.*, 2022). The aggregate-size distributions of dry soil can be used to obtain precise important of soil aggregates parameters and are helpful in manufacture soil management pronouncement and erosion predictions.

To overcome water shortage for farming is essential to increase the water use efficiency and to use marginal waters (reclaimed, saline, drainage) for irrigation. Agriculture currently uses about 70% of the total water withdraw, mainly for irrigation. Although irrigation has been practiced for millennia, most irrigated lands were introduced in the 20th century. The rigorous irrigation could provide for the growth of irrigated areas and guarantee increased food production. The global rate of increase in irrigated areas slowed considerably, mainly due to very high cost of irrigation system construction, soil salinization, the reduction of irrigation water-supplying sources, and the problems of environmental protection. Though, as the population is growing at a rapid rate, irrigation is being given an important role in increasing land use and cattle-breeding efficiency. Thus, irrigated farming is expected to expand rapidly in the future with subsequent increase of water use for irrigation. Irrigation is not sustainable if water supplies are not reliable. Especially in areas of water scarcity the major need for development of irrigation is to minimize water use. Effort is needed to find economic crops using minimal water, to use application methods that minimize loss of water by evaporation from the soil or percolation of water away from the depth of root zone and to minimize losses of water from storage or delivery systems. At the present time, during a period of dramatic changes and water resources uncertainty there is a need to provide some support and encouragement to farmers to move from their traditional high-water demand cropping and irrigation practices to modern, reduced demand systems and technologies.

Under shortage circumstances significant attempt has been devoted over time to introduce policies aiming to increase water efficiency based on the assertion that more can be achieved with less water through better management. Better management usually refers to improvement of allocative and/or irrigation water efficiency. The former is strongly related to adequate pricing, while the latter depends on the type of irrigation technology, environmental conditions and on scheduling of water application. Therefore it is hard for a farmer to tell at any certain moment whether there is a water deficit or not. Since overabundant water usually does not cause harm, farmers tend to "play safe" and increase irrigation amount, especially when associated costs are low. Over-irrigation can cause among others temporal water shortage to other farmers, water-logging conditions for the crop, favorable environment for disease development, loss of nutrients due to leaching or deep percolation, contamination of the aquifers from agrochemicals, reduction of crop yield and deterioration of the quality and increase of production cost.

1.3 Sustainable Water Management in Farming

Sustainable water management (SWM) in crop growing aims to match water availability and water needs in amount and quality, in space and time, at sensible cost and with suitable environmental impact. Its agreement involves technological harms, social behavior of rural communities, economic constrains, legal and institutional skeleton and farming practices. Under water demand management most awareness has been given to irrigation scheduling (when to irrigate and how much water to apply) charitable minor role to irrigation methods (how to apply the water in the field). Many parameters like crop growth stage and its sensitivity to water strain, climatic circumstances and water availability in the soil resolve when to irrigate or irrigation frequency. However, this frequency depends upon the irrigation process and therefore, both irrigation scheduling and the irrigation way are inter-related.

1.3.1 Localized irrigation

Localized irrigation is broadly predictable as one of the most efficient methods of watering crops. The localized irrigation systems (trickle/ drip irrigation, micro-sprayers) relate the water to individual plants by means of plastic pipes, frequently laid on the ground surface. With drip irrigation water is slowly applied through small emitter openings from plastic pipes with discharge rate ≤ 12 l/h. With micro-sprayer (micro-sprinkler) irrigation water is sprayed over the part of the soil surface occupied by the plant with a discharge rate of 12 to 200 l/h. The aims of localized irrigation are mainly the application of water directly into the root system under conditions of high availability, the avoidance of water losses during or after water application and the reduction of the water application cost (less labor). Localized tap water used by trickle irrigation intervals in fruit crops for a two litre bottle filled in the tap water and has adjust the trickle in minimum rate of 2 litre water per 4 hours was enforced in the root zone of the Kinnow mandarin/Aonla tree (Kumar *et al.*, 2022). The water is used for supplementary wet soil for dry condition, the plant can be survived in dry spell /dry seasons.

Chartzoulakis *et al.* (1991) reported that the drip irrigation's combination of water savings and higher yields typically increases at least by 50% the water use efficiency, yield per unit water (Tab. 1), and makes it a leading technology in the global challenge of boosting crop fabrication in the face of serious water constrains. Although the area of localized irrigation has expanded 50-fold over the last two decades, it still represents less than 6 percent of the world's irrigated area. The main barriers to its expansion are the high investment cost, ranging from 1,500 to 2,500 € per hectare, and the high sensitivity to clogging.

Table 1. Irrigation water, yield and WUE in kiwi irrigated with different systems

Irrigation system	Water used* (mm)	Yield (kg/tree)	WUE (kg/mm)
Drip	340	33.2 a	4.00 a
Micro-sprinkler	477	30.9 a	2.65 b
Overhead sprinkler	782	34.6 a	1.81 c
*Average of 10 years			

Development in localized irrigation systems aspiring to decrease the volumes of water useful and increase the water productivity include the use of a single drip line for a double row crop, the use of micro-sprayers in high infiltration soils, the adjustment of duration of water application and timing to soil and crop characteristics, the control of pressure and discharge variations, the use of appropriate filters to the water quality and the emitter characteristics used, the adoption of careful maintenance, automation, fertigation (efficient fertilizer application) and chemigation (easy control of weeds and soil born diseases).

1.3.2 Irrigation scheduling

Irrigation scheduling is the result making process for formative when to irrigate the crops and how much water to apply. It forms the single means for optimizing farming production and for conserve water and it is the input to improving presentation and sustainability of the irrigation systems. It necessitate good knowledge of the crops' water requirements and of the soil water characteristics that find out when to irrigate, while the adequacy of the irrigation method determines the accuracy of how much water to pertain. In most cases, the skill of the farmer determines the effectiveness of the irrigation scheduling at field level. With appropriate irrigation scheduling deep percolation and transport of fertilizers and agro-chemicals out of the root-zone is controlled, water-logging is avoided, less water is used (water and energy saving), optimum soil water conditions are created for plant growth, higher yields and better quality are obtained and rising of saline water table is avoided. In water scarce regions, irrigation scheduling is more important than under conditions of abundant water, since any excess in water use is a potential cause for deficit for other users or uses. However, in most countries these programmes do not exist since they are expensive, trained extension officers are lacking, farmer's awareness of water saving in irrigation is not sufficient and the institutional mechanisms developed for irrigation management provide low priority to farm systems. Therefore, in general, large limitations occur for their use in the farmers practice

1.4 Soil Water Approximate and Measurements

Soil water affects plant growth directly through its calculating effect on plant water status. There are two behaviors to assess the availability of soil water for plant growth: by measuring the soil water content and by measuring how strongly that water is retained in the soil (soil water potential). The correctness of the information relates to the sampling methods adopted and to the selection of locations where point explanation are performed due to the soil water unpredictability both in space and depth (Peymorte and Chol, 1992). Soil water estimates and measurements used for irrigation scheduling include: soil appearance and feel, soil water content (SMC) measurement (TDR), soil water potential measurement (tensiometers, soil spectrometers and pressure transducers), remotely sensed soil moisture. The soil moisture content (SMC) measurement by gravimetric methods in Kinnow mandarin orchards under various treatments. The SMC in 0-20 cm soil depth which was higher in 230 days after mulch materials at fruit growing orchards (Kumar *et al.*, 2022)

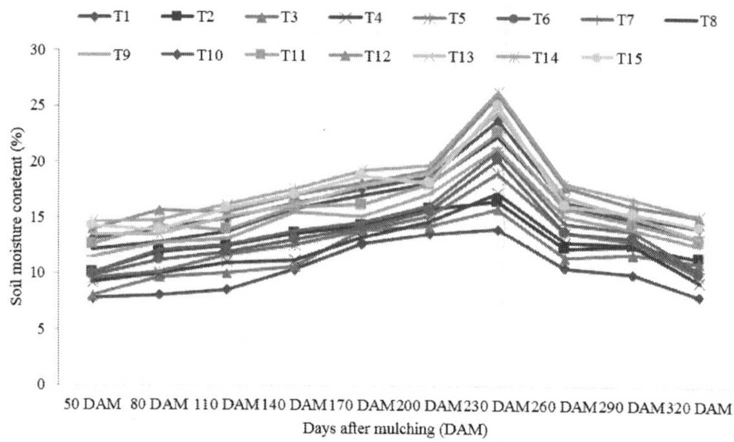

Fig.1 Soil moisture content of Kinnow mandarin during the years 2018-19 under different treatments.

1.4.1 Crop stress parameters

In its place of measuring or estimating the soil water parameters, it is possible to get a signal from the plant itself indicating the time of irrigation but not essential the irrigation depths. This significance can either come from individual plant tissues, where a correct sampling is required, or from the canopy as a whole. Therefore, crop stress parameters are useful when irrigation depths are predefined and kept constant during the irrigation season. Crop water stress parameters include leaf water content and leaf water potential, changes in stem or fruit diameter, sap flow measurement, canopy temperature, remote sensing of crop stress (Jackson *et al.*, 1981; Itier *et al.*, 1993).

1.4.2 Climatic parameters

Climatic parameters are broadly used for local or regional irrigation schemes. Weather data and empirical equations that, once they are secure by calibrated, supply accurate estimates of reference evapotranspiration (ETo) for an agreed area are used. Then, crop evapotranspiration (ETc) is probable using appropriate crop coefficients. In sequence may be processed in real time or, more often, using historical data. These techniques include evaporation measurements for ETo calculation, assessment of crop evapotranspiration using climatic data (air temperature, RH, wind speed, sunshine hours) (Allen *et al.*, 1998) and remote sensed ET.

1.4.3 Soil water balance

Soil water holding character, crop and climate data are used by complicated models to produce typical irrigation calendars. This approach can be applied from individual farms to large regional irrigation schemes. Soil water balance approach is to expect the water content in the rooted soil by way of a water conservation equation: Δ (AWC × Root depth) = Balance of entering + outgoing water fluxes, where AWC is the available water content. Though, it needs expertise, hold up by strong extension services or links with information systems. Its effectiveness is very high, but depends on farm technological development and/or support services. Examples of commercial software for irrigation scheduling based on soil-water balance approach are IMS (Hess, 1996) and SIMIS (FAO, 1999b).The water management through drip irrigation and hydrogel with black polyethylene mulch for improving the moisture retention capacity in citrus species viz. Kinnow mandarin under rainfed conditions (Kumar *et al.*, 2022)

1.5 Effective Irrigation Scheduling

The effective request of any irrigation scheduling method and effective implementation of the corresponding delivery schedule are subject to the physical capability of the collective system for delivering water according to this schedule and to the capacity of the management for operating the system properly. One of the major obstacles to effective implementation of crop-based and water-saving irrigation scheduling is the inability of most conveyance and delivery systems to deliver water at the farm gates with the reliability and flexibility required. The farmers are free to select and adopt the irrigation schedules they consider more appropriate to their crops and farming practices. However, in case of drought or limited water supply, managers can enforce restrictions to volumes delivered and/or price penalties for excess water use. It is recognized that appropriate irrigation scheduling should lead to improvements in irrigation management performance, especially at farm level. The farmer should be able to control the timing and the depth or volume of irrigation. The dependence on a collective system implies social, cultural and policy constraints. The main constrains are the lack of flexibility, either due to rigid schedules or the system limitations, the non-economic pricing of water (price covers less than 30% of the total cost), the high cost of irrigation scheduling (either for technology and/or labor), the lack of education and training of the farmers, the institutional problems, the behavioral adaptation, the lack of interactive communication between research, extension and farmers and finally the lack of demonstration and technology transfer.

1.5.1 Fertigation

The application of fertilizers with the irrigation system (fertigation) became a common practice in modern irrigated agriculture. Localized irrigation systems, which could be highly resourceful for water application, are also suitable for fertigation. Thus, the soluble fertilizers at concentrations required by crops are applied through the irrigation system to the wetted volume of the soil. Possible disadvantages include the non-uniform chemical distribution when irrigation design or operation are inadequate, the over-fertilization in case that irrigation is not based on actual crop requirements and the excessive use of soluble fertilizers.

1.5.2 Shortage irrigation practices

In the ancient times, crop irrigation necessities did not believe limitations of the available water provisions. The irrigation arrangement was then based on casing the full crop water requirements. Yet, in arid and semi-arid regions under rainfed condition increasing municipal and industrial demands for water diminish steadily water allocation to agriculture. Hence, water availability is usually limited, and certainly not enough to achieve maximum yields. Then, irrigation strategies not based on full crop water requirements should be adopted for more successful and rational use of water. Such management practices include scarcity irrigation, partial root drying and subsurface irrigation.

1.5.3 Regulated deficit irrigation

Regulated deficit irrigation (RDI) is an optimizing approach under which crops are allowed to sustain some degree of water scarcity and yield lessening. During RDI the crop is uncovered to certain level of water stress either during a particular period or all over the growing season. The main objective of RDI is to increase water use efficiency (WUE) of the crop by eliminating irrigations, and to improve control of vegetative growth (improve fruit size and quality). The resulting yield reduction may be small compared with the benefits gained through diverting the saved water to irrigate other crops for which water would normally be insufficient under conventional irrigation practices. RDI is a sustainable issue to cope with water scarcity since the allowed water deficits favour water saving, control of percolation and runoff return flows and the reduction of losses of fertilizers and agrochemicals; it provides for leaching requirements to cope with salinity and the optimization approach leads to economical viability. The adoption of deficit irrigation implies appropriate knowledge of crop ET, of crop response to water deficits including the identification of critical crop growth stages, and of the economic impact of yield reduction strategies. Therefore, appropriate deficit irrigation requires some degree of technological development to support the application of irrigation scheduling techniques.

1.5.4 Partial Root Drying

Partial root drying (PRD) is a new irrigation technique, first applied to grapevines that subject one half of the root system to dry or drying conditions while the other half is irrigated. Wetted and dried sides of the root system alternate on a 7-14 day cycle. PRD uses biochemical responses of plants to water stress to attain balance between vegetative and reproductive growth. During water stress development the vine's first line of defense is to close its stomata to conserve moisture. One of the principal compounds that elicit this response is the abscisic acid (ABA). As soil water availability falls following the cessation of irrigation, the ABA is synthesized in the drying roots and transported to the leaves through the transpiration stream (Loveys *et al.*, 1999). Stomata respond by reducing aperture, thereby restricting water loss. Improvement of WUE results from partial stomatal closure. However, an inevitable consequence is reduced photosynthesis. With PRD, switching the wet and dry sectors of root zone on regular basis, this transient response was overcome (Dry and Loveys, 1998).

1.6 Subsurface Drip Irrigation

Subsurface drip irrigation (SDI) is a low-pressure, low volume irrigation system that uses buried tubes to apply water. The applied water moves out of the tubes by soil matrix suction. Wetting occurs around the tube and water moves out in the soil all directions. The possible advantages of SDI are: a) water conservation, b) enhanced fertilizer efficiency, c) uniform and highly efficient water application, d) elimination of surface infiltration problems and evaporation losses, e) flexibility in providing frequent and light irrigations, f) Reduced problems of disease and weeds, g) lower pressure required for operation. The main disadvantages are the high cost of initial installation and the increased possibility for clogging, especially when poor quality water is used. The subsurface irrigation is suitable for almost all crops, especially for high value fruit and vegetables, turfs and landscapes. A huge variety of tubes are available in the market from PE tubes with built-in emitters or porous tubes that ooze water out the entire length of the tube. The tube is installed below the soil surface either by digging the ditches or by special device pulled by a tractor. The depth of installation depends upon soil characteristics and crop species ranging from 15-20 cm for vegetables and 30-50 cm for tree crops.

1.6.1 Farming Practices

Farming practices, such as soil management, fertilizer application, disease and pest manage are related with the sustainable water management (SWM) in farming and the protection of the environment. Agricultural practice today is characterized by the abuse of fertilizers. Farmers very hardly ever carry out soil and leaf analyses in order to clarify the proper quantity and kind of fertilizer required for each crop and they apply them empirically. These observe increases considerably the cost of agricultural production and are potentially critical for the worsening of the groundwater quality and the environment. Agrochemicals (herbicides and pesticides) are also excessively used, endangering the quality of the surface water and negatively affecting the environment. Plant-protection products (pesticides) are often used preventively, even when there is not real threat in the area.

There is a large variety of traditional and modern soil and crop management practices for water conservation (runoff control, improvement of soil infiltration rate, increase soil water capacity, control of soil water evaporation) and erosion control in agriculture, some of which apply also for weed control (Pereira *et al.*, 2002). Effort should be made in their rational use of chemicals for pest and weed control in order not to further pollute the environment. The soil management consists of:

Soil surface tillage (SST): SST which concerns superficial tillage practices to fabricate an enlarged roughness on the soil surface permitting short time storage in little depressions of the rainfall in excess to the infiltration.

Contour tillage (CT): CT is the soil cultivation is made along the land contour and the soil is left with small furrows and ridges that prevent runoff. This technique is also effective to control erosion and may be applied to row crops and small grains provided that field slopes are low.

Bed surface profile (BSP): BSP is concerns cultivation of wide beds and is typically used for horticultural row crops.

Conservation tillage (CT): CT is including no-tillage and reduced tillage, where residuals of the preceding crop are kept on the soil at planting. Mulches protect the soil from direct impact of raindrops, thus controlling crusting and sealing processes. Conservation tillage helps to maintain high levels of organic matter in the soil thus it is highly effective in improving soil infiltration and controlling erosion.

Mulching with crop residues on soil surface which shades the soil, slows water overland flow, improves infiltration conditions, reduces evaporation losses and also contributes to control of weeds and therefore of non- beneficial water use.

Improving or maintaining the amount of organic matter in the upper soil layers, since it provides for better soil aggregation, compact crusting or sealing on soil surface and improved water retention capacity of soil.

Accumulation of fine material or hydrophilic chemicals to sand/coarse soils. This technique increases the water retention capacity of the soil and manages deep percolation. Therefore, water availability in soils with low water holding capacity is amplified.

Manage of acidity by liming, likewise to gypsum application to soils with high soil pH. This treatment favors more intensive and deep rooting, better crop development and contributes to improved soil aggregation, thus producing some enlarge in soil water availability.

Adoption of appropriate weed control techniques to alleviate competition for water and transpiration losses by weeds.

1.7 Suggestion for Best Irrigation Practices

The major farming use of water is for irrigation and its supply is diminishing steadily due to competition with municipal and industrial sectors. Therefore, technological, managerial, policy innovation and human resources managing are desired to enhance the use efficiency of the available water. Sustainable water management (SWM) in agriculture can be achieved by:

Reduce of water losses in the conveyance, distribution and application networks. Water leakages should be detected via superior technologies, e.g.

telemetry systems, GIS, remote sensing. Aged water projects experiencing extensive water losses should be rehabilitated and modernized.

Improve the efficiency of irrigation system used. Improvements in sprinkler irrigation systems (efficiency up to 85%) include the adoption or correction of sprinkler spacing, the design for pressure variation not exceeding 20% of the average sprinkler pressure, the use of pressure regulators in sloping fields, the monitoring and adjustment pressure equipment, application of irrigation during no windy periods, adoption of smaller spacing and large sprinkler drops and application rates in windy areas, the approval of application rates smaller than the infiltration rate of the soil and careful system maintenance. Improvements in localized irrigation systems (Efficiency 95%) aiming to reduce the volumes of water applied and enlarge the water productivity include the use of a single drip line for a double row crop, the use of micro-sprayers in high infiltration soils. The adjustment of duration of water application and timing to soil and crop characteristics, the control of pressure and discharge variations, the use of appropriate filters to the water quality and the emitter characteristics used, the adoption of careful maintenance and automation.

Enhance water use efficiency. The obligatory use of localized irrigation systems by the farmers (with or without subsidies), the proper irrigation scheduling according to actual needs of the crops, the establishment of a system for advising farmers on their irrigation schedules, the introduction of appropriate agronomical practices and the application of salinity management techniques.

Implementation of innovative irrigation techniques. In water inadequate regions irrigation approaches not necessarily based on full crop water requirements like regulated deficit irrigation (RDI) or subsurface irrigation (SSI) must be adopted. Fertigation (efficient fertilizer application) and chemigation (easy control of weeds and soil born diseases) should also be promoted.

Water pricing strategy. The proper water pricing volumetric water metering and accounting procedures are recommended and progressive, seasonal and over-consumption water tariffs as well as temporary drought surcharges rates contribute to water savings and should be promoted. Additionally, an rising block tariff charging system, that discourages water use levels exceeding crop's critical water requirements, must be established. It will be the basis for promoting conservation, reducing losses and mobilizing resources. Furthermore, it could affect cropping patterns, income distribution, efficiency of water management, and generation of additional revenue, which could be used to operate and maintain water.

Reprocess of marginal waters (reclaimed or brackish) for irrigation. Reclaimed waters can be used under some restriction for irrigation of tree, row and fodder crops. In adding up to water they provide the soil with nutrients, minimizing the inorganic fertilizer application. They instead prefer to use surface and/or groundwater and educating farmers to accept treated sewage. In addition the tariff for this source of water should be lower than the tariff of the primary sources. This may not be difficult to achieve because the primary and secondary levels of treatment are regarded as sunk costs since they are required by the new WFD. When using low quality water for example brackish or saline water an integrating approach for water, crop (salt tolerant varieties) field management (suitable tillage) and irrigation system (adequate leaching, suitable devices) should be considered.

Wider and more effective participation of the public, NGOs and end users for the preparation of the plans, in decision-making, in monitoring the implementation and generally in the management of water. The participation of these groups in the above processes safeguards the acceptability of the plans by the general public, raises support on the part of the body politics and promotes success in possible conflict resolutions.

Capacity building: The existing "capacity building" is poor. It requires an appropriate mix of competent personnel, technologically advanced devices and facilities, legal guidelines and administrative efficient and effective processes for the sustainable management of the water resources. It includes:

- Education and training of professional, technical staff and decision makers and others, including non-public organizations, on a wide range of subjects related to sustainable water management.
- Manpower build-up. Institutions to be staffed with qualified manpower (managers, engineers, technicians, social scientists) that should be adequately compensated.
- Facilities and procedures. Water authorities at all levels of management should be equipped with technologically advanced devices and programs e.g. computers and software for the application of new techniques such as GIS, remote sensing etc. These advanced techniques facilitate the multi-sectoral information availability and use and help water managers in their decision- making.
- Legislative changes to the fragmented and antiquated legislation should be promoted. Responsibility of water resources planning and operation, especially at the overall national level, should be under one institution for proper organization and non-conflicting actions by the various authorities. Water authorities should participate fully in the formulation of agricultural policies because the development of water and land should be fully integrated. In practice, agricultural decisions are water decisions and vice versa.

References

Allen R.G., Pereira L.S., Raes D. and Smith M., 1998. Crop evapotranspiration. Guidelines for computing crop water requirements. FAO, Irrig. Drain. Paper 56, FAO Rome, 300 p.

Chartzoulakis K., Michelakis N. and Vougioukalou E., 1991. Growth and production of kiwi under different irrigation systems. Fruits 46 (1):75-81.

Dry, P.R. and Loveys, B.R. 1998. Factors influencing grapevine vigour and the potential for control with partial rootzone drying. Australian Journal of Grape and Wine Research 4: 140-148.

FAO, 1999. Land and Water Digital Media Series no 6. (www.fao.org)

Hess T. M., (1996). A microcomputer scheduling program for supplementary irrigation. Computers and Electronics in Agriculture 15: 233 – 243.

Itier B., Flura D. and Belabess K. 1993. An alternative way for C.W.S.I. calculation to improve relative evapotranspiration estimates: results of an experiment over soybean. Acta Hortic., 335, 333-340.

Jackson R.D., Idso S.B., Reginato R.J. and Pinter P.J. 1981. Canopy temperature as a crop water stress indicator.Water Resources Res. 17: 1133-1138.

Kumar V, Kumar R, Chand G, Kumar J, Singh B, Dhotra B, Khajuria S, Gupta S, Sharma N, Sinha A K, Raina V, Ahamad S. 2022. Influence of trickle irrigation and hydrogel on Kinnow mandarin (Citrus reticulate Blanco) under rainfed condition. Agricultural Mechanization in Asia 53(7):9007-9023.

Kumar V, Kumar R, Chand G, Kumar J, Singh B, Khajuria S, Dhotra B, Reena, Sharma A, Kumar M. 2022. Dynamics of soil properties and fruit productivity of Kinnow mandarin (Citrus reticulate Blanco) as affected by drip trickle irrigation and hydrogel. Agricultural Mechanization in Asia 53(3):6439-46.

Kumar V, Sharma K.R., Dhotra B, Kumar M, Khajuria S, Sharma V, Banotra M, Sharma R.2022. Soil Aggregate and Organic Carbon Fractions in the Foothill Himalayas Under Different Landuse Systems. Agricultural Mechanization in Asia. 53(4): 7477- 89.

Loveys B.R., Dry P.R. and McCarthy M.G., 1999. Using plant physiology to improve water use efficiency of horticultural crops. Acta Hort. 537: 187-199.

Pereira L.S., Cordery I., Iacovides I., 2002. Coping with water scarcity. UNESCO IHP VI, Technical documents in Hydrology No 58, UNESCO, Paris , p. 267.

Peyremorte P. and Chol P. 1992. Pilotage automatique optimisé des irrigations localisées. In: Proceedings of the Workshop on Real Time Sensing and Control of Automated Irrigation Systems, ICID, 16th European Regional Conference, Budapest (Hungary), 26 June 1992.

Chapter 17

Optimizing Agriculture: Precision Farming for Efficient and Sustainable Crop Management with Advanced Technologies and Data-Driven Practices

Bushra Rasool, Ranbir Singh, Nisar Ahmad Dar, L. V. Ravishankar, Stanzin Diskit, Amit Kumar Maurya[1]

Sher-e-Kashmir University of Agricultural Sciences and Technology, FoA, Division of Plant PathologyChatha, Jammu, Jammu and Kashmir-180009, India
[1]Intellectual Institute of Management & Technology University (IIMT), School of Agricultural Sciences, Division of Plant Pathology, Meerut, Uttar Pradesh-25001, India

Abstract

Precision agriculture revolutionizes traditional farming methods by integrating advanced technologies to optimize crop management. This approach leverages tools such as satellite imagery, GPS, sensors, and data analytics to enhance decision-making processes throughout the agricultural cycle. In precision agriculture, accurate data collection is paramount. Satellite imagery provides comprehensive insights into crop health, soil conditions, and overall field dynamics. Sensors, deployed on the ground or through drones, monitor factors like soil moisture and nutrient levels. This wealth of data is then analyzed using farm management software, empowering farmers with valuable insights. The integration of GPS technology enables precise mapping and navigation within fields, facilitating activities like planting, harvesting, and irrigation. Variable Rate Technology (VRT) allows for customized application of inputs, optimizing resource utilization and minimizing waste. Automation plays a pivotal role in precision farming, with autonomous machinery capable of executing tasks with high precision. Decision support systems provide farmers with recommendations based on real-time data, weather forecasts, and historical patterns, facilitating proactive decision-making. Precision agriculture extends beyond automation; it is a holistic approach that promotes sustainability. By fine-tuning resource application, reducing environmental impact, and enhancing overall efficiency, precision agriculture is transforming traditional farming into a technologically advanced, data-driven, and sustainable practice. This innovative paradigm represents a critical step towards meeting the growing demands of global food production while minimizing the ecological footprint of agriculture.

1. Introduction

With the rapid growth in population and the consequent demand for food, it has become increasingly important to optimize farming practices to meet the needs of the world's growing population. Many challenges impede agricultural production, which leads to a decrease in crop productivity, such as soil salinity (Said *et al.*, 2020). In addition, the climate also affects the quantity and quality of crops and may lead to an increase in soil sensitivity to desertification (El-Zeiny *et al.*, 2017). Therefore, the focus on survey land resources to use in agricultural development is necessary. In developing countries, the agricultural sector is one of the most important pillars of national income. Therefore, implementing new technologies to improve the agricultural sector is a significant issue for supporting the national economy in those countries (Nyaga *et al.*, 2021). In recent years, technological advancements have revolutionized the way we approach farming, leading to the emergence of a new approach known as "smart farming" (Sharma *et al.*, 2022a). Smart farming is an innovative approach to agriculture that integrates technology into farming practices, enabling farmers to optimize crop yields, reduce waste and improve efficiency (Said *et al.*, 2021). This approach uses a range of technologies, including sensors, drones, artificial intelligence and the internet of things (IoT), to collect data and provide real-time insights into crop health, soil quality and other key indicators (Sharma *et al.*, 2022b). Smart farming also offers numerous benefits, including increased productivity, reduced labor costs, improved crop quality, and a more sustainable approach to farming. This approach also offers greater precision, enabling farmers to target specific areas of their farms that require attention and reduce the use of chemicals and fertilizers (Raj *et al.*, 2021).The advantages of smart agriculture are summarized as follows: 1) Increasing the amount of real-time data on the crop 2) Remote monitoring and controlling of farmers 3) Controlling water and other natural resources 4) Improving livestock management 5) Accurate evaluation of soil and crops; 6) Improving agricultural production.

2. Evolution of Farming: From Traditional to Smart Farming

The evolution from traditional to smart farming traced back to the early 1990s when precision agriculture (PA) was first introduced. PA is a farming approach that involves using technology to target specific areas of the farm that require attention, such as soil moisture levels or nutrient deficiencies. This approach uses data analysis tools to optimize inputs and minimize waste, resulting in higher crop yields and reduced costs. Over the years, smart farming has evolved to include a range of advanced technologies. For example, drones equipped with cameras and sensors can provide detailed images and data on crop health and yield. Soil sensors can measure soil moisture, temperature, and nutrient levels, providing insights into the health of the soil and enabling farmers to make informed decisions about fertilization and irrigation (Boursianis *et al.*, 2022). Another technology that has revolutionized smart farming is the IoT. IoT-enabled sensors and devices can be placed throughout the farm to monitor environmental conditions, track crop growth, and optimize irrigation and fertilizer applications. This data is transmitted to a central platform, where it is analyzed and used to generate insights that can help farmers make informed decisions (Nolz *et al.*, 2013).

Technologies Involved

Global Positioning System (GPS)

GPS is a network of orbiting satellites, sending precise positional details from space back to earth. It's obtaining a birds-eye point of view, but from space. The signals are obtained by GPS receivers, to calculate the exact position and time. It has an accuracy of between 100 and 0.01 m. It enables farmers to identify field information, such as soil type, pest occurrence, weed invasion, water holes, etc based on a precise location. This becomes particularly handy when farmers need to make decisions regarding seed plantation, herbicides, pesticides, fertilizers and irrigational needs (Bakhtiari and Hematian, 2013).

Geographic information system (GIS)

It is a system designed to capture, store, manipulate, manage analyze and present spatial or geographic data. GIS comprises of hardware, software and procedures for supporting the compilation, storage, retrieval and analysis of geographical attributes and location data to produce maps. GIS binds information in one place so that it can be extrapolated when needed. Computerized GIS maps contain layers of information like yield, soil survey, crop type, nutrient levels and prone pests. A farming GIS database can provide information on field topography, soil types, surface drainage, subsurface drainage, rainfall, irrigation, rates of chemical application and crop yield. Once analyzed, this information is used to understand the relationships between the various elements affecting a crop on a specific site (Belal *et al.*, 2015).

3. Grid Soil Sampling and Variable-rate Fertilizer (VRT) Application

Grid sampling is an unbiased, simple and relatively quick method for site specific soil management. Variable-rate technologies (VRT) may be applied to numerous farming operations. They set the rate of delivery of farm inputs based on the soil type. GIS extrapolated information can control processes, such as crop determination, level of seeding, application of fertilizer, pesticide and herbicide at a variable rate in the right place at the right time. It is widely used in developed countries. Grid soil sampling, using the same principles, increases the intensity of sampling. Samples collected in a systematic grid can be mapped to specific locations thus giving us an application map. Grid soil samples are analyzed and interpreted to determine crop nutrient needs. Then using this information, a fertilizer application map is plotted. Both maps are fed into a computer which generates a detailed and systematic schedule for plantation and fertilizer needs (Ess *et al.*, 2001).

4. Sensors

Sensor technology is an integral part of precision agriculture technology and their application has been widely reported to provide essential information on soil properties, plant fertility and water status. Location Sensors are the sensors that are used to determine latitude, longitude, and altitude within feet. Optical Sensors use light to measure soil properties. Electrochemical Sensors provide key information like pH and soil nutrient levels. Mechanical Sensors measure soil compaction. They use a probe that penetrates the soil and records resistive forces through use of load cells or strain gauges. Dielectric Soil Moisture Sensors measure the dielectric constant in the soil to assess moisture levels. Airflow Sensors measure soil-air permeability (Mehta and Masdekar, 2018).

5. Rate controllers

Rate controllers control the delivery rate of chemical inputs such as fertilizers and pesticides, either liquid or granular. It monitors the speed of the tractor/sprayer traveling across the field, as well as the flow rate and pressure of the material, making adjustments in real-time to apply a targeted rate. Rate controllers are frequently used as stand-alone systems (Mehta and Masdekar, 2018).

6. Precision Livestock Farming (PLF)

Precision livestock farming (PLF) is defined as the application of precision agriculture to the management of livestock production. Processes of precision livestock farming approach focus on animal growth, egg and milk production, detection and monitoring of diseases and aspects related to animal behavior. Systems include monitoring of milk to check fat and microbial levels, helping to identify potential infections, as well as new automated feeding and weighing systems. It also focuses on automated cleaners, feed pushers and other aids. Acoustic sensors detect an increase in coughing of pigs indicating respiratory infection. Other sensors are now used to provide alerts and notifications concerning birthing and fertility (Mehta and Masdekar, 2018)

7. Internet of Things (IoT)

Internet of Things (IoT) is an intelligent and promising technology that offers unconventional and practical solutions in many areas such as smart cities, smart homes, traffic control, healthcare, smart agriculture, etc. In the agricultural field, IoT technology has made significant development in agricultural management. This technology allows all agricultural devices and equipment is to be linked together to make the appropriate decision in irrigation and fertilizer supply (Kumar and Periasamy, 2021). Smart farm management requires using ICT, ground sensors, and control systems installed on robots, autonomous vehicles, and other automated devices. The success of smart systems depends on high-speed internet, advanced mobile devices, and satellites to provide (images and positioning). James *et al.* (2019) succeed used IoT real-time to track and diagnose leaf diseases that obstruct crop growth using many satellite images and sensors positioned in farms (paddy and banana crops); this system helped to be analyzed the data and made the decision then sent back to the farmers via the web server.

8. Smart Farming Approaches in Developing Countries

The implementation of SF technology is considered very important around the globe, and the developing countries are interested in localized such technology (Arslan *et al.*, 2015). The developing countries have several challenges in implementing smart systems regarding the availability of infrastructure owned by the state and other capabilities possessed by individuals (Glaroudis *et al.*, 2020). Therefore, the obstacles to the implementation of smart agricultural technology in developing countries can be summarized as follows; 1) The availability of a suitable network of the fourth or fifth generation and is the most crucial factor in terms of data transmission between sensors via the Internet. 2) The availability of sensors as they are responsible for measuring the various phenomena and characteristics on the farm. 3) Availability of devices and equipment that can achieve agricultural operations. 4) Trained experts based on smart farms. However, their several approaches in the developing countries, in India, several factors affect the majority of farmers regarding the implementation of smart farming technology such as weak socio-economic backgrounds and face many challenges due increasing cost of cultivation. Furthermore, one of the significant natural problems influencing agricultural productivity is climate change (Srivastava and Singh, 2016).

These challenges need concrete strategies at different levels, from local to national levels. Climate-Smart Agriculture (CSA) has the potentiality to overcome these challenges (Lipper *et al.*, 2014). Doyle *et al.* (2018) innovated a methodology concerning the implementation of smart systems such as CSA to mitigate the effects of climate change in East Africa and also to reduce food insecurity at the local level, where they designed an aquaculture system to maximize the water use efficiency in irrigation-based aquatic crops. Moreover, they reuse the same water for a fish farm. This system adopts more efficient use of water for irrigation of crops compared to traditional agriculture, making it more resistant to drought and desertification, and also contributes to reducing climate impacts in rural areas in East Africa. There was another experience in South Africa to use smart irrigation based on Climate-smart agricultural technological at the farm level to address climate challenges. In Africa, smart technology faces many obstacles such as the farm areas, which their areas are ranging between 0.5 and 2 ha, climate, and environmental changes as well as water scarcity (Senyolo *et al.*, 2018).

Conclusions

The current work illustrated the importance of smart agriculture on improving and increasing agricultural production in order to contribute to reducing the food demand gap. IoT is considered the backbone of smart agricultural technology, as it connects all components of smart systems, not only in the agricultural field but also the other applications. Concerning the use of IoT in agriculture, it can be used in many practices such as farm monitoring, irrigation, pest control, harvesting, etc. IoT connects several sensors with processing units, then analyzes data, then makes appropriate decisions in real-time. The application of smart agricultural technology helps developing countries, such as some Egyptian approaches that represent the beginning toward the spread of such technology and can help develop the agriculture sector and achieve farm sustainability in those countries. Finally, these smart technologies should be supported by governments in third world countries at the level of small farms, as they aim to increase production and improve the efficient use of land and water resource.

References

Arslan, A., McCarthy, N., Lipper, L., Asfaw, S., Cattaneo, A., Kokwe, M. 2015. Climate smart Agriculture? Assessing the adaptation implications in Zambia. Journal of Agricultural Economics, 66 (3): 753-780.

Bakhtiari AA, Hematian A (2013) Precision farming technology, opportunities and difficulty. International Journal of Sciences Emerging Technologies Latest Trends, 5(1): 1–14

Belal, A. A., Mohamed, E. S. and Abu-hashim, M. S. D. 2015. Land evaluation based on GIS-spatial multi-criteria evaluation (SMCE) for agricultural development in dry Wadi, Eastern Desert, Egypt. International Journal of Soil Science, 10: 100–116

Boursianis, A. D., Papadopoulou, M. S., Diamantoulakis, P., Liopa-Tsakalidi, A., Barouchas, P. and Salahas, G. 2022. Internet of things (IoT) and agricultural unmanned aerial vehicles (UAVs) in smart farming: A comprehensive review. Internet of Things. 18: 100187

Doyle, L., Oliver, L. and Whitworth, C. 2018. Design of a Climate Smart Farming System in East Africa, IEEE. 1-6

El-Zeiny, A., El-Hefni, A., & Sowilem, M. 2017. Geospatial techniques for environmental modeling of mosquito breeding habitats at Suez Canal Zone, Egypt. The Egyptian Journal of Remote Sensing and Space Science, 20(2), 283-293

Ess, D., Morgan, M. Parson, S. 2001. Implementing site-specific management: map- versus sensor-based variable-rate application. Technical Report. Available on http://www.ces.purdue. edu/extmedia/AE/SSM-2-W.pdf.

Glaroudis, D., Iossifides, A. and Chatzimisios, P. 2020. Survey, comparison and research challenges of IoT application protocols for smart farming. Computer Networks, 168: 107037

James, A., Nair, A., Saji, D. and Joseph, D. 2019. Crop Sense–A Smart Agricultural System using IoT. Journal of Electrical and Electronic Engineering, 5 (3): 1-7

Kumar, T. U. and Periasamy, A. 2021. IoT Based Smart Farming (E-FARM)'S. International Journal of Recent Advances in Multidisciplinary Topics, 2: 85-87

Lipper, L., Thornton, P., Campbell, B.M., Baedeker, T., Braimoh, A., Bwalya, M. and Torquebiau, E.F. 2014. Climate-smart agriculture for food security. Nature Climate Change, 4(12): 1068-1072

Mehta, A and Masdekar, M. 2018. Precision Agriculture – A Modern Approach to Smart Farming. International Journal of Scientific and Engineering Research, 9: 23-26

Nolz, R., Kammerer, G. and Cepuder, P. 2013. Calibrating soil water potential sensors integrated into a wireless monitoring network. Agricultural Water Management, 116: 12-20.

Nyaga, J. M., Onyango, C. M., Wetterlind, J. and Soderstrom, M. 2021. Precision agriculture research in sub-Saharan Africa countries: A systematic map. Precision Agriculture, 1-20

Raj, E. F. I., Appadurai, M. and Athiappan, K. Precision farming in modern agriculture. 2021. In: Choudhury A, Biswas A, Singh TP, Ghosh SK, editors. Smart Agriculture Automation Using Advanced Technologies. Singapore: Springer, pp. 61-87

Said Mohamed, E., Belal, A., Kotb Abd-Elmabod, S., El-Shirbeny, M. A., Gad, A.and Zahran, M. B. 2021. Smart farming for improving agricultural management. Egyptian Journal of Remote Sensing and Space Science, 24: 971-981

Senyolo, M. P., Long, T. B., Blok, V., & Omta, O., 2018. How the characteristics of innovations impact their adoption: An exploration of climate-smart agricultural innovations in South Africa. Journal of Cleaner Production, 172: 3825-3840

Sharma V., Tripathi, A. K. and Mittal, H. 2022a. Technological revolutions in smart farming: Current trends, challenges and future directions. Computers and Electronics in Agriculture, 201: 107217.

Sharma, A., Georgi, M., Tregubenko, M., Tselykh, A. and Tselykh, A. 2022b. Enabling smart agriculture by implementing artificial intelligence and embedded sensing. Computers and Industrial Engineering, 165: 107936

Srivastava, P. K. and Singh, R. M. 2016. GIS based integrated modelling framework for agricultural canal system simulation and management in Indo-Gangetic plains of India. Agricultural Water Management, 163: 37-47

Chapter 18

Scenario of Insect Pests Management in Natural Farming

Reena[1], Manoj Kumar Jat[2], Shahid Ahmad[3], A. P. Singh[1], B.K. Sinha[4], Krishna Rolania[2] and Lomash Kumar[2]

[1]Advanced Centre for Rainfed Agriculture, Dhiansar, SKUAST-Jammu, J&K, India
[2]Department of Entomology, College of Agriculture, CCSHAU, Hisar, Haryana, India
[3]Directorate of Research, SKUAST-Jammu, Chatha, J&K, India
[4]Division of Plant Physiology, SKUAST-Jammu, Chatha, J&K, India

In India, the agriculture sector has been dominated for the last 40 years by Green Revolution had influenced the Indian economy by increasing agricultural production as well as productivity. Now a day's using conventional techniques by farmers in agriculture is decrease our soil health as well it does not only make the soil barren but eventually, the farmer goes under debt. The revolutionary impact of green revolution or modern agricultural techniques that broke away the old and outdated traditional practices. Around the world 35 million hectares of agricultural land are managed organically with India having the largest number of organic producers. Out of this cultivated area accounts for 1.08 million ha while remaining 3.4 million ha is wild forest harvest collection area. Natural farming is defined as a method of farming system which primarily aims at cultivating the land and raising crops in such a way, as to keep the soil alive and in good health by use of organic wastes (crop, animal and farm wastes, aquatic wastes) and other biological materials along with beneficial microbes (biofertilizers) to release nutrients to crops for increased sustainable production in an eco-friendly pollution free environment.

The options management of insect pests in natural farming is heavily dependent on preventive measures rather than curative practices, which are based on the ecologically safer management methods. The priority has been given to maintain the health of the ecosystem thus enabling plant to become resistance to attack by insect-pests. Broad management of ecosystem through little modification in the cultural practices such as crop rotation, soil quality management through the addition of organic amendments constitute the preliminary defense against the attack of insect-pests and diseases followed by use of the curative methods like use of predators, parasitoids, plant products and ecologically safer chemicals against management of the insect pests and diseases.

Insect Pests Management Methods in Zero Budget Natural Farming (ZBNF)

This concept was introduced by Shri Subhash Palekar, for which he was awarded India's fourth highest civilian award, Padma Shri in 2016 (Anon., 2016). The zero budget natural farming (ZBNF) is also known as zero budget spiritual farming (ZBSF). In this farming farmers are using mulching, soil protection techniques, natural pesticides, fungicides and fertilizers. The methods of ZBNF include crop rotation, green manures and compost, biological pest control, and mechanical cultivation. There are five most important methods of ZBNF i.e. Agniastra, Bramhastra, Neemastra, Jivamrita, Bijamrita.

Some unique quality of ZBNF

- In the Zero Budget Natural Farming nothing has to be purchased from the market. All things required for growth of the plant are available around the root zone of the plants.
- About 98.5 per cent nutrients are taken from air, water & solar energy. Remaining 1.5% nutrients taken from the soil are also available free of cost as it is taken from the prosperous soil which is enriched with all nutrients.
- The approach towards sustainability.
- An expense frees farming.
- The farming up to 30 acres with one native cow, farming with minimum electricity, water consumption and producing quality food.
- The techniques of farming are multi crop cultivation for higher net income and reducing external labor requirement.

Insect Pests Management methods in Zero Budget Natural Farming (ZBNF)

Sr. No.	Name of pest management methods	Composition
1	**Agniastra**	It composed of 100 liter of water, 10 liter cow urine, 1 kg tobacco, 500gm of green chilli, 500gm of garlic, 5kg neem leaves pulp.
2	**Bramhastra**	It is prepared by neem leaves, custard apple leaves, guava leaves, camellia leaves, pomegranate leaves, papaya leaves, white dhatura leaves crushed and boiled in cowurine.
3	**Neemastra**	It is made up of local cow urine (5 liter), cow dung (5kg), neem leaves and neem pulp (5kg) is fermented for 24 hrs.
4	**Jivamrita**	It is made from cow-dung (20 kg), cow urine(5-10 liter), jaggery (20kg), flour (2kg) and is applied to crops with each irrigation cycle.
5	**Bijamrita**	It is basically made up of water (20 liter), cow dung (5kg), cow urine (5 liter) and lime (50g).

- **Agniastra:** This is a mixture of chilli, garlic, neem, cow urine and used to control insect pests i.e. leaf roller, stem borer, fruit borer, pod borer.
- **Bramhastra:** It is the mixture of different locally available plants i.e. neem, guava, custard apple, pomegranate etc. with cow urine and it is used to spray on the leaves of plant for the control of all sucking pests i.e. fruit borer, pod borer.
- **Neemastra:** It is the mixture of cow dung, cow urine, neem leaves etc. and used against different leaf sucking insect pests and mealy bugs.
- **Jivamrita:** It provides nutrients, but it also acts a catalytic agent which promotes activity of different microorganisms present in soil, as well as increases activity of earthworm and it's also provides adequate protection from the insect pests and diseases during initial stages of germination.
- **Bijamrita:** This method has been used for seed treatment in protecting young roots from fungus, as well as against from soil borne insect pests control.

Insect Pests Management in Natural Farming

The goal of natural farming to control and restore a balance between pest and predator to keep pests population below down at acceptable level. During the last two decades, there has been a significant globalization of community towards environmental preservation and assuring of food quality. The natural farming is not to suppress the population of insects, as they also have play a role in the natural system. Once insect pests have started to attack a crop, the damage cannot be repaired and control becomes increasingly difficult. Where possible, use different techniques to avoid or prevent insect pests attack in the first place. Natural farming considers that it can meet both these demands and become the mean for complete development of rural and tribal areas. After almost a century of development, organic as well as natural farming is now being embraced by the mainstream and shows great promise commercially, socially and environmentally safest. While there is continuum of thought from earlier days to the present, the modern natural movement is radically different from its original form. It now has environmental sustainability at its core in addition to the founders concerns for healthy soil, healthy food and healthy people. The pest management in the natural farming should be in a such a way that it should not affect the ecological balance that existed with in an ecosystem and also enable us to obtain the greater economic yields free from insect pest attack. The series of pest management strategies should be planned well in advance to improve the overall health of the ecosystem there by making the whole ecosystem sustainable and resistance to attack by insect pests (Kumaranag *et al.*, 2013).

Insect pest management strategies for natural farming

Effective application formanagement of insect pests in natural farming strategies are classified into following categories

- The conservation practices to restore the natural enemies through provision of hedge rows, shelter belts, etc.
- Use of biological control agents such as insect predators, parasitoids, insect pathogens by applying or releasing the agents.
- Use of botanicals and organic pesticides as curative control measures.

Pest Management Options

The conservation practices to restore the natural enemies: Conservation of natural enemies involves manipulation of the environment to enhance the survival, fecundity, longevity, and behavior of natural enemies to increase their effectiveness. Such conservation efforts may be directed at mitigating harmful conditions or enhancing favorable ones. Conservation practices can be further categorized as those that focus on reducing mortality, providing supplementary resources, controlling secondary enemies, or manipulating host plant attributes to the benefit of natural enemies (Haldhar *et al.*, 2017).

Adult parasitoids feed on and depend upon nectar: Studies have shown that longevity and fecundity increase whensuitable nectar sources are available in our scenario. That can have profound effect on parasitism rates and control of insect pest's species. Flowers plants such as marigolds (Tagetes), mint (Mentha), sunflower (Helianthus annus), sunhemp (Crotalaria juncea) as well as local legumes are useful attractant plants. Hoverflies, larvae feed on greenfly are attracted to the flowers of herbs and vegetables such as fennel, celery, dill, carrots and parsnips (Umbelliferae family). The nectar and pollen that these flowers provide will help to increase the number of eggs that these insects lay. Umbellifers will also provide food to various parasitic wasps whose young live on aphids and some of caterpillars.

Adult beneficial insects often use pollen as a food source. For example, adult hoverflies use pollen and may need it to mature eggs and produce their young, which are aphid predators. Some predatory insects are able to complete their life cycle entirely on supplementary food.

Providing a constant source of food will slow emigration of beneficial insects and keep them at high population levels. Refugia is anon-crop area where beneficial insects are provided with microhabitats that contribute to their survival and persistence. They need a place to hide when they are not searching for prey. This area should also provide overwintering habitat. In general the more plant-rich an area is the more likely natural control will work. But some insect natural enemies are more effective in grass or legume monocultures than in a grass and legume mixture, so plant diversity does not universally encourage natural control. Many different ways of conserving beneficial insects have been tested like managing soil, water, and crop residue; varying cropping patterns and non-crop areas and growing plants that attract beneficial species (Haldhar *et al.*, 2017).

Field borders and hedge rows: Field borders and hedgerows represent an important component of the whole combination of habitats occupied by beneficial insects. It is unlikely that all the survival needs of beneficial insects will be met within field borders. Many beneficial insects will spend part of their time in field borders or hedge rows. These areas also serve as corridors that beneficial insects use to move from one field to another. Some consideration of all these areas is needed to develop a complete plan for encouraging natural control. Plants in hedge rows usually consist of natural vegetation that as much as practical, should be preserved. Field borders, depending on how they are managed, usually consist of annual plants. The natural complex of annual plants may provide the habitat and resources needed by beneficial insects. Certain practices can help to maintain a desired mix of plants. For example mowing at a particular time interval or height will stop the natural process of succession to woody perennial vegetation. Some plans for maintaining plant mixtures should be made. Too often, borders are left unattended until there is time to mow them. Some pest problems, such as mites, can be spread by waiting until mid summer to mow. Gather all the information that is available about maintaining field borders before deciding on a maintenance schedule. Use taller non-host border crops like maize, sorghum and pearlmillet as a barrier for whitefly (Krishina Moorthy *et al.*, 2006).

Use of biological control agents: Inundative and inoculative release or applying biological control agents such as insect predators and parasitiods will have a greater role to play in controlling the insect pests. It can be achieved on a small-scale by encouraging natural predators to live and breed in the area where pests are a problem. This can be achieved by having trees and hedges around the farm to provide a home for them. There are many insects and animals which should be encouraged because they feed on pests. These are some examples:

Frogs, toads, hedgehogs, mice, moles, bats, birds, chamelions, lizards, spiders, ants, assasin bugs, black-kneed capsids, bees, branchid wasps, parasitic wasps, dung beetles, ground beetles, earthworms, hawk moths, dragon flies, hoverflies, lacewings and stick insects.

Biological control agents for pest management in natural farming

Predator	Pest targeted	Family
	Order: Coleoptera	
Coccinella septempunctata	Aphids	Coccinellidae
Scymnus coccivora	Grapevine mealy bug	Coccinellidae
Cryptolaemus montrouzieri	Grapevine mealy bug	Coccinellidae
Rodolia cardinalis	Cottony cushion scale	Coccinellidae
Menochilus sexmaculata	Mealybugs and scales	Coccinellidae
Ground beetles	Coconut black headed	
caterpillar and rice BPH	Carabidae	
Tiger beetles	Caterpillars	Cicindellidae
Rove beetles, Paederus fuscipes	Rice leaf folder	Staphylinidae
	Order: Hemiptera	
Platymeris laevicollis	Coconut rhinoceros beetle	Reduviidae
Cyrtorhinus lividipennis	Rice brown plant hopper	Miridae
Eucanthecona furcelleta	Red hairy caterpillar	Pentatomidae
Order: Neuroptera		
Green lace wings or aphid lions: Chrysoperla zastrowsillemi	Aphids, scales and mealy bugs and boll worms	Chrysopidae
Ant lions	Soft bodied insects	Myrmeliontidae
ORDER: ODONATA		
Dragonflies and Damselflies		
(Naiads & Adults)	Caterpillars	-
ORDER: DIPTERA		
Robber flies (Adult)	Small insects	Asilidae
Hover flies (Adult)	Small insects	Syrphidae
	Insect Parasitoids	
Trichogramma chilonis	cotton bollworm, sugarcane internode	
borer and rice leaf folder	Trichogrammatidae	
Trichogramma japonicum	Rice yellow stem borer	
Chelonus blackburni	Spotted bollworm	Braconidae
Copidosoma koehleri	Potato tuber moth	Encyrtidae
Campoletis chloridae	larvae of H. armigera	Ichneumonidae

Predator	Pest targeted	Family
Bracon hebetor and Bracon brevicornis	Coconut black headed caterpillar	Braconidae
Cotesia plutella	Diamond back moth	
Isotima javensls	Pre-pupae of top shoot borer of sugarcane	Ichneumonidae
Xanthopimpla punetata	Pupae of coconut black headed caterpillar	
Encarsia formosa	Cotton whitefly	Aphelinidae
Encarsia perniciosi	Sanjose scale	
Source: Yadav. S. *et al.*, 2017		

Important Predators

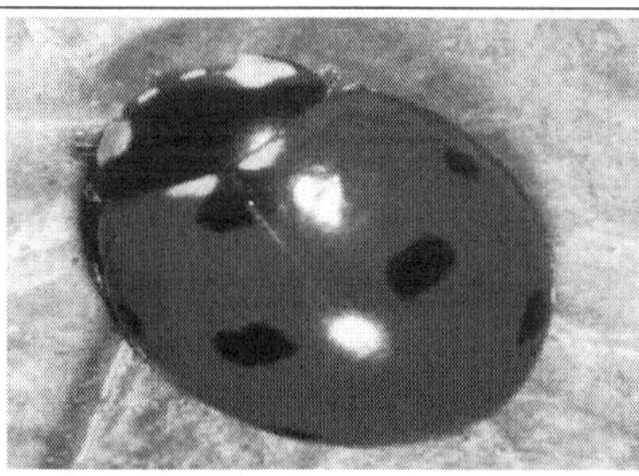

Coccinella septempunctata

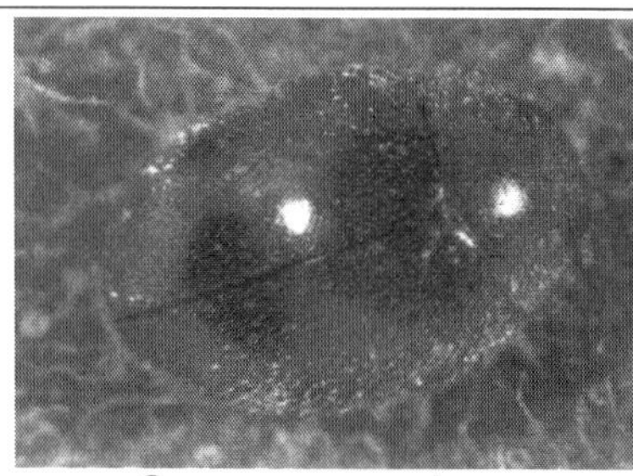

Scymnus coccivora

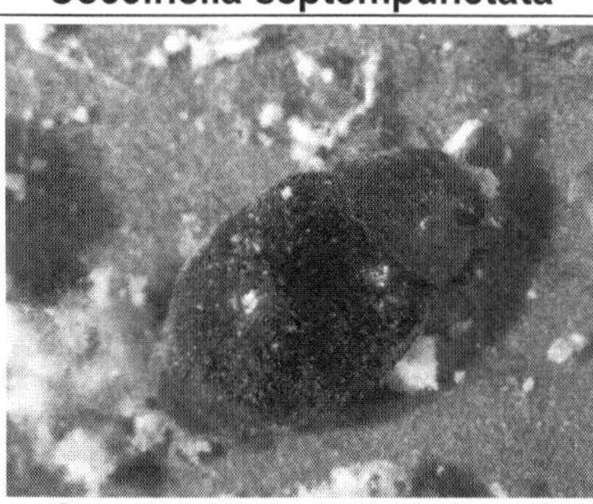

Cryptolaemus montrouzieri

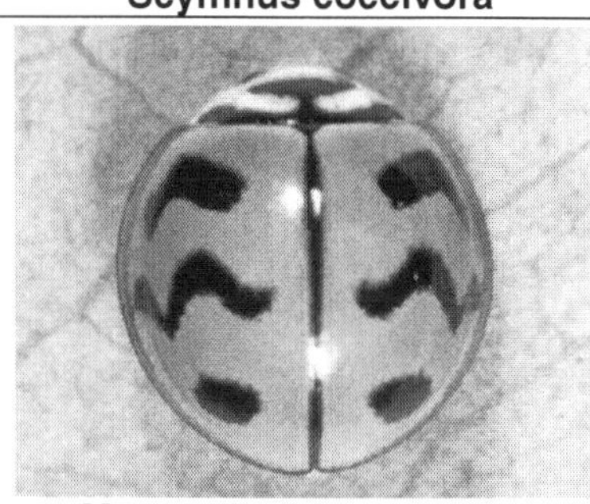

Menochilus sexmaculata

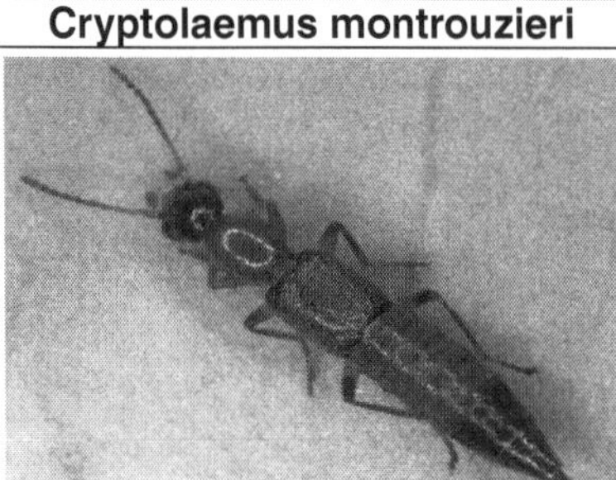

Paederus fuscipes

Cyrtorhinus lividipennis

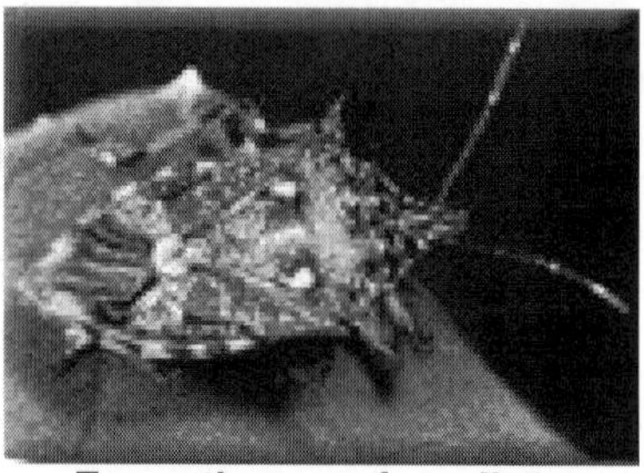

Eucanthecona furcelleta

Chrysoperla zastrowsillemi

Antlion

Hover fly

Source: http://www.nbaii.res.in/Featured%20insects/featured-insects.html

These biological control including predators and parasitoids agents will useful when there is a sudden outbreak in the pest population in natural farming unlike the earlier control measures which are to be planned well in advance. However, slow mode of action, susceptibility of these biocontrol agents to environmental conditionsand in ability to control the pest below the economic threshold level (ETL) will hinder the large scale use of biocontrol control agents in natural farming. Cotesia plutellae is the most common larval parasitoid of diamond back moth in India and causes 40-60 per cent parasitism under unsprayed conditions (Krishna Moorty *et al.*, 2006). A predatory bug, Rhinocoris fuscipes feeds on all stages of Epilachna beetles (Chandel *et al.*, 2007).

Use of botanicals and other bio-chemicals: The crude extracts as well as commercial formulations from plants like neem, pongamia, and tobacco that showed efficacy in conventional agriculture for the management of insect pests were allowed in natural farming because of their less residual action and ecological safety. The botanicals act to poison insects through their digestive systems or to repel insects with strong odours and tastes. Some interrupt life cycle stages with hormone-like substances. Crude formulation of Melia azedarach (drupes), Lantana camara (leaves), Rumex nepalensis (roots) and Artimisia brevifolia (leaves) are also highly effective in reducing lepidopterous larval complex in cabbage (Mehta *et al.*, 2005). Dashparni extract is the commonly used homemade extract used to control the insectpests in organic as well as natural farming in India. It is prepared by mixing the crushed neem leaves 5 kg,

Vitex negundo leaves 2 kg, Aristolochia leaves 2 kg, papaya (Carica papaya) 2 kg, Tinospora cordifolia leaves, 2 kg, Annona squamosa (Custard apple) leaves 2 kg, Pongamia pinnata (Karanja) leaves 2 kg, Ricinus communis (Castor) leaves 2 kg, Nerium indicum 2 kg, Calotropis procera leaves 2 kg, green chilli paste 2 kg, garlic paste 250 gm, Cow dung 3 kg and Cow Urine 5 lit in 200 lit water ferment for one month. Shake regularly three times a day.Extract after crushing and filtering. The extract can be stored up to 6 months and is sufficient for one acre. The highest per cent mortality of white grub, Leucopholis burmeisteri was noticed in the treatment of Karanj cake (44.8%) (Padmanabam, 1997). Chau and Heong (2005) reported that density of spiders was 31.1/ square feet in the treatment of chicken hogmanure compost at 7.5 tons/ ha and 31.9/ square feet in the treatment of organic fertilizer. Parthenium and neem leaves are taken in equal quantity, crushed and soaked in water for 24 hours. The extract is sprayed @ 20 ml/10 litres of water cause considerable reduction in Helicoverpa armigera damage in chilli (Kumar, 2007). Sharma *et al.* (2012) reported the minimum aphid infestation in fenugreek grown under organic management was recorded with three foliage application of neem oil (1%) at 95 DAS and 15 days interval which was significantly superior over karanj oil (1%), garlic bulb extract (5%) and neem leaf extract (5%).

Some of the commonly used animal product based concoctions in natural pest management in India are panchagavya and dasagavya. Panchagavya, an organic product hasthe potential to play the role of promoting growth and providing immunity in plant system. Panchagavya consists of nine products viz. cow dung, cow urine, milk, curd, jaggery, ghee, banana, Tender coconut and water. When suitably mixed and used, these have miraculous effects. This product is known to have a deleterious effect on the many insect pests attacking various crops when used at the dosage 3% solution. Dasagavya is a variant of panchagaya prepared by adding certain plant extracts to panchagavya. Foliar extracts of weeds such as Lantana camara, Leucasaspera, Datura metel, Phytolacca octandra, and Artemisia nilgirica, are then soaked in cow urine in the ratio 1:1 (1kg chopped leaves in 1 litre cow urine) for ten days were then added to panchagavya. Dasagavya may be sprayed once every week for all vegetable and plantation crops. Spraying dasagavya is effective in controlling diseases such as leaf spot, blight, and powdery mildew, rust of vegetables and cut flower crops and tea blister blight. Dasagavya also controls pests such as aphids, thrips, white flies, mites and also foliar caterpillars. Three times spraying dasagavya on vegetable crops @ 3% solution recorded higher yields. These animal based products have to be scientifically validated for their use in the natural insect pests management.

Conclusion

- Insect Pests management in natural farming is a challenging task without use of any insecticides. It involves careful planning in advance through a slight modification in the cultural practices as primary method of pest control. The use of eco-friendly practices such as use of biological control agents and other plant based products as second line of defense against insect pests.

- To support and develop alternative crop protection schemes that are economically, socially and environmentally sustainable, alternative lines of research, price supports, agricultural policies, and land-use practices may need to be embraced.
- To optimize crop protection in natural agriculture, research should be geared to defining and accessing suites of crop production materials and practices that work in concert as a favourable production system.

References

Anonymous. 2016. "Venkaiah Naidu congratulates farmer on winning Padma Shri". Indian Express.

Chandel R S, Dhiman K R, Chandla V K and Kumar Vinod. 2007. Intergrated pest management in potato. In: Entomology: Novel Approaches, Eds Jain, P. C and Bhargava, M. C. New India Publishing Agency, New Delhi, pp. 377-398.

Chau M L and Heong K L. 2005.Effects of organic fertilizers on insect pest and diseases of rice. Omonrice 13: pp. 26-33.

Haldhar, S. M., Jat, G. C., Deshwal, H. L., Gora, J. S. and Singh, D. 2017. Insect Pest and Disease Management in Organic Farming. Towards Organic Agriculture. 359-390.

Krishna Moorthy P N, Pandey K K, Pandey K K and Krishna Kumar N K. 2006. Status and prospects of integrated pest management strategies in selected crops: Vegetables. In: Integrated Pest Management. Principles and Applications Volume 2: Princiles, Eds. Singh, Amerika., Sharma, O. P. and Garg, D. K. CBS Publishers and Distributors, New Delhi, pp. 340-92.

Kumar S and Yadav B P. 2007. Efficacy of fungicides and phyto extracts on Colletotrichum spp. Physiology and Molecular Plant Pathology 37(2): 363-364.

Kumaranag, K. M., Kedar, S. C. and Patil, M. D. 2013 Insect Pest Management in Organic Farming Systems. Pop. Kheti, 1(4): 159-163.

Mehta P K, Sharma P C and Sood A K. 2005.Evaluation of some botanicals against lepidopterous larval complex of cabbage in Himachal Pradesh. In: First Congress on Insect Science, Punjab Agricultural University, Ludhiana, December 15-17, pp. 110-11.

Padmanabam. 1997. Evolution of oil cake against white grub. Annals of Agricultural Research 24 (3): 461-465.

Sharma S K, Trivedi A, Ameta O P, Sharma S K, Choudhary R and Hussain T. 2012. Evaluation of botanical extracts against aphid management in fenugreek under organic farming. Indian Journal of Applied Entomology 26: 88-92.

www.atimysore.gov.in/PDF/action_research1.pdf.

Chapter 19

"Breaking Ground: Prospecting Nano Fertilizers for Green Growth"

Boreddy Jayachandra Reddy, Meena Yadav, Ranbir Singh, L.V. Ravishankar and Bushra Rasool.,

Division of Soil Science and Agriculture Chemistry
Division of Plant Pathology
Sher-e-Kashmir University of Agricultural Sciences and Technology, FoA, Chatha-Jammu-180009, Jammu and Kashmir

Abstract

In order to meet the capacity for food production required by the expanding global population, traditional agricultural methods have been applied to increase crop yields. Commercially available chemical fertilizers, such as straight mixed and composite fertilizers, are crucial for providing agricultural crops with macro and micronutrients. The development of the agricultural sector by increasing resource use efficiency while minimizing damage to the production bed using modern technologies, increased concern about excessive application of these fertilizers, which results in leaching losses and contamination of the agricultural eco system, a scarcity of variable land, limited water resources, and nutrient resources. A promising new technology has been created to address the shortcomings of conventional agriculture practices. The size of the nanoparticles enables the most efficient and waste-free focused activity. Atomic and molecular aggregates with a size range of 1 to 100 nanometers are the focus of nanotechnology. Due to their fundamentally changed physiochemical characteristics brought on by their smaller size and greater surface area, these nano particles offer a wide range of applications in a number of fields. Nano fertilizers, which give essential nutrients in nano form, enhance plant development and agricultural output. This results in high fertilizer efficiency and ultra- high observations. Nano fertilizers, which give essential nutrients in nano form, enhance plant development and agricultural output. This results in high fertilizer efficiency and ultra-high observations. Application of Nano fertilizers, which boosts enzyme activity in the soil's rhizosphere, aids plant development (Tarafdar *et al.*, 2015) The increasing aluminium content of certain nanoparticles exemplifies the nonfertilizer beneficial effects. Application of silicon dioxide (SiO2) improves seed germination in Lycopersicum esculentum mill, and application of zinc oxide (ZnO), the most significant cofactor in the production of IAA, increased growth and yield in Arachis hypogea. Micronutrients, which are important for plant physiology and resistance to abiotic and biotic stresses, as well as plant growth promoting factors. The creation of nanoparticles, which calls for a high degree of input and technical proficiency, demonstrates the efficiency of the nanotechnology in terms of a variety of factors. Further study is required to determine the best use of nanotechnology in agricultural science since it has ramifications for ecological dispersion and human health.

Keywords: Nano fertilizers, conventional fertilizers

Introduction

Nanotechnology, abbreviated to "Nanotech". The definition of nanotechnology is based on the prefix "Nano" which is from the Greek word meaning "dwarf". Nanomaterials display different properties from bulk materials as a result of their size. Nanotechnology is science, engineering, and technology conducted at the Nano scale, which is about 1 to 100 nanometers. In India, the Department of Science and Technology (DST) launched the Nano Science and Technology Initiative (NSTI) in 2001 under the leadership of Prof. C. N. R. Rao. The Government of India is currently spending 1000 crores under Nano Science and Technology Mission (Nano Mission) during the Eleventh Five-year Plan period.

How it Started

The ideas and concepts behind Nano science and nanotechnology started with a talk entitled "There's Plenty of Room at the Bottom" by physicist Richard Feynman at an American Physical Society meeting at the California Institute of Technology on December 29, 1959. He described a process in which scientists would be able to manipulate and control individual atoms and molecules. Over a decade later, in his explorations of ultra-precision machining, Professor

Norio Taniguchi coined the term Nanotechnology. It wasn't until 1981, with the development of the scanning tunnelling microscope that could "see" individual atoms, that modern nanotechnology began. His and the invention of the atomic force microscope (AFM) made it possible for nanotechnology to become reality. Nanotechnology has come a long way since then and now affects many industries.

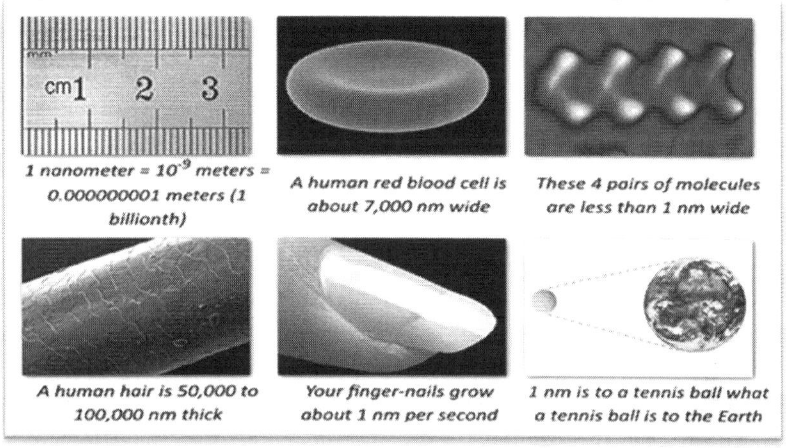

One nanometre is a billionth of a meter, or 10-9 of a meter.

- Here are a few illustrative examples:
- There are 25,400,000 nanometers in an inch.
- A sheet of newspaper is about 100,000 nanometers thick
- Dimensions between approximately 1 and 100 nanometers are known as the nanoscale.

Characteristics of Nanoparticles

- Smaller particles allow better coverage of surface area
- Nano sized particles can even pass through the cell wall in plants and animals.
- Nanotechnologists use this process to deliver at cellular level which is more effective then the conventional method.

Key focus areas for nanotechnology in agricultural research

- Nano genetic manipulation of agricultural crops
- Agricultural Diagnostics
- Nano fertilizers
- Nano-Biosensors
- Nano-pesticides
- Nano-herbicides

Why Nano fertilizers?

Nanomaterial coatings (such as a nanomembrane) may slow the release of nutrients. Formulations of Nano fertilizers can provide balanced nutrition for crops throughout the growth cycle, which in turn improves agricultural production. Additionally, several researchers have reported that a decrease in the size of nanomaterials facilitates an increase in the surface mass ratio of particles.

Why We Need Nano Fertilizers?

- Mineral fertilizers feed half of the global population (Erisman *et al.* 2008)
- Mineral fertilizers are characterized by poor nutrient use efficiency (NUE) <30%
- Nanotechnology enhances NUE and can allow minimizing input of mineral fertilizers into the biosphere
- Nano-fertilizer may be defined as the Nano particles or Nano material which can supply essential nutrients for plant growth, having higher use efficiency and can be delivered in a timely manner to rhizosphere target or at foliar site.
- After first green revolution, the agricultural production is experiencing a plateau now a days
- High ratio of NPK fertilizer i.e. 10: 2.7: 1
- Low output for per unit fertilizer use
- Increasing fertilizers requirement for producing more food grains to feed the burgeoning population of 1.4 billion in 2025
- The extent of multi-nutrient deficiencies is alarmingly increasing year by year which is closely associated with a crop loss of nearly 25–30%
- The extent of nutrient deficiencies in the country are in the order of 89, 80, 50, 41, 49 and 33% for N, P, K, S, Zn and boron respectively
- In the past few decades, use efficiencies of N, P and K fertilizers remained constant Nano fertilizers are the important tools in agriculture to improve crop growth, yield and quality parameters with increase nutrient use efficiency, reduce wastage of fertilizers and cost of cultivation.
- Nano-fertilizers are very effective for precise nutrient management in precision agriculture with matching the crop growth stage for nutrient and may provide nutrient throughout the crop growth period.
- Nano-fertilizers increase crop growth up to optimum concentrations further increase in concentration may inhibit the crop growth due to the toxicity of nutrient.

- Nano-fertilizers provide more surface area for different metabolic reactions in the plant which increase rate of photosynthesis and produce more dry matter and yield of the crop.
- It is also preventing plant from different biotic and abiotic stress.

Mechanism of Nano fertilizers

- **Size and Surface Area:** Since nano fertilizers are smaller than conventional fertilizers, their surface area can be expanded. This larger surface area improves nutrient absorption and allows for improved contact with plant roots.
- **Slow Release:** Nutrients can be released gradually over time by engineering nano fertilizers. By minimizing the chance of nutrient leaching and maximizing nutrient utilization, this controlled release guarantees a more consistent supply of nutrients to the plants.
- **Enhanced Solubility:** Poor solubility is a problem for many conventional fertilizers, which causes nitrogen loss through runoff. In order to increase nutrient availability to plants, Nano fertilizers can be engineered to increase solubility.
- **Targeted Delivery:** Specific plant tissues or cells can be the target of a nano fertilizers design. This accuracy in the distribution of nutrients ensures that they reach the places where they are most required and helps cut down on waste.
- **Enhanced Nutrient Uptake:** These fertilizers' nanoscale size allows for greater penetration of plant cell walls, which improves the uptake of nutrients by roots. Plant growth and development are promoted by this higher uptake efficiency.
- **Interaction with Soil Microorganisms:** The cycling and availability of nutrients may be favorably impacted by the interactions that Nano fertilizers have with soil microorganisms. Plant nutrition and soil health may both benefit from this.
- **Reduced Environmental effect:** Nano fertilizers have the potential to lessen the environmental effect associated with conventional fertilization techniques by increasing nutrient use efficiency and lowering nutrient runoff.

Revolutionizing Agricultural Practices: Unveiling the Potential of Nano-Fertilizers for Sustainable Crop Nutrition

Nano-fertilizers are a first step towards creating the cutting-edge fertilizers for future input-based applications. In order to provide vital plant nutrients, modern intensive farming methods use organic and mineral manures; yet, this practice has led to a major decline in the environment and ecosystems. Eutrophication and the manifestation of the effects of climate change and global warming are the results of nitrogen loss through nitrous oxide and nitrate leaching. According to reports, phosphate fertilizers have an even lower nutrient usage efficacy (NUE) of less than 20% (Lopez-Moreno *et al.*, 2010). Because of the greater nutrient uptake brought about by the reduced surface area of nanomaterials, which promotes nutrient-surface contact, nano-fertilizers have the potential to improve NUE. In addition to increasing agricultural yields sustainably, Nano fertilizers have the ability to stop fertilizer-related environmental contamination.

Nitrate leaching and denitrification were considerably decreased by slow-release fertilizers coated with nanoparticles—chemical compounds that have a small solubility in water or other solvents and are broken down progressively and slowly by the soil microbial community. In addition, controlled-releasing fertilizers which are more soluble than slow-releasing fertilizers are coated with substances that drastically cut down on the active ingredient's solvent exposure, allowing for the controlled release of nutrients through diffusion coated with nanomaterials to reduce surface area. This could make them an excellent source of plant nutrients in the future.

Important properties of Nano fertilizers which facilitate higher nutrient use efficiency are:

- Higher surface area.
- High solubility in different solvent such as water.
- Facilitates more penetration
- Improve uptake and nutrient use efficiency of the Nano fertilizer.
- Reduction of particle size results in increased specific surface area and number of particles per unit area of a fertilizer that provide opportunity to contact of Nano-fertilizers which leads to more penetration and uptake of the nutrient.

Fertilizers encapsulated in Nano-particles will increase availability and uptake of nutrient to the crop plants.

Advantages of Nano fertilizers

- Facilitate higher nutrient use efficiency: Small particle size than pore size of root and leaves leads to more penetration into the plant. Improve uptake and nutrient use efficacy of crop plants. Prevent the loss of nutrients
- Nutritional value and health: Nano fertilizers enhance growth of plant parts and metabolic process such as photosynthesis; improve the yield
- More availability of nutrients helps to increase the quality parameters of crops, such as protein, oil content, sugar content, etc.
- More availability of Nano nutrient to the plant, prevent from disease, nutrient deficiency and other biotic and abiotic stress, which result

in better yield and quality food products for human and animal consumption

- Controlled release: Nano fertilizers control the speed and dose of encapsulated nutrient/fertilizers to make more uptake by crop plant.
- Increase availability due to slow release of nutrients.
- Increase actual duration of nutrient supply
- Reduce lose and demand of fertilizers
- Nano fertilizers can take up by the plants due to slower rate of release.
- Nutrients can be taken up by plants without wastage by leaching and/or leaking.
- Reduce the demand for fertilizers
- By using foliar form of nano fertilizers in small quantities, increased growth and yield of crops were obtained compared to bulky fertilizers of the same elements
- The efficient use of nanomaterials may reduce the negative environmental impact of conventional agricultural practices
- Nano coating and nanotechnology can help in numerous ways to reduce costs and increase productivity around the farm.
- Improvement in soil aggregation, moisture retention and carbon build up.
- Nano fertilizers are made of nano sized nitrogen molecules coated in a polymer coating with biosensor, which release the particle when soil is needed.
- Nano fertilizers are being studied as a way to increase nutrient efficiency and improve plant nutrition, compared with traditional fertilizers.
- A nano fertilizers is any product that is made with nanoparticles or uses nanotechnology to improve nutrient efficiency

Nano Fertilizers Refers to a Product that Delivers Nutrients to Crops in One of Three Ways

- The nutrient can encapsulate inside Nano-materials nanotubes or Nano porous materials
- Coated with a thin protective polymer film
- Delivered as particles or emulsions of nanoscale dimensions (Wang *et al.* 2016)

Disadvantages of Nano-Fertilizers

The Catchy term 'Nanotechnology' also pose some risks and problem towards the health and also towards environment.

- **Transformation of NPs:** Owing to the property of reactivity, nanomaterials can interact with different components of environment, which leads to transformation and changes in physicochemical properties. Nanomaterials can interact with soil components and may cause toxicity
- **Accumulation of NPs:** Nano fertilizers can accumulate in plant parts, leading to growth inhibition, generation of reactive oxygen species, and cell death can accumulate in food parts and, when consumed, may cause human health problems
- **Safety concern for farm workers:** Reactivity and variability of nanomaterials have raised safety concerns for workers who may become exposed during their manufacturing and application in the field.
- Lack of production and availability of Nano fertilizers in required quantities limits the wider scale adoption of Nano-fertilizers as a source of plant nutrients.
- Lack of standardization in the formulation process brings about different results of the same nanomaterial under various pedoclimatic conditions.
- These nanoparticles produce waste toxic materials which if contacted with soil and aquatic environment can cause contamination or pollution.
- It has also ill-effect upon plant system viz; by it may plug the stomata pore, forming a toxic layer upon the stigmatic surface, which further prevent pollen tube penetration, it may enter into vascular tissue and impair translocation of water, minerals and photosynthates.
- It might heat up the soil and be toxic in full. alter the soil's texture and structure to reduce its fertility.
- Complete potential for maximal loss through application and irrigation
- Distribution that was improperly filed.

Ethical and safety issues of Nano-fertilizers application

- The major concern at world scale is whether the unknown risks of nanoparticles involving their environmental and health impact prevail over their potential benefits.
- Thus, the risks associated with the application of nanoparticles are yet to be evaluated before the application of nanoparticles is fully accepted and implemented.
- Due to the thorough quantitative analysis of the potential health impacts, environmental clearance, and safe disposal of nanoparticles, improvements in the design of further applications of nanotechnology can be anticipated.

- No direct human disease has been linked to nanoparticles so far.
- However, the interdisciplinary research of materials scientists, environmentalists, and life scientists is contributing to the identification of the true, if any, hazards of nanotechnology.

Use of Nanofertilizers in Agriculture

The key issues facing agriculture today are (a) stagnant crop yields (b) shrinking arable land as a result of land degradation and urbanization (c), low nutrient use efficiency (d), nutrient deficiencies in multiple areas of the soil (e), declining soil organic matter (e), and (e) water availability. It would be impossible to produce enough food under these conditions to support the world's expanding population, which is predicted to reach 9 billion people by 2050.

Agriculture and horticulture-related nanoscience and nanotechnology research is still in its early stages but is progressing quickly. Conventional bulk fertilizer or conventional fertilizers can be dangerous to both people and the environment in addition to being expensive for the producer. Due to this, three main problems in modern agriculture are: (a) static crop yields (b) decreasing arable land due to soil degradation and However, while nanomaterials and Nano fertilizers undoubtedly provide new opportunities for sustainable and precise farming, it is important to thoroughly assess their drawbacks before bringing them to market. Particularly, there may be a risk to the ecosystem and human health from the widespread release of nanomaterials into the environment and the food chain.

Conventional Bulk Fertilizers vs Nano Fertilizers

Conventional fertilizers are often applied to the soil by farmers or growers by surface broadcasting, subsurface placement, fertigation, or irrigation water. However, a major amount of applied fertilizer's end up in the atmosphere or in aquatic bodies, where they eventually contaminate our ecosystems. As an illustration, the 75% N of urea that was applied to the ground was lost by volatilization (as NH3, emission of N2O or NO, or through NO3 leaching or runoff to water bodies). Due to the low nitrogen consumption efficiency of the existing N fertilizer's (20 percent), eutrophication and a rise in greenhouse gas emissions result from N loss in the environment. Key macronutrient elements, such as N, P, and K, applied to the soil have been observed to be lost by 40–70%, 80–90%, and 50–90%, respectively, resulting in a significant loss of resources. The extra phosphorus is "stuck" in the soil, where it interacts chemically with other nutrients and prevents plants from absorbing it.

Environmental and Health Concern of Nano Fertilizers

Before using Nano fertilizers as a widespread agricultural practice, the use of nanostructures or nanoparticles as agrochemicals (fertilizers or pesticides) is systematically being investigated. The size, shape, solubility, crystal phase, kind of material, exposure levels, and dose concentrations of numerous nanoparticles are all thought to pose a risk to human health. Expert opinions suggest that food items containing nanoparticles that are already on the market are probably safe to eat, although more active research is needed in this area. Detail research are needed to understand the effects of nanoparticles inside the human body after exposure through nanofood in order to solve the safety problem. Researchers must evaluate and create appropriate evaluation methodologies to evaluate the effects of nanoparticles and Nano fertilizers on biotic and abiotic ecosystem components like higher plant growth after nanoscale ZnO treatment (1000 ppm), after 110 days (Prasad *et al.*, 2012)

Conclusion

- ❒ Nano-fertilizers have potential to increase crop productivity through slow or controlled delivery.
- ❒ Due to their small size and target specificity, they increase the fertilizer use efficiency.
- ❒ Nano-fertilizers have the capability to minimize soil toxicity, reduce the potential negative effects associated with over dosage and frequency of the application as well as the cost of production.
- ❒ As Nano-fertilizer exhibits a large surface area with minute size, it can easily translocate and uptake by plants and producing more photosynthate, dry matter ultimately increases the yield of a crop.
- ❒ Thus, Nano-fertilizers become more potential for achieving sustainable agriculture, especially in developing countries.
- ❒ Nano fertilizer clearly has the potential to dramatically improve agriculture production.
- ❒ Seed dressing or foliar application at lower rates get more desired results.
- ❒ Still need improvement in analysis of penetration and transport of Nano particles in plants.
- ❒ The use of nanotechnology for fertilizers is still in its infancy but is already adopted for medical and engineering applications
- ❒ In fact, the country is in need of a Second Green Revolution. Nano fertilizers are envisioned to have the potential to revolutionize agriculture.

References

Naderi, M. and Shahraki, A. (2013). Nano fertilizers and their roles in sustainable agriculture.

International Journal of Agriculture and Crop Science, 5(19): 2229-2232

Lal, R. (2008). Promise and limitations of soils to minimize climate change. Journal of Soil Water Conservation, 63: 113A-118A

Nel A, Xia T, Madler L, Li N. (2006) Toxic potential of materials at the nano level. Science.; 311(5761):622-627

Tarafdar (2015) Application of nano-fertilizer increases PGP enzymatic activities in the Rhizosphere National Symposium on Application of Clay Science: Agriculture, Environment and Industry At: NBSS&LUP, Regional Centre, Kolkata p.47-48

Lin D, Xing B. Root uptake and phytotoxicity of ZnO nanoparticles. Environ. Sci. Technol. 2008; 42:5580–5585. Doi: 10.1021/es800422x

Prasad *et al.*, (2012) of nanoscale ZnO and bulk ZnSO4 on peanut germination, and growth Lab Experiments in Petri dishes Journal of Plant Nutrition 35(6):905-927

https://www.dextrainternational.com/use-of-nanofertilizers-in-agriculture-advantages-and- safety-concerns/

Iyarin Thanka Mahil. E. and Aravinda Kumar B. N. (2019.) Foliar application of Nano fertilizers in agricultural crops – A review J. Farm Sci., 32(3): (239-249).

Debashis Dutta and Anurag Bera(2021) Nano Fertilizer on Sustainable Agriculture- A Review International Journal of Environment and Climate Change.11(8): 1-5.

Albanese, A., Tang, P. S., & Chan, C. W. (2012). The effect of nanoparticle size, shape and surface chemistry on biological systems. Annual Review of Biomedical Engineering. 14, 1-16

Benzon, H. R. L., MRU, R., Ultra, V., Jr, & Lee, S. C. (2015). Nano-fertilizer affects the growth, development and chemical properties of rice. International Journal of Agronomy and Agricultural Research, 7, 105- 117.

Mandeh, M., Omidi, M., & Rahaie, M. (2012). In vitro influences of TiO2 nanoparticles on barley (Hordeum vulgare L.) tissue culture. Biological Trace Element Research. 150, 376- 380.

Manikandan, A., & Subramanian, K. S. (2016). Evaluation of zeolite based nitrogen nano-fertilizers on maize growth, yield and quality on inceptisols and alfisols. International Journal of Plant & Soil Science. 9, 1- 9.

Mazumdar, H., & Ahmed, G. U. (2011). Phytotoxicity effect of silver nanoparticles on Oryza sativa. international Journal of ChemTech Research. 3, 1494-1500

Chapter 20

Trichoderma-Plant Interactions for Sustainable Crop Production under Organic Farming Systems

Efath Shahnaz[1], Saba Banday[2], Ali Anwar[3], Qadrul Nisa[2], Gazala Gulzar[2], Atufa Ashraf[2] and Diksha Banal[2]

[1]Dryland Agriculture Research Station, Rangreth, SKUAST-K, [2]Division of Plant Pathology, SKUAST-K, Shalimar, [3]Division of Plant Pathology, Faculty of Agriculture, SKUAST-K, Wadura, Sopore, Jammu & Kashmir, India
Corresponding author: efaths@gmail.com

Abstract

Trichoderma is one of the most versatile bio-control agents, ubiquitous in soil, multifaceted, with tremendous potential for use in organic farming systems. Since the time of its discovery, *Trichoderma* has been found to be of considerable importance in various facets of agriculture as well as industry. The industrial applications revolve around the production of various types of enzymes by different species of *Trichoderma*. However, it has more often been used as a bio-control agent, particularly for the management of soil borne diseases, though it has also been reported to be effective against foliar pathogens. It has multiple modes of action, viz., mycoparasitism, antibiosis, competition and induced systemic resistance. An important characteristic of *Trichoderma* is its bio-stimulant activity. *Trichoderma* has been found to stimulate the various growth parameters like root growth, shoot growth, flower and fruit production and total yield, in addition to disease management. In the present scenario, *Trichoderma* can be integrated with the various agricultural operations for sustainable crop production. This is particularly so because *Trichoderma* has been found efficient for soil bio-remediation and can improve the quality of soil which in turn increases the quality of yield. Thus, it is highly recommended that *Trichoderma* be incorporated in the schedule for sustainable organic farming system as it combines many benefits like management of diseases, bio-stimulant action for crop growth and soil bio-remediation.

Introduction

One of the greatest discoveries of the century, from an agriculturist point of view, is perhaps the discovery of the uses and benefits of *Trichoderma.* It was first described by Persoon in 1794 (Persoon, 1794), but due to difficulties in morphological identification, it was not until 1969 that a unified concept for the identification was initiated (Rifai, 1969; Samuels, 2006). By 2006, more than 100 phylogenetically different species had been defined. Nowadays, safe identification of *Trichoderma* can be done with the help of an oligonucleotide barcode (TrichOKEY) and a customized similarity search tool (TrichoBLAST) from online web resource www.isth.info (Kopehinskiy *et al.*, 2005). There is a wide variation in the different species of *Trichoderma,* starting from colony morphology, pigmentation, colony color, color of spores, etc. upto the branching pattern and molecular characteristics.However, most of the *Trichoderma* species are characterized by easy isolation, rapid growth, mostly green conidia of various shapes and highly branched conidiophore structure (Gams and Bissett, 1998). Currently, up to 50 new species are recognised per year with the help of revolutionary molecular methods (Cai and Druzhinina, 2021). From its first use in 1932 by Weindling (Weindling, 1932; Weindling and Fawcett, 1936), this fungus has proved to be effective against a vast number of pathogens which include fungi, bacteria, nematodes and even viruses. It produces a wide array of enzymes that are industrially important and have been commercially exploited for the production of cellulases, xylases and other enzymes (Sperandio and Filho, 2021; Passos *et al.*, 2018). It has also been used for the bioremediation of problem soils (Tripathi *et al.*, 2013; Dacco *et al.*, 2020). However, currently we shall be focussing on the bio-control aspect of Trichoderma.

Trichoderma for the management of fungal diseases

Adnan *et al.*, (2019) recently reviewed the status of management of fungal diseases by *Trichoderma* with focus on management via taxonomy, important strains, biodiversity and mode of action. In case of *Trichoderma-Fusarium* interaction, it was observed that both of these fungi recognise each other by sensing their volatile compounds (VCs) and in response produce their own VCs to inhibit the VCs produced by the other fungus (Li *et al.*, 2018). *T. Longibrachiatum* and *T. Asperelloides* produce soluble metabolites that can inhibit or kill *Magnaporthiopsis maydis,* the causal agent of late wilt of maize, besides improving the seedlings' wet biomass and total yield improvement through increase in various growth parameters (Degani and Dor, 2021). Mendez *et al.,* (2020) found that *T. Asperellum* induces systemic defences against *Sclerotium cepivorum* in onion plants under tropical climatic conditions.The endophytic *Trichoderma* reduces colonization of *Phaeoacremonium minimum* and protects the plant by limiting the development of grapevine trunk disease (Carro *et al.,* 2020). Based on in vitro diagnostic assay and analysis of metabolite fractions, Tomah *et al.,* (2020), deduced that *T. virens* HZA14 could cause colony collapse and degradation of *Phytophthora capsici.* Besides, different species of *Trichoderma* were found effective against foliar blight of onion, powdery mildew and black spot of rose, tomato root rot, postharvest anthracnose of chilli, peanut brown root rot, common bean damping off, tomato vascular wilt and early blight (Shahnaz *et al.,* 2013; Amin *et al.,* 2018; Kashyap, *et al.,* 2020; Boat *et al.,* 2020; Carino *et al.,* 2020; Ruangwong *et al.,* 2021; Erazo *et al.,* 2021) and many other diseases to name a few.

Trichoderma for the management of bacterial diseases

Khan *et al.,* (2020) found that the secondary metabolites (SMs) from different *Trichoderma* isolates produced severe morphological changes, such as rupturing of the bacterial cell walls, disintegration of cell membrane and cell content leaking out of phytopathogenic bacteria *Ralstonia solanacearum* and *Xanthomonas compestris.* Konappa *et al.,* (2018) also found that the application of *T. asperellum*isolates delayed wilt caused by *R. solanacearum,* decreased the disease incidence, increased fruit yield, and improved plant growth promotion of tomato. However, Yan and Khan (2021) suggested that the secondary metabolites of *T. Harzianum* could be utilized as low cost and environment friendly alternative for sustainable management of bacterial wilt of tomato. *Trichoderma* was also found beneficial for the management of bacterial spot of tomato caused by *Xanthomonas perforans* (Chien and Huang, 2020). A consortium of *Bacillus subtilis* and *T. Harzianum* effectively suppressed common scab of potato caused by *Streptomyces* spp. and increased tuber yield as a result of primary colonization (Wang *et al.,* 2019). The growth of *Erwinia carotovora* was reduced by *Trichoderma* spp. resulting in reduced disease incidence of potato tuber soft rot (Sulaiman *et al.,* 2020). However, Morán-Diez *et al.* (2020) reported that *Trichoderma* spp. are not able to control the fully pathogenic strain of *Pseudomonas syringae* pv. *tomato.*

Trichoderma for the management of viral diseases

The management of viral diseases by *Trichoderma* spp. is mostly through the induction of systemic resistance. *T. hamatum* was found to promote plant growth, induce resistance and boost innate immunity against tobacco mosaic virus (TMV) infection (Abdelkhalek *et al.,* 2022). The release of salicylic acid by *Trichoderma* bioagents may result in the suppression of pepper leaf curl virus (PeLCV) (Rochal *et al.,* 2021). In some cases, metabolites like trichorzins may inhibit viral infections like cucumber mosaic virus infection of cowpea. *T. Asperellum* could also control cucumber mosaic infections in cucumber by inhibiting the virus and induction of systemic resistance (Tamandegani *et al.,* 2021). *T. harzianum* can also been used as a bio-control agent for the management of papaya ringspot virus (Etim*et al.,* 2022). Jaddawi *et al.,* (2019) suggested that tomato seedlings are protected from cucumber mosaic virus through stimulation of plants to produce pathogenesis related proteins, whereas, Taha *et al.,* (2020) evaluated the ability of 6-pentyl-□-pyrone (6PP) isolated from *Trichoderma koningii*for the induction of systemic resistance in tobacco against tobacco mosaic virus. Nigirpexin E, an azaphilone, derived from *T. Afroharzianum* was found to have anti-tobacco mosaic virus activity (Xie *et al.,* 2022).

Modes of Action

Until recent past, it was believed that *Trichoderma* induces its action by way of mycoparasitism or by the production of antibiotics. It has now been shown that it, in fact, exerts a multipronged strategy for the management of plant pathogens. *Trichoderma* has multiple modes of action and attacks its target through many tactics. These may be direct like competition for nutrients or space, parasitisation of the pathogen, release of antibiotics or production of a cascade of cell wall degrading enzymes that may directly affect the pathogenic population. There is also an indirect mechanism, that is induction of systemic resistance and plant growth promotion. A brief account of the various mechanisms employed by *Trichoderma* species is provided in Table 1.

Table 1. Modes of action employed by *Trichoderma* against different plant pathogens

S. No.	Title	Pathogen	Mode of Action	Reference
1.	*Trichoderma viride* cellulase induces resistance to the antibiotic pore-forming peptide alamethicin associated with changes in the plasmamembrane lipid composition of tobaccoBY-2 cells	-	Antibiosis	Aidemark *et al.,* 2010
2.	Appraisal of Combined Applications of *Trichodermavirens* and a Biopolymer-Based Biostimulant on Lettuce Agronomical, Physiological, and Qualitative Properties under Variable N Regimes	-	Biostimulation	Rouphael *et al.,* 2020

S. No.	Title	Pathogen	Mode of Action	Reference
3.	The 4-phosphopantetheinyl transferase of *Trichoderma virens*plays a role in plant protection against *Botrytis cinerea* through volatile organic compound emission	**Botrytis cinerea**	Volatile Organic Compounds	Cornejo *et al.*, 2014
4.	*Trichoderma harzianum*T6776 modulates a complex metabolicnet work to stimulate tomato cv. Micro-Tom growth	-	Plant growth stimulation	Fiorini *et al.*, 2016
5.	Biological control of foliar pathogens by means of *Trichoderma harzianum* and potential modes of action	Botrytis cinerea, Pseuperono-sporacuben-sis, Sclero-tinia sclero-tiorumand Sphaerothe-cafusca	Local and systemic induced resistance, production of enzymes	Elad, 2000
6.	*Trichoderma hamatum*: Its hyphal interactions with *Rhizoctonia solani* and *Pythium* spp.	*Pythium* spp. and*Rhizocto-nia solani*	Mycoparasitism	Chet *et al.*, 1981
7.	Inhibition of plant pathogenic fungi by endophytic *Trichoderma* spp. through mycoparasitism and volatile organic compounds	*Sclerotinia sclerotiorum*-TSS, *Sclerotium rolfsii*-CSR and *Fusarium oxysporum*-CFO	Volatile Organic Compounds	Rajani *et al.*, 2021
8.	Inhibitory Mechanism of *Trichoderma virens* ZT05 on *Rhizoctonia solani*	*Rhizoctonia solani*	Hyperparasitim and antibiosis	Halifu *et al.*, 2020
9.	Biological characteristic and biocontrol mechanism of *Trichoderma harzianum*T-A66 against bitter gourd wilt caused by *Fusarium oxysporum*	*Fusarium oxysporum*	H2 O2 burst and callose deposition, as well as increasing antioxidant enzymes activities	Zhang *et al.*, 2020

S. No.	Title	Pathogen	Mode of Action	Reference
10.	Mechanism antagonism of *Trichoderma viride* against several types of pathogens and production of secondary metabolites	*Alternaria solani, Fusarium oxysporum, Rhizoctonia solani,* and *Sclerotium rolfsii*	Competition and mycoparasitism	Muhibbudinet *al.*, 2021
11.	*Trichoderma asperellum* T42 induces local defense against *Xanthomonas oryzae*pv. *oryzae*under nitrate and ammonium nutrients in tobacco	*Xanthomonas oryzae* pv. *oryzae*	oxidative burst-mediated defence	Singh *et al.*, 2019
12.	Systemic inducing resistance against late blight by applying antagonist *Trichoderma viride*	*Phytophthora infestans*	Systemic Induced resistance	Purwantisari *et al.*, 2018
13.	Bioactive Secondary Metabolites from *Trichoderma* spp. against Phytopathogenic Fungi	-	Secondary metabolites	Khan *et al.*, 2020
14.	*Trichoderma viride* Controls *Macrophomina phaseolina* through its DNA disintegration and Production of Antifungal Compounds	*Macrophomi-naphaseolina*	DNA disintegration and secondary metabolites	Khan *et al.*, 2021
15.	*Trichoderma asperellum* T76-14 Released Volatile Organic Compounds against Postharvest Fruit Rot in Muskmelons (*Cucumis melo*) Caused by *Fusarium incarnatum*	*Fusarium incarnatum*	Volatile Organic Compounds	Intana *et al.*, 2021
16.	Biological control of *Fusarium oxysporum* f. sp. *ciceri*and *Ascochyta rabiei* infecting protected geographical indication Fuentesaúco-Chickpea by *Trichoderma* species	*Fusarium oxysporum f.* sp. *ciceri*	Directly and by induction of local and systemic resistance	Poveda, 2021
17.	Morphological and protein alterations in *Sclerotinia sclerotiorum* (Lib.) de Bary after exposure to volatile organic compounds of *Trichoderma* spp.	*Sclerotinia sclerotiorum*	Volatile Organic Compounds	da Silva *et al.*, 2020

S. No.	Title	Pathogen	Mode of Action	Reference
18.	*Trichoderma* Isolates Inhibit *Fusarium virguliforme* Growth, Reduce Root Rot, and Induce Defence-Related Genes on Soybean Seedlings	*Fusarium virguliforme*	mycoparasitism and induction of defence-related genes in plants	Pimentel *et al.*, 2020
19.	Mycelial Inhibition of *Sclerotinia sclerotiorum* by *Trichoderma* spp. Volatile Organic Compounds in Distinct Stages of Development	*Sclerotinia sclerotiorum*	Volatile Organic Compounds	da Silva *et al.*, 2021
20.	In vitro biocontrol potential of *Trichoderma pseudokoningii* against *Macrophomina phaseolina*	*Macrophomi-naphaseolina*	Disintegration of DNA	Khan and Javaid, 2020

Formulations of Trichoderma

For successful biological control, the bio-control agent used must be highly efficient; ease of mass production of highly effective and viable propagules; delivery systems should aid the BCA giving it a competitive edge without loss of viability; economical; cost effective; protection against microbial contamination; long shelf life (Harman, 1991; Harman *et al.*, 1991). Besides, it should have high rhizosphere competence and added advantage of promotion of plant growth for increased acceptance among the farming community.Most of the times BCAs are formulated as liquid formulations or solid formulations. Both have their own advantages and disadvantages. Liquid formulations allow for optimization of biomass production and quality with ease of preparation but their acceptability among the farming community is low. Solid formulations can be prepared on agricultural waste materials like sawdust, wheat, rice or maize straw, spent compost, vermicompost, etc. but they are bulky and need larger space for production and packaging. Further, popularization of BCAs is very low among the farming community and only 2 % biopesticides are available in the market (Kumar *et al.*, 2014). Out of these 60% are *Trichoderma* based products (Pintaric, 2019). The different types of formulations available in the market are based on talc, vermiculite, pesta granules, alginate prills, press mud, coffee husk based, oil based or based on different low cost agricultural waste materials.

Table 2: Some important reviews and research on mass production and formulation of *Trichoderma*in the last ten years

S. No.	Title	Year	Reference
1.	*Trichoderma:* Mass production, formulation, quality control, delivery and its scope in commercialization in India for the management of plant diseases	2014	Kumar *et al.,* 2014
2.	Mass production, formulation, quality control anddeliveryof Trichoderma for plant diseasemanagement	2010	Ramanujam *et al.*, 2010
3.	Advances in Formulation of *Trichoderma* for Biocontrol	2014	Cumagun, 2014
4.	*Trichoderma harzianum* based novel formulations: potential applications for management of Next Gen agricultural challenges	2018	Fraceto *et al.*, 2018
5.	*Trichoderma* as biological control agent: Scope and prospects to improve efficacy	2021	Ferreira and Musumeci, 2021
6	Trichoderma-based products and their widespread use in agriculture	2014	Woo *et al.*, 2014
7.	Development of *Trichoderma* sp. formulations in encapsulated granules (CG) and evaluation of conidia shelf-life	2018	Locatelli *et al.*, 2018
8.	Mass production of *Trichoderma harzianum* for managing fusarium wilt of banana	2004	Thangavelu*et al.*, 2004
9.	Prospects of indigenous mass production and formulation of *Trichoderma.*	2006	Jeyarajan, 2006
10.	Mass production of *Trichoderma* spp. and application	2012	Panahian*et al.*, 2012
11.	Optimization of culture conditions for mass production and bio-formulation of *Trichoderma* using response surface methodology	2018	Sachdev *et al.*, 2018
12.	Isolation, identification and mass multiplication of *Trichoderma* an important bio-control agent	2013	Babu and Pallavi, 2013
13.	Mass multiplication and shelf life of liquid fermented final product of *Trichoderma viride* in different formulations	2011	Khan *et al.*, 2011
14.	Advances in Formulation of *Trichoderma* for Biocontrol	2014	Cumagun, 2014
15.	Optimizing mass production of *Trichoderma asperelloides* by submerged liquid fermentation and its antagonism against *Sclerotinia sclerotiorum*	2020	Rezende *et al.*, 2014
16.	Lowcost carrier material for mass production of *Trichoderma* inoculants	2018	Rini*et al.,* 2018

17.	Mass production and determination of shelf-life of two *Trichoderma* sp. in compost formulation	2022	Fernando and Shehani, 2022
18.	Evaluation and assessment of shelf life of liquid substrates and talc formulation for mass production of native *Trichoderma* spp.	2020	Boblina *et al.*, 2020
19.	Optimization of Culture Conditions and Production of Bio-Fungicides from *Trichoderma* Species under Solid-State Fermentation Using Mathematical Modeling	2021	Malatu *et al.*, 2021
20.	Isolation, identification and mass production of five *Trichoderma* spp. on solid and liquid carrier media for commercialization	2018	Hewavitharana *et al.*, 2018
21.	Mass Multiplication of *Trichoderma* in Bioreactors	2020	Prakash *et al.*, 2020
22.	Shelf life studies of different formulations of *Trichoderma harzianum*	2019	Komala *et al.*, 2019
23.	Mass Production and Formulation of Antagonists	2021	Rajeshwari and Appanna, 2021
24.	Development, Production, and Storage of Formulations for Agricultural Applications	2022	Prasad *et al.*, 2022
25.	Quality Control of Fungal Biocontrol Agents with Emphasis on *Trichoderma*	2021	Correa *et al.*, 2021
26.	Effect of physiological parameters on mass productionof *Trichoderma* species	2018	Ghazanfar *et al.*, 2018

Method of Application

The various formulations of *Trichoderma* are applied in various ways. The most common is by seed treatment or soil application. Seed biopriming is the preferred method of control of various soil borne diseases. For treatment of seedlings, seedling dip method for about 20 minutes is followed. Liquid *Trichoderma* is often formulated as pellets, granules or dusts and can be applied to the soil directly or through *Trichoderma* enriched FYM (Ramanujam*et al.*, 2010). Recently, *Trichoderma* has been used for the production of different types of nanoparticles which show high antifungal activity.

Conclusion

The uses of *Trichoderma* are manywith diverse modes of action, methods of application, production processes, production media and above this is perhaps the most studied and least understood and least utilized fungus. A lot of research has been going on but still lots more needs to be done on the potential of this important fungus.

References

Passos Douglas de França, Nei Pereira, Aline Machado de Castro,2018. A comparative review of recent advances in cellulases production by *Aspergillus, Penicillium* and *Trichoderma* strains and their use for lignocellulose deconstruction,*Current Opinion in Green and Sustainable Chemistry,* Volume 14, , Pages 60-66,

Abdelkhalek, Ahmed, Abdulaziz A. Al-Askar, Amr A. Arishi, and Said I. Behiry. 2022. *Trichoderma hamatum* Strain Th23 Promotes Tomato Growth and Induces Systemic Resistance against Tobacco Mosaic Virus*Journal of Fungi* 8, no. 3: 228. https://doi.org/10.3390/jof8030228

Aidemark Henrik Tjellström, Anna StinaSandelius, Henrik Stålbrand, Erik Andreasson,Allan G Rasmusson, Susanne Widell.2010.*Trichoderma viride* cellulase inducesresistance to theantibiotic pore-forming peptide alamethicin associatedwith changes in the plasma membrane lipid composition of tobaccoBY-2 cells. *BMC Plant Biology* 10:274.

AL-Jaddawi, A.A., Elbeshehy, E.K. and Mattar, E.H., 2019. Effects of different species from *Trichoderma* spp. on inducing systematic resistance in tomato plants infected by Cucumber mosaic virus. Formerly The Philippine Agriculturist.

Amin, Faheem, Qazi, N. A., Banday, S. Dar, S. H. and Efath Shahnaz. 2018. Biological control of powdery mildew and black spot diseases of rose. *Journal of Pharmacognosy and Phytochemistry.* 7(3): 2826-2828

Anton Muhibbudin, EndahMulyaningtyas Setiyowati, Antok Wahyu Sektiono. 2021. Mechanism antagonism of *Trichoderma viride* against severaltypes of pathogens and production of secondarymetabolites *Agrosaintifika :JurnalIlmu-IlmuPertanian* Volume 4 No. 1

Babu, K.N. and Pallavi, P.N., 2013. Isolation, identification and mass multiplication of *Trichoderma* an important bio-control agent. *Int. J. Pharm. Life Sci,* 4(1), pp.2320-2323.

Boat, Marie Amperes Bedine, Modeste Lambert Sameza, Beatrice Iacomi, Severin NguemeziTchameni& Fabrice FekamBoyom. 2020. Screening, identification and evaluation of *Trichoderma* spp. for biocontrol potential of common bean damping-off pathogens, *Biocontrol Science and Technology,* 30:3, 228-242, DOI: 10.1080/09583157.2019.1700909

Boblina, B., Beura, S.K., Panda, A.G. and Mishra, M.K., 2020. Evaluation and assessment of shelf life of liquid substrates and talc formulation for mass production of native *Trichoderma* spp. *Journal of Pharmacognosy and Phytochemistry,* 9(3), pp.911-915.

Cai, F., Druzhinina, I.S. In honor of John Bissett.2021. Authoritative guidelines on molecular identification of *Trichoderma.* Fungal Diversity 107, 1–69. https://doi.org/10.1007/s13225-020-00464-4

Cariño Ramírez-, H.F., Guadarrama-Mendoza, P.C., Sánchez-López, V. *et al.*2020. Biocontrol of *Alternaria alternata* and *Fusarium oxysporum* by *Trichoderma asperelloides* and *Bacillus paralicheniformis* in tomato plants. Antonie van Leeuwenhoek 113, 1247–1261. https://doi.org/10.1007/s10482-020-01433-2

Carro-Huerga G, Compant S, Gorfer M, Cardoza RE, Schmoll M, Gutiérrez S and Casquero PA. 2020. Colonization of *Vitis vinifera* L. by the Endophyte *Trichoderma* sp. Strain T154: Biocontrol Activity Against *Phaeoacremonium minimum*. Front. *Plant Sci.* 11:1170. doi: 10.3389/fpls.2020.01170

Chet, I., Harman, G.E. & Baker, R. 1981. *Trichoderma hamatum*: Its hyphal interactions with*Rhizoctonia solani* and*Pythium* spp. MicrobEcol 7, 29–38. https://doi.org/10.1007/BF02010476

Chien, YC., Huang, CH. 2020. Biocontrol of bacterial spot on tomato by foliar spray and growth medium application of *Bacillus amyloliquefaciens* and *Trichoderma asperellum*. *Eur J Plant Pathol* 156, 995–1003. https://doi.org/10.1007/s10658-020-01947-5

Cornejo Hexon Angel Contreras- & Lourdes Macías-Rodríguez &Alfredo Herrera-Estrella & José López-Bucio The 4-phosphopantetheinyl transferase of *Trichoderma virens*plays a role in plant protection against *Botrytis cinerea*through volatile organic compound emission. *Plant Soil* 379: 261-274.

Correa, M.A., Pinto, M., Molinatti, M., Valbuena, O. and Pavone, D., 2021. Quality Control of Fungal Biocontrol Agents with Emphasis on *Trichoderma*. In Biopesticides in Organic Farming (pp. 281-284). CRC Press.

Cumagun, C. J. 2014. Advances in Formulation of Trichoderma for Biocontrol. http://dx.doi.org/10.1016/B978-0-444-59576-8.00031-X

da Silva, L.R., Muniz, P.H.P.C., Peixoto, G.H.S., Luccas, B.E.G.D., da SILVA, J.B.T. and de MELLO, S.C.M., 2021. Mycelial inhibition of *Sclerotinia sclerotiorum* by *Trichoderma* spp. volatile organic compounds in distinct stages of development.

da Silva, L.R., Valadares-Inglis, M.C., Moraes, M.C.B., Magalhães, D.M., Sifuentes, D.N., Martins, I. and de Mello, S.C.M., 2020. Morphological and protein alterations in *Sclerotinia sclerotiorum*(Lib.) de Bary after exposure to volatile organic compounds of *Trichoderma* spp. *Biological Control,* 147, p.104279.

Daccò, Chiara, Lidia Nicola, Marta Elisabetta Eleonora Temporiti, Barbara Mannucci, Federica Corana, Giovanna Carpani, and Solveig Tosi. 2020. *Trichoderma*: Evaluation of Its Degrading Abilities for the Bioremediation of Hydrocarbon Complex Mixtures*Applied Sciences* 10, no. 9: 3152. https://doi.org/10.3390/app10093152

de Rezende, L.C., de Andrade Carvalho, A.L., Costa, L.B., de Almeida Halfeld-Vieira, B., Silva, L.G., Pinto, Z.V., Morandi, M.A.B., de Medeiros, F.H.V., Mascarin, G.M. and Bettiol, W., 2020. Optimizing mass production of *Trichoderma asperelloides* by submerged liquid fermentation and its antagonism against *Sclerotinia sclerotiorum*. *World Journal of Microbiology and Biotechnology*, 36(8), pp.1-14.

Degani, Ofir, and Shlomit Dor. 2021. Trichoderma Biological Control to Protect Sensitive Maize Hybrids against Late Wilt Disease in the Field *Journal of Fungi* 7, no. 4: 315. https://doi.org/10.3390/jof7040315

Elad Y.,2000. Biological control of foliar pathogens by means of *Trichoderma harzianum* and potential modes of action,*Crop Protection*,Volume 19, Issues 8–10,2000,Pages 709-714,

Erazo J.G., S.A. Palacios, N. Pastor, F.D. Giordano, M. Rovera, M.M. Reynoso, J.S. Venisse, A.M. Torres, 2021. Biocontrol mechanisms of *Trichoderma harzianum* ITEM 3636 against peanut brown root rot caused by *Fusarium solani* RC 386, *Biological Control*, Volume 164, 104774,

Etim, D.O. and Onah, D.O., 2022. *Trichoderma harzianum* as Biocontrol Agent and Molecular Characterisation of Papaya ringspot virus (PRSV) on *Cucumeropsismannii* in Calabar, Cross River State, Nigeria. *Asian Journal of Research in Botany* 7(4): 26-34

Fernando, T.H.P.S. and Shehani, K.L.K., 2022. Mass production and determination of shelf-life of two *Trichoderma sp.* in compost formulation.

Ferreira, F.V. and Musumeci, M.A., 2021. *Trichoderma* as biological control agent: Scope and prospects to improve efficacy. *World Journal of Microbiology and Biotechnology*, 37(5), pp.1-17.

Fiorini, L., Lorenzo Guglielminetti& Lorenzo Mariotti& Maurizio Curadi&Piero Picciarelli& Andrea Scartazza& Sabrina Sarrocco& Giovanni Vannacci. 2016. *Trichoderma harzianum* T6776 modulates a complexmetabolicnetwork to stimulate tomato cv. Micro-Tom growth. *Plant Soil* 400: 351-366

Fraceto, L.F., Maruyama, C.R., Guilger, M., Mishra, S., Keswani, C., Singh, H.B. and de Lima, R., 2018. *Trichoderma harzianum*□based novel formulations: potential applications for management of Next□Gen agricultural challenges. *Journal of Chemical Technology & Biotechnology*, 93(8), pp.2056-2063.

Ghazanfar, M.U., Raza, M. and Raza, W., 2018. Effect of physiological parameters on mass production of *Trichoderma* species. *Pakistan Journal of Phytopathology*, 30(1), pp.61-67.

Halifu, Saiyaremu, Xun Deng, Xiaoshuang Song, Ruiqing Song, and Xu Liang. 2020. Inhibitory Mechanism of Trichoderma virens ZT05 on Rhizoctonia solani*Plants* 9, no. 7: 912. https://doi.org/10.3390/plants9070912

Harman, G.E., 1991. Seed treatments for biological control of plant disease. *Crop Prot.* 10, 166–171.

Harman, G.E., Jin, X., Stasz, T.E., Peruzzotti, G., Leopold, A.C., Taylor, A.G., 1991. Production of conidial biomass of *Trichoderma harzianum* for biological control. *Biol. Control* 1, 23–28.

Hewavitharana, N., Kannangara, S.D.P. and Senanayake, S.P., 2018. Isolation, identification and mass production of five *Trichoderma* spp. on solid and liquid carrier media for commercialization. *International Journal of Applied Sciences and Biotechnology*, 6(4), pp.285-293.

Inhibition of plant pathogenic fungi by endophytic Trichoderma spp. through mycoparasitism and volatile organic compounds,*Microbiological Research*,Volume 242,2021,126595,

Intana, Warin, SuchawadeeKheawleng, and Anurag Sunpapao. 2021. *Trichoderma asperellum* T76-14 Released Volatile Organic Compounds against Postharvest Fruit Rot in Muskmelons (*Cucumis melo*) Caused by *Fusarium incarnatumJournal of Fungi* 7, no. 1: 46. https://doi.org/10.3390/jof7010046

Jeyarajan, R., 2006. Prospects of indigenous mass production and formulation of *Trichoderma*. In Current status of biological control of plant diseases using antagonistic organisms in India. Proceedings of the group meeting on antagonistic organisms in plant disease management held at Project Directorate of Biological Control, Bangalore, India on 10-11th July 2003 (pp. 74-80). Project Directorate of Biological Control, Indian Council of Agricultural Research.

Kai, K., Mine, K., Akiyama, K., Ohki, S. and Hayashi, H., 2018. Anti-plant viral activity of peptaibols, trichorzins HA II, HA V, and HA VI, isolated from *Trichoderma harzianum* HK-61. *Journal of Pesticide Science*, pp.D18-039.

Kashyap, P.L., Solanki, M.K., Kushwaha, P. Sudheer Kumar & Alok Kumar Srivastava. 2020. Biocontrol Potential of Salt-Tolerant *Trichoderma* and *Hypocrea* Isolates for the Management of Tomato Root Rot Under Saline Environment. *J Soil Sci Plant Nutr* 20, 160–176. https://doi.org/10.1007/s42729-019-00114-y

Khan, I.H. and Javaid, A., 2020. *In vitro* biocontrol potential of *Trichoderma pseudokoningii* against *Macrophominaphaseolina*. *International Journal of Agriculture and Biology*, 24(4), pp.730-736.

Khan, I.H., Javaid, A. and Ahmed, D., 2021. *Trichoderma viride*controls *Macrophomina phaseolina* through its DNA disintegration and Production of Antifungal Compounds. *Int. J. Agric. Biol*, 25(4), pp.888-894.

Khan, R.A.A., Najeeb, S., Hussain, S., Xie, B. and Li, Y., 2020. Bioactive secondary metabolites from *Trichoderma* spp. against phytopathogenic fungi. *Microorganisms*, 8(6), p.817.

Khan, Raja Asad Ali, Saba Najeeb, Zhenchuan Mao, Jian Ling, Yuhong Yang, Yan Li, and BingyanXie. 2020. Bioactive Secondary Metabolites from *Trichoderma* spp. against Phytopathogenic Bacteria and Root-Knot Nematode *Microorganisms* 8, no. 3: 401. https://doi.org/10.3390/microorganisms8030401

Khan, S., Bagwan, N.B., Iqbal, M.A. and Tamboli, R.R., 2011. Mass multiplication and shelf life of liquid fermented final product of *Trichoderma viride* in different formulations. *Advances in Bioresearch,* 2(1), pp.178-182.

Konappa, N., Krishnamurthy, S., Siddaiah, C.N. Niranjana Siddapura Ramachandrappa & Srinivas Chowdappa 2018. Evaluation of biological efficacy of *Trichoderma asperellum* against tomato bacterial wilt caused by *Ralstonia solanacearum. Egypt J Biol Pest Control* 28, 63. https://doi.org/10.1186/s41938-018-0069-5

Kopchinskiy A, Komon M, Kubicek CP, Druzhinina IS. 2005.TrichoBLAST: a multilocusdatabase for Trichoderma andHypocrea identifications. *Mycol Res* 109:658–660

Kumar, S., Thakur, M and Rani, A. 2014. *Trichoderma:* Mass production, formulation, quality control, delivery and its scope in commercialization in India for the management of plant diseases. *African Journal of Research* 9(53): 3838-3852

Li N, Alfiky A, Wang W, Islam M, Nourollahi K, Liu X and Kang S. 2018. Volatile Compound-Mediated Recognition and Inhibition Between *Trichoderma* Biocontrol Agents and *Fusarium oxysporum. Front. Microbiol.* 9:2614. doi: 10.3389/fmicb.2018.02614

Locatelli, G.O., dos Santos, G.F., Botelho, P.S., Finkler, C.L.L. and Bueno, L.A., 2018. Development of *Trichoderma* sp. formulations in encapsulated granules (CG) and evaluation of conidia shelf-life. *Biological control,* 117, pp.21-29.

Méndez William Rivera-, Miguel Obregón, María E. Morán-Diez, Rosa Hermosa, Enrique Monte.2020. *Trichoderma asperellum* biocontrol activity and induction of systemic defences against *Sclerotium cepivorum* in onion plants under tropical climate conditions, *Biological Control,* 141, 104145

Morán-Diez, María E., Eduardo Tranque, Wagner Bettiol, Enrique Monte, and Rosa Hermosa. 2020. Differential Response of Tomato Plants to the Application of Three *Trichoderma* Species When Evaluating the Control of *Pseudomonas syringae* Populations *Plants* 9, no. 5: 626.

Mulatu, A., Alemu, T., Megersa, N. and Vetukuri, R.R., 2021. Optimization of Culture Conditions and Production of Bio-Fungicides from *Trichoderma* Species under Solid-State Fermentation Using Mathematical Modeling. *Microorganisms,* 9(8), p.1675.

Panahian, G.R., Rahnama, K. and Jafari, M., 2012. Mass production of *Trichoderma* spp. and application. *International Research Journal of Applied and Basic Sciences,* 3(2), pp.292-298.

Persoon CH. 1794.*Dispositamethodicafungorum. Römer's Neues Mag Bot* 1:81–128

Pimentel, M.F., Arnão, E., Warner, A.J., Subedi, A., Rocha, L.F., Srour, A., Bond, J.P. and Fakhoury, A.M., 2020. *Trichoderma* isolates inhibit *Fusarium virguliforme* growth, reduce root rot, and induce defence-related genes on soybean seedlings. *Plant Disease*, 104(7), pp.1949-1959.

Pintaric, S. T. 2019. *Trichoderma:* Invisible Partner for Visible Impact on Agriculture. In: IntechOpenDOI: http://dx.doi.org/10.5772/intechopen.83363

Poveda, J. 2021. Biological control of *Fusarium oxysporum* f. sp. *ciceri*and *Ascochyta rabiei* infecting protected geographical indication Fuentesaúco-Chickpea by *Trichoderma* species. *Eur J Plant Pathol* 160, 825–840 (). https://doi.org/10.1007/s10658-021-02286-9

Prakash, V. and Basu, K., 2020. Mass multiplication of *Trichoderma* in bioreactors. In *Trichoderma:* Agricultural Applications and Beyond (pp. 113-126). Springer, Cham.

Purwantisar Susiana, AchmadiPriyatmojo, Retno Peni Sancayaningsih, Rina Sri Kasiamdari and KadarwatiBudihardjo. 2018. Systemic inducing resistance against late blight by applying antagonist *Trichoderma viride Journal of Physics*: Conference Series 1025 012053.

Rajani P., C. Rajasekaran, M.M. Vasanthakumari, Shannon B. Olsson, G. Ravikanth, R. Uma Shaanker,

Rajeshwari, R. and Appanna, V., 2021. Mass Production and Formulation of Antagonists. In Biopesticides in Horticultural Crops (pp. 216-226). CRC Press.

Ramanujam, B., R. D. Prasad, S. Sriram and R. Rangeswaran. 2010. Mass production, formulation, quality control and delivery of *Trichoderma* for plant disease management. Journal of Plant Protection 2(2): 1-8

Rifai MA. 1969. A revision of the genus Trichoderma. *Mycol Pap*116:1–56

Rini, C.R., Ramya, J., Jayakumar, G. and Shajan, V.R., 2018. Low cost carrier material for mass production of *Trichoderma* inoculants. *Journal of Tropical Agriculture*, 56(1).

RochalKepngopKouokap Lanvin, Eke Pierre, YoumbiYimta Diane, KuleshwarPrasard Sahu, Nya Dinango Vanessa, KamdemWankeu Teddy Herman, GhomsiTamghe Pierre Gilbert, KekeunouSévilor, Kansci Germain & Nana Wakam Louise. 2021. Biological elicitor potential of endospheric *Trichoderma* and derived consortia against pepper (*Capsicum annuum* L.) leaf curl virus, *Archives of Phytopathology and Plant Protection*, 54:19-20, 1926-1952, DOI: 10.1080/03235408.2021.1957412

RouphaelYoussef ,PetroniaCarillo, Giuseppe Colla, Nunzio Fiorentino,Leo Sabatino, Christophe El-Nakhel, Maria Giordano, Antonio Pannico,Valerio Cirillo, Edris Shabani,Eugenio Cozzolino, Nadia Lombardi,Mauro Napolitano and Sheridan L. Woo. 2020.Appraisal of Combined Applications of *Trichodermavirens* and a Biopolymer-Based Biostimulant onLettuce Agronomical, Physiological, and QualitativeProperties under Variable N Regimes. *Agronomy* 10: 196

Ruangwong, On-Uma, ChaninunPornsuriya, KitsadaPitija, and Anurag Sunpapao. 2021. Biocontrol Mechanisms of *Trichoderma koningiopsis* PSU3-2 against Postharvest Anthracnose of Chili Pepper *Journal of Fungi* 7, no. 4: 276. https://doi.org/10.3390/jof7040276

Sachdev, S., Singh, A. and Singh, R.P., 2018. Optimization of culture conditions for mass production and bio-formulation of *Trichoderma* using response surface methodology. *3 Biotech*, 8(8), pp.1-8.

Samuels GJ. 2006.*Trichoderma*: systematics, the sexual state, andecology. *Phytopathology* 96:195–206

EfathShahnaz , V. K. Razdan , S. E. H. Rizvi, T. R. Rather, S. Gupta and M. Andrabi. 2013. Integrated Disease Management of Foliar Blight Disease of Onion: A Case Study of Application of Confounded Factorials. *Journal of Agricultural Science*.5(1): 17-22

Singh Bansh Narayan,Padmanabh Dwivedi, Birinchi Kumar Sarmac and Harikesh Bahadur Singh. 2019. *Trichoderma asperellum* T42 induces local defence against *Xanthomonas oryzae*pv. *oryzae* under nitrate and ammonium nutrients in tobacco. *Royal Society of Chemistry Advances* 9: 39793-39810

Sperandio, G.B., Filho, E.X.F. 2021. An overview of *Trichoderma reesei* co-cultures for the production of lignocellulolytic enzymes. *Appl. Microbiol Biotechnol* 105, 3019–3025 (). https://doi.org/10.1007/s00253-021-11261-7

Sulaiman Mohammed Mahmood, Saad Tariq Abdulmalek Yass, Ammar Amgad Aish, Lara Basheer Yasir, Saja Jamal Abdullah and Sahar A. Youssef. 2020. Activity of *Trichoderma* spp. against *Erwinia carotovora* causal agent of potato tuber soft rot. *Plant Archives* 20 (Supplement 1): 115-118.

Taha, M.A., Ismaiel, A.A. & Ahmed, R.M. 2021. 6-pentyl-□-pyrone from *Trichoderma koningii* induces systemic resistance in tobacco against tobacco mosaic virus. *Eur J Plant Pathol* 159, 81–93 (). https://doi.org/10.1007/s10658-020-02142-2

Tamandegani, P.R., Sharifnabi, B., Massah, A. and Zahravi, M., 2021. Induced reprogramming of oxidative stress responses in cucumber by *Trichoderma asperellum* (Iran 3062C) enhances defense against cucumber mosaic virus. *Biological Control*, 164, p.104779.

Thangavelu, R., Palaniswami, A. and Velazhahan, R., 2004. Mass production of *Trichoderma harzianum*for managing Fusarium wilt of banana. *Agriculture, ecosystems & environment*, 103(1), pp.259-263.

Tomah Ali Athafah, Iman Sabah Abd Alamer, Bin Li, Jing-Ze Zhang, 2020. A new species of Trichoderma and gliotoxin role: A new observation in enhancing biocontrol potential of *T. virens* against *Phytophthora capsici* on chilli pepper, *Biological Control*, Volume 145, 104261,

Tripathi, P., Singh, P.C., Mishra, A. Puneet S. Chauhan, Sanjay Dwivedi, Ritu Thakur Bais& Rudra Deo Tripath. 2013. *Trichoderma:* a potential bioremediator for environmental clean up. Clean Techn Environ Policy 15, 541–550. https://doi.org/10.1007/s10098-012-0553-7

Wang Zhenshuo, Yan Li, Lubo Zhuang, Yue Yu, Jia Liu, Lixia Zhang, Zhenjiang Gao, Yufeng Wu, Wa Gao, Guo-chun Ding, Qi Wang, 2019. A Rhizosphere-Derived Consortium of Bacillus subtilis and Trichoderma harzianum Suppresses Common Scab of Potato and Increases Yield, Computational and Structural Biotechnology Journal, Volume 17, Pages 645-653,

Weindling, R. & O. H. Emerson. 1936. The isolation of a toxic substance from the culture filtrate of *Trichoderma. Phytopathology* 26, 1068–1070.

Weindling, R. (1932) *Trichoderma lignorum* as a Parasite of Other Soil Fungi. *Phytopathology*, 22, 837-845.

Wong, Ken K. Y. & John N. Saddler. 1992.*Trichoderma* Xylanases, Their Properties and Application, *Critical Reviews in Biotechnology*, 12:5-6, 413-435, DOI: 10.3109/07388559209114234

Woo, S.L., Ruocco, M., Vinale, F., Nigro, M., Marra, R., Lombardi, N., Pascale, A., Lanzuise, S., Manganiello, G. and Lorito, M., 2014. *Trichoderma*-based products and their widespread use in agriculture. *The Open Mycology Journal*, 8(1).

Xie, X., Zhao, Z., Yang, H. *et al.*2022. Nigirpexin E, a new azaphilone derivative with anti-tobacco mosaic virus activity from soil-derived fungus *Trichoderma afroharzianum* LTR-2. J *Antibiot*75, 117–121 (). https://doi.org/10.1038/s41429-021-00485-4

Xu XM, Jeger MJ. 2013. Combined use of two biocontrolagents with different biocontrol mechanisms most likelyresults in less than expected efficacy in controlling foliarpathogens under fluctuating conditions: a modelling study. *Phytopathology* 103:108–116

Yan, L., Khan, R.A.A. 2021. Biological control of bacterial wilt in tomato through the metabolites produced by the biocontrol fungus, *Trichoderma harzianum. Egypt J. Biol. Pest Control* 31, 5 (). https://doi.org/10.1186/s41938-020-00351-9

Zhang, F., Liu, C., Wang, Y. *et al.* 2020. Biological characteristic and biocontrol mechanism of *Trichoderma harzianum* T-A66 against bitter gourd wilt caused by *Fusarium oxysporum. J. Plant Pathol.* 102, 1107–1120 (). https://doi.org/10.1007/s42161-020-00573-8

Chapter 21

Extrusion Technology in Food Processing and its Impact on Physical and Chemical Properties of Protein Rich Products

Anjum Ayoub[a], Mohd Naser[b], Shahid Ahamad[c],*

**a,b,c Sher-e-Kashmir University of Agricultural Science and Technology Jammu – 180009*
*Corresponding author, E- mail:anjumparay77@gmail.com

Abstract

Extrusion is a thermodynamic process that combines several unit operations such as mixing, cooking, kneading, shearing, shaping, and forming. This technology is gaining increasing popularity in the global agro-food processing industry, particularly in the food and feed sectors. Extrusion cooking methods are widely used in cereal and protein processing. Today, their capabilities extend to conveying, mixing, shearing, separating, heating or cooling, molding, co-extrusion, venting, controlling volatiles and moisture, flavor generation, encapsulation, and sterilization. These processes can also operate at relatively low temperatures, as seen in the production of pasta, spaghetti, noodles, and similar products. In light of this, the current article provides a detailed explanation of the different types of extruders and their effects on the physical and chemical properties of various protein-rich products.

Keywords: Extrusion, Protein rich product, Twin, Extruder, Protein digestibility, Physical properties.

Introduction

Extrusion is defined as the process of pushing mixed ingredients through a small opening, known as a die, to form and shape materials (Launay and Lisch, 1983). The resulting products are referred to as extrudates (Berk, 2009). The first extruders used in food processing were cylinder and piston-type extruders, initially developed for preparing meats and sausages (Harper, 1981). Although still functional today, cylinder extruders have undergone minimal advancement but continue to be useful in specific applications. In the 1930s, cylinders were replaced by single-screw extruders, which allowed for a continuous process rather than the previous batch method. Single-screw extruders were particularly applied to pasta production and revolutionized the food industry with the speed and efficiency they offered (Ainsworth, 2011). Later, in the 1960s, twin-screw extruders were introduced, and their increased capacity led to a broader range of food extrusion options, expanding into a variety of new snack and cereal products during the 1980s (Mercier *et al.*, 1989).

Modern extruders now offer an incredible range of versatility and functionality. They are widely favored due to their ability to produce fast, continuous processes that can be applied to create a variety of food products, such as snacks, breakfast cereals, pellets, pet foods, and pre-gelatinized flours, among others (Singh *et al.*, 2007). This system integrates several unit operations—such as mixing, kneading, cooking, shaping, and cutting—into a single piece of equipment. As a result, the process is relatively simple, highly efficient, and cost-effective compared to other processing methods (Fellows, 2009).

Typical extrusion processing line

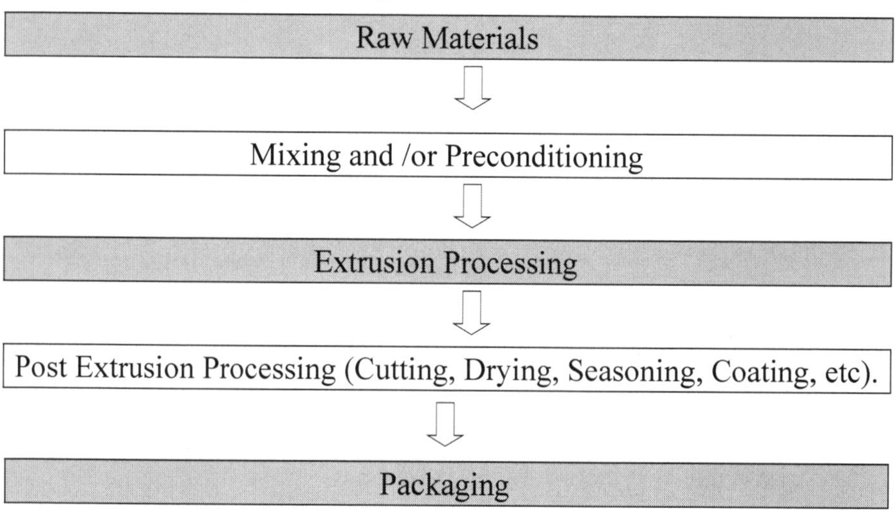

Figure 1. Flow chart of a typical extrusion processing line

Figure 1 provides a broad overview of a typical food extrusion processing production line. The process begins with identifying and obtaining the raw ingredients, which play a crucial role in determining the final product's consistency. The raw materials then undergo mixing and, in some cases, preconditioning using equipment such as ribbon blenders and preconditioners to ensure uniformity before entering the extruder. However, mixing and preconditioning are optional for certain products, depending on their specific requirements.

The next step in the process is extrusion, which serves as the main cooking phase where raw ingredients are transformed into cooked and shaped products. After extrusion, post-processing tasks follow, such as cutting the extruded products into appropriate sizes, drying them to achieve the desired moisture levels, and adding flavoring or coatings to enhance taste and flavor before packaging. Additionally, there may be further processing steps depending on the type of product being produced and its intended application.

Extruders

Extruders come in a few structures, depend upon their application. Some extruders are structured just to pass on the raw materials, while others are intended to blend and knead them; most, in any case, are intended to confer mechanical and warm vitality to the raw materials to achieve wanted physico-chemical changes.

Extruders can be mainly classified on the basis of:

Method of operation

- Cold Extruders and
- Hot Extruders (Extrusion Cooking)

Method of construction:

Single-or twin- screw extruders.

The most commonly used extruders in food processing are single-screw and twin-screw extruders. While extruders with more than two screws have been employed in the plastics industry, they have not been widely adopted in food processing.

Extruders are composed of five main parts:

1. Pre-conditioning system
2. Feeding system
3. Screw or worm
4. Barrel
5. Die and the cutting mechanism

In addition, extruders can vary in terms of screw, barrel, and die configurations. The selection of each component depends on the raw materials used and the desired final product (Riaz, 2000).

Pre-conditioning

Pre-conditioning with steam or water has always been a crucial part of the extrusion process, although it is not applied to all extrusion methods. Generally, this step is used when the moisture content is around 20 to 30% and the material requires extended residence times. Pre-conditioning helps achieve uniform particle hydration, reduces retention times within the extruder, and increases throughput. It also extends the equipment's lifespan by reducing wear on the barrel and screw components, while simultaneously lowering the energy costs associated with the process.

Feeding system

Ensuring a consistent and uninterrupted feeding of raw materials into the extruder is crucial for the efficient and uniform operation of the extrusion process.

Screw

The screw of the extruder is arguably its most crucial component, as it plays a key role in determining the cooking degree, gelatinization and dextrinization of starch, as well as protein denaturation, which all impact the final product quality. Screws can be either monolithic (formed from a single piece) or multipart (composed of various components). The number and shape of screw components can vary, with each section designed for a specific function. Some components are responsible for conveying raw or pre-conditioned material into the extruder barrel, while others are involved in compressing and degassing the feedstock.

Barrel or sleeves

The barrel of the extruder is divided into sections such as feeding and kneading zones. The sleeves surrounding the screw can be solid, but they are often jacketed to allow the circulation of steam or superheated oil for heating, or water and air for cooling. This design enables precise control of the temperature across different areas of the extruder.

Die

The die serves two main functions: shaping the final product and improving the resistance to material flow within the extruder, which increases the internal pressure. It can have various designs and numbers of orifices, depending on the desired product characteristics.

Mechanism of Cutting

The cutting mechanism must be capable of producing end products of uniform size. The size of the products is determined by the rotational speed of the cutting edges. This mechanism can be either vertical or horizontal, depending on the design and requirements of the extrusion process.

Single screw extruder

Single-screw extruders feature a single rotating screw within a metal barrel and come in various designs. The most commonly used single screws have a consistent pitch. Raw materials, fed in granular form into the hopper located in the feed area, are moved by the rotating screw into the transition section. In this section, the screw channel becomes shallower, causing the material to be compacted. A significant amount of mechanical energy is transferred here, leading to an increase in the material's temperature. This process causes starch to gelatinize and the material to become more solid. It is then transported further by the metering section and pushed through the die opening. The barrels of single-screw extruders typically have helical or axial grooves on their inner surfaces, which aids in more effective and efficient conveying and mixing of the material (Caldwell, 2000).

Twin screw extruder

Twin-screw extruders consist of two interlocking screws that rotate within a single barrel, which usually has a smooth inner surface. Although less commonly used in the food industry, twin-screw extruders offer more efficient processing properties. When the material enters the barrel, the ingredients are thoroughly mixed before proceeding to other stages of the extruder. In the initial stage, the screw is designed with a deep screw channel to provide ample space between the screw base and the barrel, allowing for effective mixing. Often, the screws are designed with reverse threads to facilitate intense mixing and longer residence times before the material is conveyed further. In the subsequent section, the diameter of the screw root increases rapidly while the channel depth becomes shallower, which enhances the cooking process. This adjustment increases the pressure on the product, leading to the gelatinization of starch and the denaturation of proteinaceous materials (Chang, 2001).

Effect of Extrusion on Physical Properties of Protein

The hardness of extruded products is significantly influenced by the protein content of the raw material. High protein content results in a less extensible product with a more rigid structure, leading to higher resistance to cutting. However, factors such as feed moisture and barrel temperature also affect the hardness of the extrudate. Hardness is measured as the maximum force required for a test to penetrate the extrudate. Typically, higher density and lower expansion result in a harder extrudate. Increasing the moisture content in the raw material improves the breaking quality of extruded products, while higher temperatures decrease it.

Expansion occurs both radially and axially, depending on the viscoelastic properties of the melt. Moisture vaporization and cooling transition the product from a liquid to a rubbery state, and further drying usually produces a brittle, fracture-prone surface. Products processed at high temperatures and short extrusion times generally have a porous, open structure, giving them a "crunchy" texture (Barrett, 2003).

An increase in raw material moisture content and a simultaneous decrease in barrel temperature reduce the expansion of extruded products, resulting in higher density and hardness. Temperatures below 90°C delay expansion and layer formation (Cheftel *et al.*, 1992). The presence of wheat proteins also affects the expansion of wheat starch. A decrease in expansion was observed as gluten content increased to 11%, followed by an increase in expansion as protein content rose to 16%. At a given temperature, higher moisture content leads to milder and less texturized extrudates due to reduced protein-protein interactions and lower viscosity. Increased feed moisture sharply raises extrudate density, while higher screw speeds and barrel temperatures cause a slight decrease in extrudate density.

Changes in color can occur due to nonenzymatic browning, specifically the Maillard reaction between proteins and reducing sugars, which is triggered by high temperatures (Petitot *et al.*, 2009). The extrusion cooking process, which involves high barrel temperatures and low feed moisture, further promotes the Maillard reaction and can reduce the nutritional availability of lysine (Camire, 2000). Increasing protein content while keeping feed moisture constant leads to greater hardness, crispness, and brittleness, but decreases color intensity.

Water absorption serves as a qualitative measure for assessing the cross-link density of protein biopolymers. Water retention decreases and tensile properties improve as molding temperature rises, up to 130°C. Beyond this temperature, water absorption increases slightly, accompanied by a minor decline in tensile strength and elongation at break, indicating protein degradation. Water absorption is influenced not only by covalent cross-linking but also by the protein's ability to interact with water. Exposing hydrophobic groups, whether through chain rearrangement or chemical additives, can also affect water absorption.

Feed moisture and temperature have a significant impact on the water absorption index (WAI) and water solubility index (WSI) of wheat extrudate. However, feed rate and screw speed do not notably affect these indices. Increasing feed moisture and barrel temperature both significantly decrease the WAI of wheat extrudate. Conversely, higher feed moisture and barrel temperature lead to a substantial increase in the WSI of the extrudate (Ding *et al.*, 2006). The water absorption capacity of the mixture largely depends on the flour composition and increases with higher levels of protein, pentosans, and damaged starch (Martinez *et al.*, 2013).

Extruded whey protein isolate (WPI) under alkaline conditions can develop a fibrous texture, making it suitable for meat applications. Higher barrel temperatures (140 to 180°C) produce better textures even at relatively low moisture contents. However, at higher moisture levels, the temperature should be reduced to prevent significant water loss due to moisture flash-off if a cooling die is not used (Thiebaud *et al.*, 1996).

Effect of Extrusion on Chemical Properties of Protein

Protein solubility, both before and after processing, is often used as an indicator of cross-link formation during extrusion. The denaturation of proteins during extrusion leads to reduced protein solubility, as cross-linking reactions and the formation of covalent bonds occur at high temperatures. Increasing the temperature from 140 to 180°C results in a relative decrease in disulfide linkages in extruded soy protein isolates (Areas, 1992).

High temperatures, screw speeds, and moisture levels lead to more insoluble extruded products, indicating increased cross-linking. At elevated temperatures, hydrophobic and electrostatic interactions contribute to the formation of insoluble aggregates, similar to the fibrous structure found in meat analogs (Sluimer, 2005). Protein denaturation at high temperatures during extrusion enhances digestibility and deactivates antinutritional factors, such as antitrypsin factors and lectins, while reducing bitter taste and unwanted volatile compounds in soy protein (Kitabatake and Doi, 1992).

Extrusion of whey protein isolate (WPI) at 35 or 50°C improves gel quality, but gel strength is lost when WPI is extruded at 75 or 100°C. However, these high temperatures have minimal impact on WPI's foaming and absorbability properties. Extruding WPI under acidic or alkaline conditions affects its functional properties significantly. Alkaline conditions greatly increase solubility and pasting properties, while acidic conditions enhance solubility but reduce pasting properties of WPI. The nutritional quality of protein is dependent on the quantity, digestibility, and availability of essential amino acids (Singh *et al.*, 2007).

At the molecular level, extrusion induces only minor changes in starch and protein components. The drying step primarily affects the protein fraction. High drying temperatures can preserve the protein network even after prolonged cooking. Higher drying temperatures lead to greater protein aggregation and insolubilization. Very high temperatures may promote the formation of protein aggregates stabilized by irreversible interactions, such as inter-peptide cross-linking or Maillard reactions. These structural modifications, especially at high drying temperatures, can strengthen the protein network and better protect starch from enzymatic degradation in cooked pasta. Variations in alveogram characteristics are mainly related to changes in gluten and starch, both of which influence the extrusion process. Increased treatment intensity leads to greater degradation of the gluten matrix (Gomez *et al.*, 2011).

Conclusion

Extrusion processing offers significant flexibility in producing low-cost, highly nutritious, and convenient food products. Texturized proteins have been successfully manufactured and marketed worldwide through extrusion technology. Key factors influencing the physical and chemical properties of the final products during extrusion include high barrel temperature, feed moisture, the composition of raw materials (primarily protein and starch), and screw speed. Additionally, extrusion reduces antinutritional factors present in the raw materials and enhances the digestibility of the end products. The extrusion process modifies ingredients such as protein, starch, salt, and fiber, and affects various process variables and product qualities, including surface appearance, texture, structure, expansion, and sensory characteristics.

References

Ainsworth, P. (2011). Extrusion. In Food Processing Handbook, 429–453. Wiley-VCH Verlag GmbH & Co. KGaA.

Areas JA (1992). Extrusion of food proteins. *Critical Reviews in Food Science and Nutrition*, 32: 365-392.

Barrett A (2003). Characterization of macrostructure in extruded products. In: Characterization of cereal and flours: properties, analysis and applications (G Kaletunc and Breslauer (Eds.), *CRC Press, Boca Raton, United States of America*, 369-386.

Berk, Z. (2009). Chapter 15 – Extrusion. In Food Process Engineering and Technology, 333–350. San Diego: Academic Press.

Caldwell, E. F.; Fast, R. B.; Ievolella, J.; Lauhoff, C.; Levine, H.; Miller, R. C.; Slade, L.; Strahm, B.S. Whalen, P. J. (2000). Unit operation and equipment. I. Blending and cooking. In: *Breakfast cereals and how they are made*, (2nd ed.) , R. B. Fast & E. F. Caldwell, (Eds.), pp.165-216, American Association of Cereal Chemists, ISBN 978- 189-1127-15-1, Saint Paul, United States of America.

Camire ME (2000). Chemical and nutritional changes in food during extrusion. In: Extruders in food applications, *CRC Press, Boca Raton, United States of America*, 127-147.

Chang, Y. K.; Hashimoto, J. M.; Moura-Alcioli, M. & Martínez-Bustos, F. (2001). Twin-screw extrusion of cassava starch and isolated soybean protein blends. *Molecular nutrition and food research*, Vol.45, No.4, (August 2001), pp.234-240, ISSN 1613-4133.

Cheftel JC, Kitagawa M and Queguiner C (1992). New protein texturization processes by extrusion cooking at high moisture levels. *Food Reviews International*, 8: 235-275.

Ding QB, Ainsworth P, Plunkett A, Tucker G and Marson H (2006). The effect of extrusion conditions on the functional and physical properties of wheat-based expanded snacks. *Journal of Food Engineering*, 73: 142-148.

Fellows, P.J. 2009. 15 – Extrusion. In Food Processing Technology, 3rd ed., 456–477. Woodhead Publishing.

Gomez M, Ruiz-Paris E and Oliete B (2011). Influence of wheat milling on low-hydration bread quality developed by sheeting rolls. *Food Science and Technology International*, 17: 257-265.

Harper JM (1981). Extrusion of Foods (Vol. 2, RJ Alexander and HF Zobel (Eds.)), *CRC Press, Boca Raton, Florida*, 37-64.

Kitabatake N and Doi E (1992). Denaturation and texturization of food protein by extrusion cooking, In: Food extrusion: Science and Technology (JL Kokini, CT Ho and MV Karwe (Eds.)), *Marcel Dekker, New York, United States of America*, 361- 371.

Launay, B., and J. Lisch. 1983. Twin-Screw Extrusion Cooking of Starches: Flow Behaviour of Starch Pastes, Expansion and Mechanical Properties of Extrudates. *Journal of Food Engineering*, 2(4): 259–280.

Martinez M, Oliete B and Gomez M (2013). Effect of the addition of extruded wheat flours on dough rheology and bread quality. *Journal of Cereal Science*, 30: 1- 6.

Mercier, C., P. Linko, and J.M. Harper. 1989. Extrusion Cooking. American Association of Cereal Chemists.

Petitot M, Abecassis J and Micarda V (2009). Structuring of pasta components during processing: impact on starch and protein digestibility and allergenicity. *Trends in Food Science and Technology*, 20: 521- 532.

Riaz, M.N. 2000. Extruders in Food Applications. Technomic Pub. Co.

Singh S, Gamlath S and Wakeling L (2007). Nutritional aspects of food extrusion: a review. *International Journal of Food Science and Technology*, 42: 916- 929.

Singh, S., S. Gamlath, and L. Wakeling. 2007. Nutritional Aspects of Food Extrusion: A Review. *International Journal of Food Science & Technology* 42(8): 916–929.

Sluimer P (2005). Principles of bread making: functionality of raw material and process steps, *American Association of Cereal Chemists, Saint Paul, United States of America*.

Thiebaud M, Dumay E and Cheftel JC (1996). Influence of process variables on the characteristics of a high moisture fish soy protein mix textured by extrusion cooking. *Journal of Food Science and Technology*, 29: 526-535.

Chapter 22

Biotechnological Interventions in Agriculture and Food Processing

Anjum Ayoub[*1] and Shahid Ahamad[2,] Mahital Jamwal*

[1]Division of Processing and Food Engineering, [2]Directorate of Research Sher-e-Kashmir University of Agricultural Sciences and Technology – Jammu Chatha -180009

**Corresponding author: anjumparay77@gmail.com*

Abstract

Technological advances and transformation have increasingly risen over the past few decades. Biotechnology developments were for instance used as a method for increasing the production of food. Specifically, advancement in genetic manipulation has made it possible to control crops to maximize production, meaning that the world's rising population has food resources. Many of the citizens of community, however, did not accept transgenic crops well and social gains and the future impact on human safety remain unknown. However, only industrialized countries embraced the gains of agricultural development, while those residing in emerging and underdeveloped countries endured from chronic hunger, starvation and deprivation. Policies will also be developed to insure that advancement in biotechnology is converted into improved agricultural practices capable of satisfying growing demand for food. The modernization of the agriculture sector, however, must take into account the importance of sustainable development in modern societies and an increasing desire for the use of so-called organic food, based on the notion that these foods are of higher quality and stimulate regional agricultural production. This study explores the role of biotechnology in farm production and the growth of the food industry throughout the course of history.

Keywords: Agriculture, growth, agriculture, organic food, genetically modified food, education.

1. Introduction

The rise of the world's human population has been a significant challenge to food health. Growth of populations will lead to an increase of 70% of food production by 2050 (Delaney, 2015), particularly in the developing economies, and a significant increase in productivity will become a priority for the coming decades. Biotechnology activities have centered in this respect on improving techniques capable of growing crop yields.

While application of biotech may be traced to 6000 BC in the 1970s, advancement of the fields of biotechnology brought a new impetus that allowed a number of new potential applications in genetic instruments and cellular and tissue engineering (Awais *et al.*, 2010).

The growing recognition of biotechnology as an economic and social growth factor has encouraged governments in many countries to provide financial support to their local biotechnology companies to promote research, production, and commercialization of ideas and products (Awais *et al.*, 2010);Furthermore, the predictions for hunger, drought and unprocessable diseases have been improved as well as enhancing the efficiency and service rendered by these firms through biotechnological progress (Boccia and Sarnachiaro, 2015; Dash *et al.*, 2016). The production of genetically modified (GM) crops for insect and weed management that result in increased crop yields is a significant example of biotechnological innovation (Dash *et al.*, 2016).

The attempts to reduce hunger and disparities, mainly in rural areas, have been inadequate given the huge biotechnological developments in agricultural products over recent 50 years. In different regions and social strata, food security is managed by different approaches (FAO, 2012) and people still suffer from starvation and diseases related to food safety in fully-developed countries. Taking into account the advances in technology, hunger is an unparalleled injury.

Quality and safety have influenced consumer behavior on food products over the last decade (Boccia and Sarnachiaro 2015). There are currently no acceptable population for highly processed foodstuffs with a high chemical content and there is a clear tendency towards fresh food consumption. Via genetic engineering of crops, biotechnology has developed a way of ensuring food quality and enhancing crop safety (Fernández-Suaréz, 2009).

Nevertheless, genetically engineered organisms were the focal point of controversy as there were conflicts of opinion on the effect on human wellbeing with the usage of such kinds of foods. This study discusses biotechnological developments at various historical times and their attempts to tackle food scarcity and related malnutrition. This study analyzes emerging developments in healthy eating and sustainable growth.

The purpose of this analysis is to debate the dispute over the usage of such biotechnological instruments to increase agricultural production.

Agricultural Biotechnological Development

Directly and indirectly, agriculture was the backbone of human food supply. Farmland is limited; however, as the world population has increased and has declined over time. Global agricultural production will also improve to satisfy the increasing demand for food (FAO , 2013).

Agriculture is regarded as one of the oldest human activities. At first, agriculture was practiced manually; later on, primitive plow-and-harrow technology allowed for increased productivity in agriculture.

In the years 1875 to 1885, the technological revolution contributed to rapid economic development that was a geopolitical driver for the increasing of countries (FAO, 2012), which contributed to the relocation of rural citizens to urban cities. The usage of automation in farms has been crucial, and substantial changes have been made to boost farm productivity and to satisfy the needs of customers and growers.

With the introduction of chemicals, crop protection from disease and higher yields were allowed around the same period. Nitrogen was one of the first chemicals used, but its application was restricted due to high prices.

Agriculture in England became a science with new experiments aimed at improving farm methods (Cubero, 1993), ultimately leading to significant innovations such as crop rotation and new findings such as the ability of some vegetables to convert atmospheric nitrogen into nitrates (Overton, 1996). Cotton Mather's 1716 report was one of the earliest records of plant hybrids. The following significant data was given by the quarter in 1776 for the production of plant hybridization. The postulates of Gregor Mendel on the passage of different features from one generation to the next were the start of new techniques for improving species of vegetables, and his paper "Experiments on Plant Hybridization" published in 1866 (Persley, 1991).

Yet hybridization of plants took place well before Mendel 's experiments; most possibly accidentally at first and in some step of the plant domestication cycle (Vasey, 1992). Indeed, for the next 100 years , many researchers increased the crosses for both floral morphology and hybrid vigour (Teranishi, 1978). In 1970, the genetics of plants made for more competitive plant species that were immune to some pests.

In 1960, a new movement changed agriculture: The Green Revolution which, through the introduction of crop species of high yield and modern agricultural techniques, was used to speed up food production , especially for underdeveloped and developing countries (FAO 2004).

In this era that usually consists of bioengineered seed, which operates with chemical fertilizers as well as high levels of irrigation, the Green Revolution has a huge impact on three main grains (maize, wheat and rice). The Green Revolution has been the result of a global food shortage which was threatening in the post-World war period.

The finding, which culminated in the genetic code breaking down in the 1960s, contributed to the transition of genetic information from one person to another through genetic engineering (Dash *et al.*, 2016). In the 1960s, the genetic code was breaking. The key influence in the development of GMOs(Table 1) was the convergence of genetic manipulation and biotechnology.

The International Maize and Wheat Improvement Center (CIMMYT), founded in Mexico in the 1960s, was one of the first agricultural research centres, and supported by the Rockefeller Foundation (FAO, 2000; Ceccon, 2008). Mexicans and foreign scientists partnered in this institute to develop genetic manipulation and achieve excellent results: in only 2 decades,The production of wheat grew from 750 kg ha-1 in 1950 to 3200 kg ha-1 in 1970. Today, CIMMYT research-produced wheat and maize is planted around the world on millions of hectares. One limitation of fundamental and applied agricultural research, however, is that it is normally carried out in developed countries with specific climate conditions, and novelties and improved crop yields can only be achieved under similar conditions.

For this cause, the latest technology and innovations needed to be adjusted to colder or tougher conditions in underdeveloped countries.

Table 1. Summary of the main events in the development of biotechnology

Period	Events	Implementation by
1664	Discovery of microorganisms	Robert Hooke and Antoni Van Leeuwenhoek
1884	Discovery of bacteria.	George Gaffky
1857	Microbiology of lactic fermentation.	Justus Von Liebig
1860	End of the spontaneous generation theory	Louis Pasteur
1866	Mendel's postulates of classical genetics.	Gregor Mendel
1875	Beginning of the use of machinery in farming	JAGDISH REDDY
1941	One gene-one enzyme hypothesis	George Wells Beadle
1946	Bacterial conjugation	Joshua Lederberg and Edward Tatum
1953	DNA structure.	James Watson
1959	Gene regulation.	Jacques Lucien Monod

Period	Events	Implementation by
1960	Green Revolution	Monkombu Sambasivan Swaminathan
1966	Genetic code decoding	Francis Crick
1970	High specificity of restriction enzymes.	Werner Arber, Hamilton O. Smith, and Daniel Nathans
1973	Recombinant DNA.	Herbert Boyer
1978	Human proinsulin gene isolation	Graeme Bell
1985	Polymerasechainreaction.	Mullis
1992	Beginning of the Golden Rice project	Peter Beyer
1996	Full-fledged commercialization of GM crops	Wandy Craig
2000	Computer simulation-based biology	Judith Rae Lumb
2004	Nature Biotechnology	Peter Brown
2008	Advance nature biotechnology	Timothy Denison,
2012	Waste water treatment	Kenneth S Watson
2016	The microbiome	Antonie van Leewenhoek
2018	Drought Tolerance	Boyer J. S.

The MON87701 soya vine, made with the brand name of Itanka and expressed crystalline insecticide CytAc1 from Bacillus thuringiensis (Bt toxin), was one of the first genetically modified plants in the world to guard against food-borne harm from certain lepidopteran pests (EFSA, 2011).Monsanto has also introduced MON89788, the gene for the soil bacterium cp4-epsps, which encodes the EPSPS enzyme, which gives the herbicide glyphosate resistance. Monsanto then designed a stacked soybean with an insecticide Bt toxin, Cry1Ac, to combine its resistance to glyphosate into a GMO, MON89788 x MON87701 (Then and Bauer-Panskus, 2017).

In the last few decades, the production of GM soybean has advanced rapidly. Currently genetic modifications have been effective in plantations of maize, tomatoes, rice, and cotton, particularly when exogenous (transgene) genes are introduced, which covers a protein that is toxic to different pests. (Sanahuja, 2011; Tabashnik *et al.*, 2011).

It was noticed that some bio fertilizer-like behaviors of the agricultural sector have been demonstrated by some micro-organisms including plant growth-promoting microbes, fungi and cyanobacteria (Mahanty et al . 2016). Microbes are actually to a degree thought to be a form of enhancing crop production, by means of bio fertilizer alternatives. The use of nanotechnology can promote agriculture and is becoming one of the most important instruments in modern farming. Nanotechnology provides new agrochemicals, new delivery mechanisms and promises to decrease pesticide use through agrochemical Nano formulation. The use of nana-sensors / Nano bio sensors in crop protection is also possible with nanotechnology (Bhupinder, 2014).

Food Biotechnology Engineering

For years, the biotechnology sector has been used in the food industry by additive and component processing and the advancement of more effective and cheaper food manufacturing operations. Moreover, biotechnological developments centered on the alteration or enhancement of the flavor, scent, shelf life, shape and food products' nutritional value, oxidation, enzyme processing, nanotechnology and molecular biology.

Technology was considered to provide a way to address the issues connected with the worldwide food supply. But it didn't happen. However, the great advances which have been produced by the discovery of micro-organisms and evolving biological , biochemical and molecular techniques that have provided for the development of the fields of science are worth noting (Table 1).

The microbiology of lactic fermentation and the function of yeast in alcoholic fermentoes were discovered by Luis Pasteur in 1857 and 1860. The most common method to acquire food remains such fermentation (Madigan, 2007). These procedures, such as cocoa fermentation, may be done naturally. Foods such as yogurt and cheese may be developed utilizing inocula.

The production of primary metabolites (alcohol and organic acids) and of secondary metabolites (enzymes and Polysaccharides) can be fermented in previously developed crop mediums, such as in beer processing and development.

Fermentation is a extremely effective food production method at present. The sustainability of rural and urban citizens has been significantly influenced by fermented foods coming from varied sources such as milk, rice, bananas, vegetables and meat.

The fermentation species are also subject to research. For example, the task of successful cell manufacturers for the development of functional biomolecules and food ingredients in improving the quality of grain-based beverages can be fulfilled by lactic acid bacteria (Waters et al . 2015).

Enzyme technology became a method used in the area of food biotechnology as it helped to automate and speed up bioprocesses. Enzyme as biocatalysts play a crucial role in biotransformation processes in various food processing fields, such as baking, juice production, starch, dairy and other similar industries.

For the food sector a wide variety of enzymes are used. For instance, proteases work on wheat flour protein in bakery, rising the elasticity of gluten and, subsequently, decreasing dough shrinkage.

Rennet has been used in the dairy industry for the processing of cheese since ancient times. The essential milk use is the use of lactase for the processing of low-lactose milk. The use of enzymes such as cellulose and pectinases in the processing of fruit juice will increase the consistency of the juice. Papain and bromeline are both proteases that are used as softeners for foods.

Conventional methods for food processing are well known, but sometimes they contribute to lower nutritional value (Nakitto *et al.*, 2015). Biofortification is the usage of biotechnology to improve with additional supplementations and modern methods have been used for biofortification, such as bread and milk injection of vitamin D to resolve deficiencies.

A second landmark in the advancement of biotechnology has been the emergence of classical genetics with the research performed by Mendel. Nectarines, pear flavored apples and other synthetic forms have been developed and customers are embracing such goods.

The progress of biotechnology has been driven by the discovery of the genetic manipulation findings in the mid-1920s. GM crops are made from genetically engineered species. Transgenic plant materials are typically GM foods. The first industrial GM milk, FlavrSavr, was updated by California's Calgenecorporation to mature without softening (Bagwan *et al.*, 2010).

Consumers who bought the fruit with a significant premium on standard tomatoes were welcome to do so. Nevertheless, manufacturing problems and rivalry with a historically produced variety of long existence impeded the productivity of the drug.Purdue University and the US have grown a cancer-fighting tomato containing 3 times as much lycopene (an antioxidant) as conventional varieties. Agricultural Study Program (Awais *et al.*, 2010) of the Department of Agriculture. Lycopene is believed to bind toxic molecules that damage human body tissue, and may reduce the likelihood of cancer of the breast and prostate.

The enhancement of protein, vitamin A and vitamin E as well as iron and zinc components through gene insertion is also essential in genetic engineering. The scheme for Golden Rice began in 1992. Golden rice is a type of Oryza sativa and is engineered to produce the endosperm, which is an edible component of rice, beta-carotene, a precursor of vitamin A.

It was produced with the collaboration of Peter Bayer from the University of Freiburg by Ingo Potrykus from the Institute of Plant Sciences of the Swiss Federal Institute of Technology (Bagwan *et al.*, 2010). Golden Rice has been established as a fortified food to be used in places where dietary vitamin A is missing.The 2005 launch of a new variety named Golden Rice 2. Golden Rice 2 is 23 times stronger than the original edition, containing beta-carotene.

Bt maize is a GM maize variant which represents the European borer Bt bacterial toxin. Moreover, a variety of herbal GM foods (glyphosate) and/or insect-resistant (using the Bt toxin) are included in these foods including crops, such as soya, canola, sweet maize and sugar beet (Bagwan *et al.*, 2010).

Bt maize is a genetically modified maize that expresses bacterial Bt toxin to European cornea borer. There are, however, many foods developed by genetically modified crops that are herbicide (glyphosate), and/or insect-resistant (Bt toxin), including crops including soybean, canola, sweet maize or sugar beet (Bagwan et al . , 2010).

Nevertheless, the use of transgenic products remains correlated with the lack of awareness of their environmental and human-health consequences given the various biotechnological achievements. (Sarnacchiaro and Boccia, 2015).

Biological and Food Safety Problem

Biomechanics are loosely described as "a technique that uses live or animal organisms to generate or change a drug, enhance plants or animals, or grow micro-organisms for a specific application" (Persley, 2000). (Persley, 2000). It also deals with the construction by recombinant DNA techniques, culture of tissue, transfer of embryo and other methods of useful microorganisms, cells, plants or animals aside from traditional techniques of genetic breeding.

While biotechnology covers a number of fields, farm biotechnology seems to be most important for African countries and, in particular, for farmers with a poor population whose sole livelihood depends on farming. The biotechnology methodology alone cannot address all issues relating to agriculture, but it has the potential to tackle particular problems such as increasing seed sustainability, crop diversification, improving food's nutritional quality, growing the environmental effects of farm development and promoting competition in the industry.

In several areas of the developing world, crop yields have risen slowest, in particular in Africa. Cereal yields in Africa are estimated to have risen by almost half of the Latin American rate since 1970 (World Bank, 1993). Food safety in many parts of Africa continues to be undermined by poor soil, low rainfall, high temperatures and the prevalence of pests. The high costs of imported agricultural inputs compound these challenges. Better investment analysis and dependence on new technology would be required to boost the situation.

Another important aspect of food security is to enhance the nutritional value of crops. The modification of rice to improve its vitamin A content is a positive example in this area. The UN projections reveal that the numbers for Africa will increase dramatically, although chronic malnutrition will decrease in Asia and Latin America in the next decades. Biotech makes it easier, while enhancing the nutritional value, to maintain traditional diets.

Modern biotechnology could help to boost the competitiveness and integration of agricultural products of developing countries in the global economy. In addition to promoting food food safety in those areas as well as adding new crops to the world food market, efforts to diversify agricultural production in the developing world.

Biotechnology and Agriculture Sustainables

Sustainable agriculture, as laid down in the 1990 'Farm Bill,' implies the 'integrated program of plant and animal development activities with a site-specific implementation that will satisfy human food and fiber needs, (ii) improve environmental sustainability and the foundation of natural capital on which agricultural economies rely, (iii), exploit r-non-renovable energy as efficiently as practicable.

The cluster of biotechnology techniques can transform farms by reducing post-harvest loss and enhancing crop resistance to drought, if effectively harnessed and deployed. The key restricting factor for developing countries to profit from developments in contemporary biotechnology is their lack of science and technology and their weak entrepreneurial growth. Such countries are responsible for the implementation of policies and plans for the greater usage of biotechnology.

To order to encourage sustainable agriculture to the developing world, international collaboration and collaborations are therefore necessary. African discourse on biotechnology and sustainable farming poses a number of core questions:

- How are biotechnology transferred and enhanced technology skills in African countries to gather, assimilate, develop and implement improved agricultural production technologies efficiently?
- What policy and institutional arrangements should be established to allow rural farmers in the area to access the technology and its products?

Biotechnology alone might not be the panacea for the global food shortage issue. However, genetic engineering offers an outstanding potential to increase the efficiency and sustainability of the improvement of the crop as well as animal production.

This can be achieved once the whole technology is integrated into the traditional farming systems of small farmers. Sustainable agriculture would allow developing countries to choose prudently and not necessarily to use today's available technologies. For these choices, access to a wider range of technologies , in particular those arising from molecular biology advances, is needed.

In ensuring that access to potential benefits of new technology by poor people and environment conservation is ensured, the international public research system has a critical role. The need for greater public interest in biotechnology and to support private sector work needs to be recognized, openness and accountability must be ensured and a large variety of public goods testing encouraged, even as consumers are that for outcomes of private goods science. All concerned stakeholders need win-win scenarios and creative efforts to identify and implement mechanisms enabling developing countries to benefit from the gene revolution.

In order to win the battle for food security and sustained agriculture, enhanced crop productivity should be given priority, giving the majority of the population easy access to food.

Potential of Biotechnology in Crop Improvement

The emphasis must be placed on the fact that biotechnology should not replace traditional tools for improving crops, but the integration of recombinant technology into conventional breeding programmes, which can significantly improve the efficiency of agricultural research and development. While conventional crucifixion is restricted to the sharing of genetic material throughout some plant organisms, recombinant techniques allow useful genes to be transmitted between plants, even kingdoms. Traditional methods of breeding plants are limited by the limit of sexual incompatibility.

Moreover, plants do not present many valuable features such as herbicide tolerance and insect tolerance. The new techniques of DNA manipulation are practical for overcoming the constraints of plant breeders. Genes can be isolated from and expressed in plants by bacteria , viruses , fungi, or animals. For the future agricultural needs, the following fields of research are especially important:

- Susceptibility to particular diseases and rising plagues and mosquitoes.
- Those reducing fertilizers' environmental burden,
- Those who reduce the irrigation water requirement and
- Those that continue to improve crop production per hectare.

The key goals of transgenic breeding would therefore be to achieve:

Agronomic Traits

The agronomic traits are input traits that cover all genetically modified plants which are capable of stabilizing or improving farmers' outputs. Mechanisms that are usually just a single gene (monogenic traits) are prevalent input characteristics in plagues and disease tolerance. Various transgenic plagues and resistance to diseases have already been marketed.

In evaluating its potential value, the global cultivation loss due to biotic stress factors must be taken into account at a level of 25 – 30 per cent (Oerke *et al.*, 1994). Without the need for pesticide applications, biotechnology could significantly reduce these losses. Other desirable agricultural crop characteristics include enhanced genetic output potential and mechanisms for tolerance of abiotics, such as drought, cold and soil nutrient deficiencies.

Different genes (polygenic traits) also decide certain characteristics. Molecular mapping and functional genomics have recently shown that similar biotech technologies are very feasible in the close to medium term (Abelson and Hines, 1999). This is a very exciting development. Improved plant varieties could thus be adapted to the marginal agro-ecological regions that the green revolution largely neglected.

Quality Traits

The definition of quality marks here applies to the production characteristics of the crop commodity, such as appearance and chemical composition. It also includes enhanced macro and micro nutrient densities, which are essential for healthy human diets. Such features can benefit particularly the low revenue population, which frequently lack buying power to provide higher-value and more nutritious foods, by incorporating them into staple food crops. For example, transgenic rice varieties with greatly improved vitamin A content have been produced by researchers and are now used in rice breeding programmes.A number of additional important vitamins and minerals have also reported promising advances in biotechnological research to improve micronutrient density in plants (Potrykus *et al.*, 1999).

There have been limited human , financial and infrastructural capacity, poorly defined or undefined institutional arrangements for research and development in biotechnology and poor / poor biotechnology policy as a key constraint on the ability of biotechnology to create food security in Africa.

Risks of Biotechnology

In addition to the significant biotechnological potential for improved food output and farm profitability, the threats must not be overlooked. Often there were fears that the dire manipulation of organisms might give rise to a new dimension to risk to the environment and human health. Since human knowledge is limited, there can be no absolute confidence in the existence of unknown risks for both genetically modified crops and any other technology.

There is no evidence to suggest that GM crops are more risky than traditionally breeded species (Cooks, 2000). According to current scientific knowledge. This does not mean there is no risk whatsoever. The expected risks are not, however, correlated with the biotechnological method, but are linked to the end result. Therefore, risk evaluation studies for each individual technology product were carried on a case-by - case basis.The possible loss of biodiversity, adverse impacts on natural food chains and the development of aggressive pathogen populations are environmental risk issues that need to be addressed. Safety danger involves possible incidence of unintended harmful by-products in foods, transfer of antibiotic resistances (used as marker genes) to human digestive microorganisms and unexplained food consumers' allergic reactions.

In general, both the developed and developed countries are affected by the individual risk aspects. However, it is important to note that in the developing world, there is a greater chance of transgenes escaping to the wild via cross-pollination. In tropics and subtropics, the main center of agricultural biodiversity is found (Virchow, 1999), and in the vicinity of the farmland, more wild relative to domesticated species are grown. If transgenes contain characteristics that provide a major competitive edge, any natural vegetation is likely to be lost.

Sustained agricultural development requires a responsible management of biotechnology and it calls for effective bio-safety and food safety regulation wherever transgenic crops are developed and released. However, responsible management of technology must not be confined solely to the risk side.Risks should always be juxtaposed with potential benefits and some of the residual risks are considerably more likely to be admissible. It should be kept in mind that biotechnology may also have a substantial environmental positive effect, such as lower concentrations of traditional pesticides or minimize extension of agricultural areas into an environmental vulnerable climate.

In addition to the above-mentioned risk to health and the environment of transgenic crops, there is also a socioeconomic concern. For instance, biotechnology technologies may, to the disadvantage of many developed countries, alt foreign trading flows; trade in agricultural raw products can be substituted by tropical plants or their goods generated in the laboratory, as for vanilla and other useful substances is anticipated. In this regard, innovation can worsen existing income disparities if biotechnology research and development benefit the rich only while neglecting the need of the poor. Such risks must not be underestimated, but they are not biotechnologies-inherent. Any new technology of far-reaching importance can have unwanted equity impacts under adverse social and economic conditions.

Policy Challenges for Developing Countries

- Specific goals for innovation in biotechnology are desperately required. That requires the countries concerned to determine certain areas or technological paths in which to invest in order to achieve specified objectives and make optimum use of available skills and resources.
- The accessibility of funds for research and development of biotechnology is indeed an important field of need. There is not enough current investment in this field. This could be done by forging strategic partnerships with the private sector to ensure that the public good of the use of food is not affected by motivation for profit.
- As a pre-condition for sustainable agricultural production, responsible management of biotechnology should be established and effective regulations on biological health and food protection where transgenic crops are produced and released should be created.
- The role and impact of intellectual property rights on the acquisition, development and dissemination of biotechnology should be taken into consideration in developing countries too.

Conclusion

One of the most important advances in contemporary science is modern agricultural biotechnology. It can increase crop productivity , enhance pest and disease resistance, develop tolerance for adverse weather conditions, improve the dietary values of some foods and improve the durability of the product during harvest or shipping. Used in collaboration with traditional and conventional breeding methods.This can be made available for small farmers with little or no risk to human health and the environment by means of reasonable biosafety regulations and appropriate policies. Therefore, we should not ignore any part of the possible solution, including agricultural biotechnology, in a world where thousands of children die from inaction.

References

Andreassen M., E. Rocca, T. Bøhn, O.G. Wikmark, J. van den Berg, M. Løvik, T. Traavik, U.C. Nygaard. 2014. Humoral and cellular immune responses in mice after airway administration of Bacillus thuringiensis Cry1Ab and MON810 Cry1Ab-transgenic maize. Food AgricImmunol 26(4):521–537.

Awais, M., A. Pervez, A. Yaqub, R. Sarwar, F. Alam, and S. Siraj. 2010. Current status of biotechnology in health. American Eurasian J. Agric. & Environ. Sci 7(2):210–220.

Bagwan, J.D., S.J. Patil, A.S. Mane, V.V. Kadam, and S. Vichare. 2010. Genetically modified crops: Food of the future. Intl. J of Adv Biotech and Research.1(1):21–30.

BaltáArandes, A., J. BaróBasora, and V. Blanco Saiz. 2013. Alimentostrangenicos:larealidad no siempresupera la ficción. Universidad dé Barcelona, España.

Bhupinder, S.S. 2014. Nanotechnology in Agri-food production: an overview. Nanotechnology, Science and Applications. 7:31-53.

Boccia, F., and P. Sarnacchiaro.2015. Genetically modified foods and consumer perspective. Recent Pat Food Nutr Agric.7:28–34.

Carman, J.A., H.R. Vlieger, L.J. VerSteeg, V.E. Sneller, Robinson Garth W., Clinch-Jones, C.A; Haynes, Julie I; Edwards, John W. 2013. A longterm toxicology study on pigs fed a combined genetically modified (GM) soy and GM maize diet. J OrgSyst 8(1):38–54.

Chamas, A. 2000. Alimentostransgénicos.Invenio3(4–5):149–159.

Dash, A., D.Kundu, M. Das, D. Bose, S. Adak, and R. Banerjee. 2016.Food biotechnology: A step towards improving nutritional quality of food for asian countries. Recent Pat Biotechnol 10:43–57.

Ewen, S.W., and A. Pusztai. 1999. Effects of diets containing genetically modified potatoes expressing Galanthusnivalislectin on rat small intestine. Lancet 354:1314–1315.

EFSA - European Food Safety Authority. 2011. Scientific opinion on the application (EFSA- GMO-BE-2010-79) for the placing on themarket of insect resistant genetically modified soybean MON 87701 for food and feed uses, import and processing under Regulation (EC) No 1829/2003 from Monsanto. EFSA Journal 9(7):2309.

EFSA - European Food Safety Authority. 2012.Guidance on the risk assessment of food and feed from genetically modified animals and on animal health and welfare aspects. EFSA Panels on Genetically Modified Organisms (GMO) and Animal Health and Welfare (AHAW). EFSA Journal 10(1):2501.

FAO (1999). The state of food insecurity in the world.Food and Agriculture Organisation of the United Nations, Rome.

FACTA (1990). "Food, Agriculture, Conservation and Trade Act of 1990. Public Law 101-624, Title XVI, Subtitle A, Section 1603. The National Agricultural Law Center. http://www.nationalaglawcenter.org/assets/farmbills/1990-6.pdf.

FAO - Food and Agriculture Organizations of the United Nations. 2000. El Estado Mundial de la Agricultura y la alimentación. La alimentación y agricultura en el mundo: enseñanza de los 50 últimosaños. Roma.329 p.

Hines PJ (1999). Abelson PH. The revolution of the plants.Bordeaux, 295; 367-275.RJ (2000). Cook R.J. Scientific risk assessment in agricultural and other environments for the approval and use of plants In: persley GJ, Lantin MM (eds): agri-based and poor biotechnology. International Agricultural Research Consultative Group, Washington, D.C. 123-130. Pages 123 to 130.

McCalla AF (1999). Prospects for food security in the 21st century with special emphasis on Africa. In: Agric. Econ., 20(2): 95-103.

Oerke EC, Dehne HW, Schonbeck F, Weber A (1994). Crop production and crop protection: Estimated losses in major food and cash crops. Elsevier, Amsterdam.

Persley GJ (2000). Agricultural Biotechnology and the Poor.Promethean Science. In: Perseley GJ, Lantin MM (eds): Agricultural Biotechnology and the Poor. Consultative Group on International Agricultural Research, Washington D.C. pp. 3-21.

Pinstrup-Andersen P, Pandya-Lorch R, Rosegrant M (1999). World food prospect: critical issues for the early twenty-first century. 2020 Vision, Food Policy Report, International Food Policy Research Institute.

PRB (2006). "Population Reference Bureau".2006 World Population Data Sheet.Economic Information Bulletin Number 18. Washington, DC: USDA Economic Research Service;August2006. http://www.ers.usda.gov/publications/eib18/eib18.pdf.

Potrykus I, Lucca P, Xudong E, Al-Babilli S, Hurrel RH, Beyer P (1999). Contribution to food security by genetic engineering with rice.paper presented in the general meeting of the international programme on rice biotechnology, Rockefeller Foundation, Phuket Thailand.

United Nations Administrative Committee on Coordination SubCommittee on Nutrition (2000). Fourth report on the world nutrition situation Geneva, Switzerland: ACC/SCN in collaboration with the International Food Policy Research Institute.

Virchow D (1999). Conservation of genetic resources; Costs and implications for a sustainable utilization of plant genetic resources for food and agriculture. Springer, Berlin.

World Bank (1993). World development report: investing in health. Washington DC, Oxford University Press, as Cited in United Nations Children's Fund. The State of the World's Children. New York, NY: Oxford University Press.

Chapter 23

Frontier Technologies for Sustainable and Profitable Vegetable Production: A New Paradigm

Anil Bhushan, Sonali Sharma Sandeep Chopra and B. Singh

Division of Vegetable Science, FoHF, Main Campus Chatha SKUAST-Jammu (J&K) 180009
Email: anilbhushan2008@gmail.com

Introduction

Agriculture is a crucial sector for Indian economy to grow sustainably. The idea of an abundance of grains, roots, and tubers—the basic crops that offer accessible amounts of nutritional energy has long been linked to food security. But as the idea of food security and nutrition security have merged, and as the value of dietary diversity for optimum health has emerged, this picture is shifting (Schreniemachers *et al.*, 2018). Fruits and vegetables are the main components of foods high in nutrition. Among these, vegetables could be a preferable choice among them due to their short growing season, less area requirement and ease of availability for combating nutritional issues (Bhallan *et al.*,2018). However, the vegetable production falls short of the demand if balanced diet is provided to every individual. For the responsible regulatory organisations and the production sectors that are focused on the market, maintaining a steady supply of vegetables through continuous production to satisfy the always shifting market demand is a significant problem. The necessity for ongoing technological advancements that are responsive to dynamic market conditions is felt in the industrial sector and value-chain management (Weerakkody and Mawalagedera, 2020). Hence, bridging the gap between the demand and supply must involve novel technologies that would resolve the current constraints in production, postharvest, and marketing of vegetables. Vegetable growing could be promoted to every single home of the country through research and extension programmes. Even these could be made available to those who have less or no farm areas. Sustainability and profitability could be achieved using modern high-tech techniques such as precision farming, nanotechnology, soil-less cultivation (hydroponics, aeroponics, aquaponics), vertical gardening, etc. In addition, the government is promoting programmes to set up vegetable gardens in schools to give students a healthy diet. Small booklets or pamphlets on cultivation practices in soil-less cultivation, nutritional value and medicinal value of every single vegetable must be introduced to every home as well as school for the betterment of nation. It will also guide the farmers to adopt hi-tech horticulture strategies for raising their income through increased productivity and low cost of production. Worldwide, the vegetable industry is now in a transition period, changing from expanding quantity to increasing quality and efficiency. This is being achieved by the introduction of innovative production technology and improvement of market competitiveness. Consumers in developing countries have been able to purchase a broader range of relatively expensive vegetable commodities such as off-season produce, new or exotic vegetables, and organic produce due to rapid growth in mean per capita incomes of the emerging educated middle class, desire for variety, and awareness of nutritional benefits, resulting in the expansion of the consumption and domestic vegetable markets, particularly in urban and peri-urban areas. The frontier technologies improve crops, reduce production costs and protect the environment and can be incorporated into traditional farming to improve production practices (Ampatzidis, 2020). In the last few decades, several "smart" technologies have been developed for vegetable production and processing. However, growers are confronted with a variety of challenges when considering adopting new technology or adjusting existing technology. Growers are being offered solutions that might not work in their specific production system or might not be economically feasible. Thus, it is clear that technological innovations are vital in local vegetable cropping to increase yield productivity, production efficiency, food quality, and food safety. This chapter presents examples of state-of-the-art technologies that may be used in vegetable production today or in the near future.

Brief Scenario of Vegetable Production in Indian Context

India's diverse climate ensures the availability of all varieties of fresh vegetables. Vegetable production activity is booming in many regions with mild climates, favourable to vegetable production since it is usually more lucrative compared to field crops, have high added value and high income generation potential (Dias, 2011). India has witnessed increased vegetable production during the past few years. Area under vegetable cultivation, production and productivity have all advanced significantly. The area under vegetable cultivation has increased from 5593 thousand hectare in 1991-92 to 11065.06 thousand hectare in 2021-22. The production in India has increased manifolds from less than 20 million tonnes in 1947 to 199882.49 million tonnes in 2021-22 due to gradual increase in productivity and area under vegetable cultivation over the years, which maintains its 2nd ranking in global vegetable production after China (Anonymous, 2021). This has led to availability of 357 grams/person/day of vegetables in the country. India has first rank in pea and okra production, while it ranks 2nd in tomato, cauliflower, potato, onion and brinjal production (Mohan *et al.*, 2021). Onion production is estimated to be 31.70 Million Tonne against 26.64 Million Tonne in 2020-21. Potato production is expected to be 53.58 Million Tonne, compared to 56.17 Million Tonne in 2020-21. Tomato production is expected to be 20.34 Million Tonne, compared to 21.18 Million Tonne in 2020-21. The vast production base offers India tremendous opportunities for export. During 2021-22, India exported fresh vegetables worth Rs. 5745.54 crores/ 767.01 USD Millions. By 2025, the demand for vegetables will be 225 million tonnes and 350 million tonnes by 2050. There is an urgent need to increase vegetable production in India in order to ensure food and nutritional security.

What is the Need of New Technological Interventions?

Ever-increasing population and shrinking land holdings are the major problems need to be addressed. Birth rates are almost double than the death rates which means it is difficult to stop the growing population. Further, area is also limited and is further shrinking for agricultural purposes due to urbanization. Vegetables play a key role in food and nutrition security as the country's food system shifts from food quantity to diet quality and health benefits. Traditional locally available underutilized vegetable crops as well as exotic vegetable crops could be utilized to improve nutrition and health. Most vegetable production is carried out by smallholder farmers who use basic farm implements on their farms. Production takes place predominantly on raised beds, ridges, or bare ground in open fields. Nursery establishment for those vegetables that do not need to be planted at stake is mostly on nursery beds created close to the site of cultivation. The vegetables are often cultivated on a piece of land owned or rented by these farmers as is mostly the case. Manual labor is used for most farm activities by the farmer with the help of family members or hired farm hands. For example, weeding is by use of hoe, cutlass, and/or spraying of chemicals using knapsack sprayers. Vegetable growers face a variety of challenges, including pests and diseases, labor shortages, and climate change. One of the strategies for promoting vegetable production is the development and adoption of innovative and modern technologies to address major challenges impeding the advancement of vegetable production However, in recent times some farmers who have larger farms are incorporating some level of mechanization such as the use of tractor-drawn tools (plough and harrow) for land preparation. Nevertheless, most of the other practices such as seeding, weeding, and harvesting are still carried out manually (Osei *et al.*, 2021). The demand for vegetable crops is continuously increasing. Sustaining this demand requires the infusion of technologies for efficient utilization of sources for deriving higher result per product of inputs with a superb quality of produce . It will be possible only through the deployment of modern technologies for enhanced sustainability and profitability.

Various Frontier Technologies which can be Used for Increasing Vegetable Production on Sustainable Basis are

1. Precision farming

Precision vegetable farming is a contemporary practise that makes use of the most recent technologies, including cloud computing, information technology (IoT), wireless sensor networks (WSN), artificial intelligence (AI), and machine learning, to increase crop output (ML).The majority of research points to the positive effects PA-based techniques have on productivity and sustainability. Prescription maps play a crucial role in enabling farm owners to quantify the information required for useful plants at any specific development stage since the primary objective of precision agriculture is producing surplus yield by optimising the materials, such as water, pesticide sprays, plant foods, etc., for source (Kaushik, 2021). Following are some aspects of how precision farming technology is being used in many sectors of vegetable production:

I. Simplify Surveying

Field surveys for disease/pest scouting and to assess plant stress are expensive, labour intensive and time consuming. Since labour shortage is a major issue in vegetable production, small unmanned aerial vehicles (UAVs) equipped with various sensors (remote sensing) can simplify surveying procedures, reduce the labor cost, decrease data collection time and produce critical and practical information. For example, recently UAVs and remote sensing have allowed growers to constantly monitor crop health status, estimate plant water needs and even detect diseases. The precision agriculture team (@PrecAgSWFREC) at the University of Florida Institute of Food and Agricultural Sciences (UF/IFAS) Southwest Florida Research and Education Center (SWFREC) developed a cloud-based application called Agroview. Agroview can process, analyze and visualize data collected from UAVs and other aerial platforms (e.g., small planes and satellites). This technology utilizes machine learning (an application of artificial intelligence) to detect single plants and assess plant size and stress (Ampatzidis, 2020). Agroview and UAVs were initially used to create plant inventories in citrus and to detect specific diseases in vegetables with high accuracy. Early detection and eradication of infected plants is crucial to controlling disease and pest spread throughout the field.

II. Smart Sprayers

Majority of conventional sprayers apply agrochemicals uniformly, despite the fact that pests and diseases are patchily distributed, resulting in wastage of valuable compounds, increased costs, crop damage risk, pest resistance to chemicals, environmental pollution and contamination of products. Contamination can be related to run-off after application, discharge from drainage and off-target deposition of spray due to wind (spray drift). This pollution can be significantly reduced by improving spraying techniques. Each application of an agricultural pesticide results in spray drift, which can result in a loss of up to 50% of the chemical. A major global challenge is reducing the harmful effects of agrochemicals (including spraying technology). Several smart technologies have been created in recent years for insect detection and optimising spraying procedures. By incorporating the most recent developments in electronics, artificial intelligence (AI), and automation, these new spraying technologies have demonstrated a significant improvement in efficiency and safety. One example is the See & Spray machine developed by Blue River Technology for weed control in arable crops (Ampatzidis, 2020). See & Spray applies pesticide just to the weeds after using computer vision and AI to recognise and identify specific plants and weeds. Compared to conventional broadcast sprayers, this machine can reduce the amount of herbicide used by more than 90%. The UF/IFAS team has created yet another low-cost smart sprayer for targeted weed control in vegetables. In the early evaluation studies, intelligent technology was capable of accurately detecting and differentiating weeds from crops and applying herbicides only to specific weeds, thereby avoiding crops and areas without weeds. This innovative robotic sprayer was created to be easily deployable in the field and to be self-reconfigurable and self-adjustable. This architecture allows the robot to adapt to managing weeds in various vegetable fields (e.g., with different row spacing and raised bed sizes).

III. Robotic Harvesting

Rapid population increase has put a great deal of strain on the agricultural industry. A paradigm shift from conventional to smart agriculture systems, which increase precision and production through sensor technologies, various types of technology, and robotics, has been taking place this decade as a result of technological improvements.As a result, important advances have been made in the planting, weeding, harvesting, diagnosis of plant diseases, evaluation of vegetable damage and defects, grading of their quality, identification of insects and pests, etc. Harvesting is a crucial stage in gathering crops from the fields, and robots are helping to add precision, efficiency, and effectiveness to this process. With the use of robots, some typical vegetable crops, including tomatoes and sweet peppers, can be harvested (Islam *et al.*, 2022). Fresh market produce is quickly perishable, and almost all of it is hand-harvested. The cost of harvesting increases as a result of the manpower shortage that vegetable growers are experiencing. For vegetable growers, mechanical and robotic harvesting methods could simultaneously reduce their reliance on manual labour, cut harvesting costs, and boost market competitiveness overall.

IV. Role of global positioning system (GPS) in precision vegetable farming

The most important element in precision vegetable gardening is the GPS component. GPS's acceptance is due to the higher, dependable precision with which position, time, and direction can be identified, in addition to its ease of use and global all-weather process. Understanding the most important action to do at the right time and place may need the use of various high-tech tools and intricate statistics or other analysis. Making the perfect point, however, begins with superb supervisors and new operators using common resources as planters, fertiliser applicators, and harvesters regardless of what else is required. In this situation, GPS will be an integral part of agricultural perfection. Applications for GPS-based precision farming have been utilised for yield mapping, soil sampling, tractor steering, harvest scouting, and farm preparation. GPS permits farm owners to function during low visibility field problems, for instance, rainfall, particles, fog, and darkness. Growers can achieve additional benefits by combining better utilization of soil amendments and fertilizers, identifying the economic threshold for combating pest and weed infestations and preserving the organic energy sources for future use. Many growers now use GPS-derived products to streamline operations within their farming businesses. GPS devices collect location data to map region boundaries, highways, irrigation systems, and plant disease or weed infestation hotspots (Addicott, 2020). By using GPS, farmers are able to return year after year to certain locations in the region to gather soil samples or even check on crop conditions. . Growing companies and agricultural service providers should anticipate more adjustments as GPS technology is updated frequently.

V. Role of sensor technologies in precision vegetable farming

Optical and thermal sensors will be the most preferred sources in plant phenotyping. The classification of sensors is based on the amount and wavelength selection of measured wavelengths. Even though certain indices will probably be steadier when estimated in broader bands, the measurement reliability decreases the wider the band is examined around a particular wavelength (Yuan, 2019). Due to the the reconstruction of expanding constructions in 3D, the employment of assessment and stereo camera rigs by computer programmes of images taken from various angulations enables drawing complex versions. However, sensors that rely on reactance measurement can offer much more useful data to create physiological research. By evaluating leaf reactance at various wavelengths, multispectral imaging provides information needed to calculate crucial vegetation indices. Numerous scientific research suggest hyperspectral sensing tools for the early detection of various symptoms related to biotic stressors in the Solanaceae family. These investigations show how hyperspectral objects can be used to choose quality-related criteria and plant diseases, increasing synthetic horticultural compounds across the supply chain. Spectrophotometry was also used to measure the content of chlorophyll. These two occurrences are connected to the moderate penetration and development of free water in the tissue. (Tripodi *et al.*, 2018). Smallholder farmers, numerous other stakeholders, and extension workers would benefit from updated knowledge about their plant life through data collected by sensors, which will increase productivity. Improved data and information will optimise the time and resources used by farming extension staff and other players on the ground, ultimately resulting in great electrical power usage and cost savings. Lastly, information is required to enable the private sector and many other intermediaries to make decisions that could result in smallholder farmers receiving applications for financial and insurance support.

VI. Role of geographic information system (GIS) in precision vegetable farming

GIS is a key component of automated field operations, known as precision agriculture or possibly satellite farming. GIS has a hardware-software repository unit to collect, store, enter, influence, analyze, screen, map as sort, and spatially referenced geographical information. GIS maps are energetic. Farmers have improved their decision making ability to control growing of crops in order to increase their yields. This information is gathered via distant sensors and sensors put on farm equipment. To make information consumption easier, advanced GIS technologies provide image analysis and picture inspection. Furthermore, the mobile version will allow farmers to keep informed about all their plots through information on soil analysis, the crop's problem, and the managing of evapo-transpiration (Burrough *et al.*, 2015). On the computer display screen, map computer users can check out a GIS chart in almost any program, zoom in or out, and modify the characteristics of the information discovered within the chart. The capability of GIS to evaluate and imagine farming environments and work has shown to be genuinely beneficial to each of those connected with the agriculture business. With this knowledge, they may create and put into practise control strategies that guarantee the ideal utilisation of inputs to raise output and profits. Because of this, geospatial technologies give farmers access to information resources that they can use to make informed decisions that will ensure effective and efficient management of the farm and increase its productivity. Therefore, in order to maximise the benefits of their farms, farmers must comprehend and put these methods into practise together with their knowledge and experience (Afrooz *et al.*, 2018). In order to properly prepare for actual farming, it is most helpful to consider the previous crop yields, surface characteristics, natural material content, pH, moisture content, and nutrient levels of the soil. Crop yields and quality measures like water content and chlorophyll levels can be assessed by combine harvesters equipped with GPS tracking devices in real time and at the precise location in the region from which they are collected. Adjustable velocity engineering (VRT) makes it possible for the information to be applied correctly. In order to apply precise amounts of growing inputs at certain times or locations, it combines farming equipment, management techniques, and application products. VRT-enabled precision farming has advantages for the environment and the economy. Applying nutrients, seed, fertiliser, and perhaps insecticides just where and when they are required could result in significant cost savings for the farmer and an increase in profits. Additionally, negative environmental effects from excessive usage of several synthetic materials are reduced, and data analysis suggests that the use of particular chemical compounds may be completely phased out. The farmer can get the right length between too much and too little help with persistent problems like nitrogen programmes.

2. Nanotechnology

Nanotechnology has been known for its immense potential as one of the most advanced, progressive and key technologies of this century (Ma *et al.*, 2017). Nano is basically derived from the Greek word which means dwarf or very small and the word technology means manufacturing, utilising knowledge of various machines, tools and techniques in order to perform a specialised function. It is the study of manipulating and rearranging matter on an atomic and molecular scale or study of making small microscopic things. Nanoagriculture involves the use of nano-sized particles with different and specialised properties to boost crop productivity. Nanomaterials are basically the entities viz., nanoparticles, nanoclusters, nanocrystallites that are developed within the nano-metre scale range of 1–100 nm. These materials are available as aerosols, powder form, suspension, emulsions or dispersions. Nanoparticles possess distinctive properties viz., physical, biological and chemical properties according to their atomic strength, high reactivity, high surface area, improved catalytic activity and quickly dispersible leads to better uptake of nutrients by the plants (Shang *et al.*, 2019). The different ways in which nanotechnology is being used to produce vegetable crops is described. The current state of nanotechnology in vegetable crop production over time for several factors such seed germination, plant growth, and development has improved vegetable yield and quality, and biotic and abiotic stress detection and management are discussed.

I. Seed germination

The process of seed germination and seedling development is positively impacted by nanoparticles' ability to penetrate the seed coat. It was discovered that applying nanoparticles has a beneficial impact on the process of seed germination because it aids in the seeds' ability to absorb water, promotes their antioxidant system, reduces antioxidant stress by lowering the level of peroxides and superoxide radicles, increases enzyme and catalase activities, and ultimately increases the germination rate (Almutairi & Alharbi, 2015). Additionally, the use of nanoparticles enhanced the quality of seeds by promoting oxidation-reduction reactions by lowering the superoxide ion radical during the germination phase, which quenches free radicals in the seeds as they sprout (Anandaraj & Natarajan, 2017). A method of seed planting preparation known as "nanopriming" involves pre-soaking seeds in a solution containing nanoparticles. The prime objective of nanopriming is to increase germination rates, several seedling growth parameters and seedling vigour. It is a cutting-edge technique for priming seeds with nanoparticles (such as zinc oxide, iron oxide, titanium dioxide, silver nanoparticles, etc.), and when applied to plants, it causes both abundant and restricted nutrient uptake at the targeted location (Pawar & Laware, 2018). When compared to other seed priming agents, such as polyethylene glycol (PEG), water, and vitamins, it has been observed that using nanoparticles results in the highest binding proportion between the seeds and the priming agent (Anand *et al.*,2019). Nano priming has been found to have favourable and significant effects on a number of vegetable crops, as shown in Table 1.

Table 1: Impact of nano-particles application on seed parameters and seed priming

Crop	Mode of application	Impact	Reference
Tomato	Seed treatment with ZnO nanoparticles (250 mg/l)	At lower concentrations, seed germination increased	Basith *et al.*,2018
French bean	Seed treatment with zinc oxide nanoparticles (**500** mg/l)	Seed germination & seedling development enhanced	Savassa *et al.*, 2018
Carrot & beetroot	Seed priming with nano-ironpyrite (100 µg/mL)	Increase in crop yield	Das *et al.*, 2016
Cucumber	Seed treatment with nano silica suspension (200 mg/l)	Seed germination, fresh and dry weight, relative water content increased	Alsaeedi *et al.*, 2019
Watermelon	Seed priming with silver nanoparticles (250 mg/l	Enhanced germination, growth yield and quality (glucose, sucrose content etc.) improved	Acharya *et al.*, 2020

II. Plant growth and development

Vegetables are voracious nutrient mining crops which have a very high requirement of nitrogen and phosphorus so crreation of nano- form of these will be appropriate for the different vegetable crops to improve the nutritional quality (Chhipa, 2017). Depending on the type of nanoparticle, dosage, timing, plant species, developmental stage, and many other aspects, nanoparticles' effects on plant growth and development can be either beneficial or negative. With the aid of nanotechnology, nanofertilizers are essentially synthetic or advanced versions of conventional fertilisers or extracted from a variety of vegetables or plant reproductive parts using a variety of methods (chemical, physical, mechanical, or biological), which has enormous potential to improve soil fertility, yield, and crop quality (Zulfiqar *et al.*, 2019). Numerous methods, including foliar, soil, seed priming, root dipping, aerosol dusting, and fertigation using a drip irrigation system, can be used to apply these nano fertilisers. Contrary to conventional fertilisers, nanomaterial-based fertilisers have a number of benefits for crops, including improved crop production, more precise targeting, a threefold increase in nutrient utilisation efficiency (NUE), gradual nutrient release, and lower ecotoxicity. Numerous studies on the effects of nanoparticles and nanofertilizers on growth parameters in vegetable crops are presented in Table 2.

Table 2: Effect of nano-fertilisers on growth traits in vegetable crops

Crop	Nanoparticle	Improved trait	Reference
Okra	Foliar spray with lithovit nano fertiliser (0.75 g/ litre)	Increased plant height, number of branches, pod length, diameter , weight, number of pods per plant and total yield	El-Hamd & Abd-Elwahed, 2017
Onion	Soil application of zinc nano-particles (0.5, 1.5 and 3 mg/litre)	Increased plant height, number of leaves and reduced incidence of fungus *Phythium aphanidermatum*	Taimooz,2018
Potato	Foliar application of nano chitosan NPK fertiliser (10%)	Increased yield & modification of chemical composition	Elshamy *et al.*, 2019
Cucumber	Foliar application of liquid nano NPK (6 mL and 9 mL/plant)	Plant height, number of leaves/plant, total yield and qualitative parameters viz., TSS, chlorophyll content and firmness of the fruits increased	Merghany *et al.*, 2019
Brinjal	Soil application of nano potassium (1.5 g/l)	Improved yield & yield contributing traits	Al-Fahdawi & Allawi, 2019

III. Disease detection and disease management

With the advancement of nanotechnology, diseases can be detected by the fluorescent silica nano probes and nano-based disease diagnostic kits . Quantum dots, which are closely packed semiconductor crystals made of hundreds or thousands of atoms, nano biosensors, are used to assess crop health and identify microbial or chemical contamination, and nanofabrication, which is helpful for examining the pathogen infection process and behaviour inside host plants, can all be used to detect diseases (Alghuthaymi *et al.*, 2015). Nano-based products such as nano fungicides and nano bactericides have been utilised effectively for the management of diseases as it enables the smaller quantities of the fungicides to be used over a given period of time for the application of lower amounts of fungicides (Abd- Elsalam & Alghuthaymi, 2015). Silver nanoparticles were described as innovative antibacterial agents because they may effectively and efficiently manage diseases to a great extent without leaving any hazardous residues (Franci *et al.*, 2015). The use of various nanoparticles and nanofungicides on diverse vegetables had been reported to produce the best outcomes.

Table 3: Effect of nano-particles on disease management in vegetable crops

Crop	Nanoparticle	Mode of action	Disease managed	Reference
Cauliflower	*In vitro* evaluation of silver nanoparticles (100ppm and 110 ppm) in potato dextrose agar medium (PDA)	Antifungal	Black spot	Gupta & Chauhan,2016
Brinjal	Foliar application of zinc nanoparticles (200 µg/mL)	Antiviral	Cucumber mosaic virus	El- Sawy *et al.*, 2017
Potato	Foliar application of silver nanoparticles (50ppm) in 40 days old seedlings	Antiviral	Tomato spotted wilt virus	Shafie *et al.*, 2018
Tomato	Root immersion and foliar shoot application with silver (Ag) nanoparticles (60ppm) for 10 minutes	Antifungal	Fusarium wilt	El-Waseif *et al.*, 2019
Sweet potato	Disk diffusion assay of zinc oxide nanoparticles (50ppm) under *in vitro* conditions	Antifungal	Rhizopus sot rot	Nafady *et al.*, 2019

IV. Insect-pests management

Nano pesticides are successfully employed for the management of insect-pests in vegetable crops since nano materials may effectively raise the water solubility of the active ingredients as accompanying substances increased the efficacy, durability and decreased the amount of pesticide used. Insect-pest management employs a variety of nanoformulations, including nanoparticles, nanoemulsions, and nanoencapsulations (Hayles *et al.*, 2017). The packaging of small/nano-scale active ingredients in a tiny "envelope or shell" is known as nano encapsulation. Compared to conventional pesticides, nano encapsulation has several benefits, including controlled and slow pesticide delivery, improved pest control effectiveness, and protection from photolysis and degradation (Urkude, 2019). The utilisation of zinc oxide nanoparticles in the nano electronic nose, an artificial intelligent nose system, was proven to be a highly quick and efficient method for controlling insect pests in vegetable crops. Researchers recently reported a significant impact of various nanopesticides to control insect pests in vegetable crops (Table 4).

Table 4: Effect of nano-pesticides on insect-pest management in vegetable crops

Crop	Mode of application of nano-pesticide	Mode of action	Insect pest	Reference
Brinjal	Foliar application of green silver nanoparticles (12.75 mg/100 mL)	Nematicidal	*Meloidogyne incognita*	Abdellatif *et al.*, 2016
Okra	Foliar spray of iron nanoparticles (120 mg/mL) derived from Stenotrophomonas maltophilia KBS 2.4	Nematicidal	*Meloidogyne incognita*	Sharma *et al.*, 2017
Cauliflower	Titanium dioxide nanoparticles LC25 (20.81 µg/mL), LC50 (37.57 µg/mL), LC75 (67.83 µg/mL) and LC95 (158.69 µg/mL)	Larvicidal	*Plutella xylostella*	Preetha *et al.*, 2018
Tomato	Nematodes exposed to silver nanoparticles (0.0005 µg) suspended in 0.05 mg chitosan	Nematicidal	*Meloidogyne incognita*	Bernard *et al.*, 2019

V. Abiotic stress management

Plants are exposed to various types of environmental stresses throughout their entire life cycle. They evolve defence mechanisms against those stresses at various levels by activating physiological, biochemical, and molecular pathways. Different plant activities evolved abiotic stress tolerance as a result of changes in the characteristics of nanomaterials and nanoparticles (Saxena *et al.*, 2016). Utilization of nanofertilizers has been reported to be particularly successful against soil toxicity and other associated challenges because plants use less chemical than is actually delivered, and the remaining chemical remains unused and enters the soil to enhance soil toxicity (Kalteh *et al.*, 2014). Nanotechnology is the novel and promising emerging science that helps in overcoming the stresses by enhancing the antioxidant defence system and less toxic fertilisers. Numerous researches have demonstrated the efficient use of nanoparticle/nano fertiliser in the treatment of abiotic stress in vegetable crops (Table 5).

Table 5: Effect of nano-particles on abiotic stress management in vegetable crops

Crop	Stress	Nano-particle	Improved trait	Reference
Okra & Cowpea	Salinity	Foliar spray of zinc oxide nanoparticles (10 mg/litre) after 20, 40 and 60 days from sowing	Enhanced growth parameters and induced salinity tolerance	Alabdallah & Alzahrani, 2020
Tomato	Salinity	Foliar application of selenium nanoparticles (10 mg/ litre combined with NaCl)	Increased enzymatic compounds in the leaves and fruits, promote growth, yield and aerial biomass	Morales-Espinoza *et al.*, 2019
Cucumber	Salinity	Foliar application of manganese tetroxide nanoparticles (1 mg/ plant) applied to 3-week old plants for 7 days	Increased leaf photosynthetic rate, net photosynthesis, biomass, reduced oxidative stress and alleviate the salt stress	Lu *et al.*, 2020
Cabbage	Drought	Treatment with yttrium dopping-stabilised Fe_2O_3 nanoparticles (40 mL/pot/day) through irrigation in a nutrient solution under controlled conditions	Enhanced growth rate of leaves, chlorophyll content, catalase activity and increased drought tolerance levels	Palmqvist *et al.*,2017
Cucumber	Drought	Treatment with carbon nanotubes	Maintained plant health with dark green leaves, regulation of drought stress marker genes and enhanced drought tolerance when watering withheld for 21 days	Kim *et al.*,2019

VI. Improvement of the yield and quality

Nanotechnology exhibited enormous potential of improving the growth, yield and quality of vegetable crops in the form of nano fertlisers (boron, zinc, ZnO, chelate, Fe and its compounds and chitosan) when sprayed at lower concentrations (Zahedi *et al.*,2019). Nano-urea can reduce the amount of nitrogen that must be applied to the soil by up to 12.4–41.7% while increasing yield by 10.2% and agronomic efficiency of nitrogenous fertilisers by 44.5% compared to regular urea (Huang *et al.*, 2015). Treatment with 70% nano nitrogen fertiliser resulted in yields that were 11.6% greater than those from conventional fertilisation. Application of nanoparticles raised the firmness of edible fruits, prolonged their shelf life and quality, and also induced the buildup of numerous beneficial substances as ascorbic acid, lycopene, carotenoids, and

TSS and antioxidant capacity of vegetable crops (Vargas *et al.*, 2018). Numerous researchers have documented the significant effects of nanoparticle treatment on the growth and yield of vegetable crops (Table 6).

Table 6: Effect on yield & quality of vegetable crops

Crop	Nanofertilizer	Trait improved	Reference
Sweet pepper	Foliar application of nanofertilizer Nano active forte (1% and 4 kg/ha)	Increased total yield, marketable yield, dry matter, total soluble solid, phosphorus and potassium	Wolska *et al.*,2018
Carrot	Foliar application of grapehene oxide and zinc oxide nanoparicles (0.10 mg/mL)	Increased plant growth, chlorophyll, carotenoids and proline contents	Siddiqui *et al.*, 2019
Broccoli	Foliar application of nano fertilisers viz., nano iron + nano zinc (200 mg/L + 100 mg/litre)	Increased plant height, leaf area, shoot dry weight, dry matter, flower disc diameter, flower disc weight & total flower disc yield	Al-Tameemi *et al.*, 2019
Chilli	Seed invigouration with zinc oxide nanoparticles (1000 mg/kg)	Increased germination percentage	Kumar *et al.*, 2019
Lettuce	Exposed to 50% Hoagland nutrient's solution ammended with titanium dioxide nanoparticles (50–200 mg/litre) for one week	Increased root and shoot dry biomass, total chlorophyll and carotenoids content	Hu *et al.*, 2020

3. Soil-less vegetable production

Soil-less culture, in its simplest form, is the practise of growing plants without soil. Recently, a variety of novel methods for cultivating plants in soilless media have been created. Plants are cultivated on organic or inorganic substrates such fertiliser solutions, cocopeat, vermiculite, perlite, etc. instead of soil. Modern cultivation techniques such as aeroponics, hydroponics, and aquaponics use nutrient-rich water instead of soil to grow plants (Waiba *et al.*, 2020). Compared to conventional agricultural systems, these advanced methods use less water and area. In controlled environment settings, productivity is also stable throughout the year with less pesticide use. This technique aids in managing the production system for effective use of natural resources and the reduction of malnutrition, as well as assisting in overcoming the difficulties of unfavourable climatic conditions. The various strategies are discussed as under:

I. Hydroponics

It is simply the process of growing vegetables in water. For optimal crop growth and development, plants require the vitamins and minerals that soil can offer them in the form of light, H2O, CO2, and O2. Inert materials such as pebbles, coco coir fibre, peat moss, vermiculite, etc. are utilized as a growing medium. They are fed a solution containing macro- and micronutrients. Practically any vegetable can be effectively hydroponically produced. Farmers and producers utilise it extensively around the world due to a number of benefits, including the fact that crops can grow closer together and use less space to produce more (Jones *et al.*, 1991). The nutrient solution also ensures that a consistent supply of nutrients is always accessible, resulting in healthy crop growth. Hydroponic crops are more productive than crops grown in soil because of this combination of factors. As a result, many farmers in different nations are starting to grow vegetables hydroponically. One of the main benefits of using hydroponics is that it considerably conserves water. It uses 10% less water than soil-based agriculture, which uses more than 80%.

II. Aeroponics

An example of a closed soilless vegetable cultivation system is this one. The r oots of the crops are suspended in the air. Sealing root chambers wrapped in polystyrene or another material are utilised as nutrient solution reservoirs. Typically, a water pump is used to sprinkle nutritient solution around the plant's root zone several times per minute. As with other types of hydroponic technique, a timer is installed at the edge of the tank to manage the nutrient pump for a brief cycle that runs the pump for a few seconds every couple of minutes (Waiba *et al.*, 2020). Roots will quickly dry out if the misting cycles are interrupted since they are exposed to the air. The initial framework in this system, which consists of three different types, has high pressure and doesn't typically utilise a water pump. Soakaponics, a low pressure framework, is the second type. Using common submersible water pumps, the water and nutrient solution easily pour out of the sprinkler, or mister heads (higher water pressure). Ultrasonic foggers that create a fog make up the third framework. In this technique, virtually all vegetable crops can be cultivated with ease.

III. Aquaponic system

This technique combines hydroponics and recirculating aquaculture, and it is used to produce both fish and vegetables in a mutually beneficial setting. Water is pumped from the fish tank to the plants using a water pump. The crop doesn't need to be supplemented with additional nutrients because fish waste is enough to support plant growth and development. Because fish excreta contains a lot of ammonia, bacteria turn nitrite and ammonia into nitrate. The fish tank receives the extra water. Due to the fact that it serves as a bio-integrated paradigm for sustainable fresh and nutritious vegetable production, it has recently attracted greater attention. The combination of hydroponic and aquacultural practises is becoming more and more popular (Goddek *et al.*, 2015). It ensures food security in urban areas where conventional vegetable farming is not possible. Likewise, resource scarcities such as decreasing fertile land, soil degradation, lack of freshwater supplies for the crop, and soil nutrient depletion add additional difficulties for soil-based vegetable farming. For mitigation of such a challenges, aquaponic systems can be good solutions. Various research workers have grown different vegetable crops under soilless culture conditions which have been presented in Table 8.

Table 8: Different soil-less modern cultivation systems used in vegetable crop production

System used	Vegetable Crop	Recommendation	Reference
Hydroponics			
(a)	Tomato (Interaction effects of silicon application on tomato physiology and quality using low-quality water	Si application both as pre- and post harvest could increase physiological and quality aspects	Costan *et al.*, 2020
(b)	Lettuce (Different treatments under hydroponic system)	Static pipe with 10.1 cm diameter gave highest yield per plant while it was lowest in geoponics	Qadeer *et al.*,2020
(c)	Water spinach (use of pretreated swine wastewater (PSW)	Fresh stem-leaf production got elevated by 38.9% as compared to control	Jin *et al.*, 2020
Aeroponics			
(a)	Potato (use of aeroponics system for production of potato in more sustainable manner)	Besides providing food security, it will help farmers to get rid of soil-borne diseases and pests.	Tunio *et al.*, 2020
(b)	Lettuce (effect on growth, total polyphenols content and antioxidant activity by varying aeroponics atomizers (droplet sizes)	Airless atomizers were identified as best with spraying interval of 20 min on and 3hr off.	Lakhiar *et al.*, 2019
Aquaponics			
(a)	Cucumber (effect on nutrient content and antioxidant activity by growing it under aeroponics systems)	Aquaponic system had highest content of Cd, Cr, Fe, Ni, Pb, and Zn while it had lower B, Cu, and Mn contents	Guzel *et al.*, 2018

System used	Vegetable Crop	Recommendation	Reference
(b)	Butter head lettuce (growth response and tissue elemental concentration of butterhead lettuce under aeroponics conditions)	Aquaponics operated at pH 7.0 negated negative biomass and elemental effects compared with high or lower pH.	Anderson *et al.*, 2017

4. Vertical gardening

The urban population in the world will double in 30 years, leading to the challenges for the mankind to meet food and nutritional security, besides the problem of environment. In cities, environmental benefits can be achieved when horticulture is planned as a part of the urban landscape, including safe recycling of solid waste and water. Vertical farming is a new technology that aims to boost crop yield per unit land area in response to rising pressure on aagriculture (Beacham *et al.*, 2019). Many scientists have worked on improving agricultural methods, and some are still doing so, in an effort to maximise productivity from a small amount of land while minimising costs. The concept of utilization of the vertical area is not a recent discovery instead it is known to everybody from old ages, in a sense that the people build two or more storied houses. Growing crops vertically in a controlled environment utilising technology such as LED lighting, HVAC systems, sensors and intelligent software, drones, and mobile apps to maintain complete environmental control is known as vertical farming. It is also pressing need to grow more & more crops in a sustainable way to provide food to everyone. Vertical farming and hydroponics are two of the greatest choices that may be used. As vegetables are thought of as the best source of various types of vitamins, minerals, bioflavonoids, and other compounds that aid in maintaining the metabolism of the organism, vertical farming enables us to grow a wide variety of vegetables, crops, and condiments, etc. that would not only provide good returns but also help in minimising the major problem of malnutrition (which is more prone among younger generations). The results of vertical farming are far better when hydroponics and controlled environmental conditions are employed together than when they are used alone. Currently, hydroponics is viewed as a viable method for producing a variety of short-lived crops, such as vegetables, year-round in relatively small places with little labour and making efficient use of the available resources (Sharma *et al.*, 2018).The use of vertical farming and other methods, such as hydroponics, can be utilised to produce some vital vegetables, such as onions, all year long while increasing productivity and preventing sudden rise in price. Additionally, compared to open-air farming, vertical farming facilitates earlier crop harvesting. The vegetable crops suitable to be grown in vertical farming include Tomato, Chilli, Brinjal, Green Bean,

Bell Pepper, Potato, Cabbage, Cauliflower, Lettuce, Sweet Basil (Salad crop), Parsley, Onion, Cucumber etc.

Urban and peri-urban vegetable farming

Urban and peri-urban vegetable farming generally refers to vegetable production in open spaces or fields mostly within and along the outskirts of major urban areas. In addition, due to urbanization, there have been a rise in commercial urban and peri-urban farming to meet the needs of the ever-increasing population within the urban centers. Nevertheless, the actual number of farming sites keeps fluctuating due to the loss of some farming sites as a result of development projects and the establishment of new or previously unidentified areas. Most of the farmers in these urban and peri-urban sites cultivate different crops and vegetables all year round. This is due to the ability to find ways of providing supplementary irrigation in the absence of rain with the aid of watering cans, buckets, small motorized pumps, and occasionally furrow and sprinkler systems (Keraita *et al.*, 2014). This ensures an almost regular supply of vegetables to consumers all year long. Nevertheless, with most being small-scale farmers, production is not sufficient to prevent the shortage of particular vegetables during certain times of the year. Most of the vegetables grown by farmers involved in peri-urban and urban farming include exotic vegetables such as cabbage, lettuce, carrot, spring onions, and bell pepper. Other exotic vegetables such as cauliflower, radish, and traditional vegetables including okra, tomato, hot pepper and spinach can also be cultivated.

Conclusions

The expanding population and decreasing land holdings are the main issues that need to be addressed. Investments in soil and water management as well as the creation of institutions to supply better technology, inputs, information, and services are necessary to reduce the yield gap. The application of frontier technologies along with the conventional technologies can significantly improve productivity, provide nutritional security and encourage the sustainable use of natural resources. Frontier technologies like remote sensing, GIS and GPS have advanced significantly in India, but their potential has thus far gone unrealized due to a variety of policy and non-policy causes. Additionally, these technologies can be used profitably to strengthen the resilience of agricultural production systems by warning farmers about unpredictable events like changing weather and pest outbreaks and infestations. Soilless vegetable farming may not seem significant at current situation but it has huge potential in the coming generation. The progressive loss of arable land due to the emergence of uncontrolled human civilization makes it imperative to look for innovative alternative technologies and techniques for the mass production of vegetables in order to feed the world's fast expanding population. If the advanced new technologies and techniques are applied and inspect properly it will open a doorway to produce different vegetables crops in outer space. Successful adoption, nevertheless, comprises at least three phases as execution, evaluation, and exploration. In order to operationalise the growth potential of modern technologies, there is a need to prioritise efforts and investments on a continuum, and create a level playing field for generation and dissemination of agricultural research through appropriate policies and institutions.

References

Abd- Elsalam, K.A., & Alghuthaymi, M.A. 2015. Nanobiofungicides: Are they the next- generation of fungicides? Journal of Nanotechnology and Materials Science, 2, 1–3

Abdellatif, K.F., Hamouda, R.A., & El-Ansary, M.S.M. 2016. Green nanoparticles engineering on root-knot nematode infecting eggplant and their effect on plant DNA modification. Iran Journal of Biotechnology, 14, 250–259.

Acharya, P., Jayaprakasha, G.K., Crosby, K.M., Jifon, J.L., & Patil, B.S. 2020. Nanoparticle-mediated seed priming improves germination, growth, yield, and quality of watermelons (Citrullus lanatus) at multi-locations in Texas. Scientific Reports, 10, 1–16.

Addicott JE (2020) The Precision Farming Revolution. In: The Precision Farming Revolution. Springer, pp 1-35

Addicott JE. 2020. The Precision Farming Revolution. In: The Precision Farming Revolution. Springer, pp 1-35

Afrooz AE, Lowe R, Leao SZ, Pettit C. 2018. 3D and virtual reality for supporting redevelopment assessment. Real Estate and GIS: The Application of Mapping Technologies 151

Alabdallah, N.M., & Alzahrani, H.S. 2020. Impact of zinc oxide nanoparticles on growth of cowpea and okra plants under salt stress conditions. Biosciences Biotechnology Research Asia, 17, 329–340.

Al-Fahdawi, A.J.J., & Allawi, M.M. 2019. Impact of biofertilizers and nano potassium on growth and yield of eggplant (Solanum melongena L.). Plant Archives, 19, 1809–1815.

Alghuthaymi, M.A., Almoammar, H., Rai, M., Galiev, E.S. & Abd-Elsalam. 2015. Myconanoparticles synthesis and their role in phytopathogens management. Biotechnology and Biotechnological Equipment, 29, 221–236.

Almutairi, Z.M., & Alharbi, A. 2015. Effect of silver nanoparticles on seed germination of the crop plants. Journal of Advances in Agriculture, 4, 280–285.

Alsaeedi, A.H., Elgarawany, M.M., El-Ramady, H., Alshaal, T., & Al-Otaibi, A.O.A. 2019. Application of silica nanoparticles induces seed germination and growth of cucumber (Cucumis sativus). Environment and Arid Land Agriculture, 28, 57–68.

Al-Tameemi, A.J.H., Al-Aloosy, Y.A.M., & Al-Saedi, N.J.J. 2019. Effect of spraying chelated and nano of both iron and zinc on the growth and yield of broccoli (Brassica oleracea var. italica). Plant Archives, 19, 1783–1790.

Ampatzidis,Y. 2020. Technology to Improve Vegetable Production. In: Speciality Crop Industry Ag Net.

Anand, A., Kumari, A., Thakur, M., & Koul, A. 2019. Hydrogen peroxide signalling integrates with phytohor mones during the germination of magnetoprimed tomato seeds. Scientific Reports, 9, 8814.

Anandaraj, K., & Natarajan, N. 2017. Effect of nanoparticles for seed quality enhancement in onion [Allium cepa (Linn) cv. CO (On)] 5. International Journal of Current Microbiology and Applied Sciences, 6, 3714–3724.

Anderson, T. S., De Villiers, D., & Timmons, M. B. 2017. Growth and tissue elemental composition response of butterhead lettuce (Lactuca sativa, cv. Flandria) to hydroponic and aquaponic conditions. Horticulturae, 3(3),43.

Anonymous.(2021). First Advance Estimates of 2021-22, Area and Production of Vegetable Crops. Department of Agriculture and Farmers Welfare.

Balafoutis A, Beck B, Fountas S, Vangeyte J, Wal TV der, Soto I, Gómez-Barbero M, Barnes A, Eory V (2017) Precision agriculture technologies positively contributing to GHG

Basith, M.A., Rao, K.P., & Ramteke, P.W. 2018. Phytotoxic effect of zinc oxide nanoparticles on seed germination in tomato (Lycopersicon esculentum L.). Plant Archives, 18, 1791–1794.

Beacham Vickers LH, Monaghan JM. 2019. Vertical farming: a summary of approaches to growing skywards. The Journal of Horticultural Science and Biotechnology, 94(3):277-283.

Bernard, G.C., Fitch, J., Min, B., Shahi, N., Egnin, M., Ritte, I. and Bonsi, C. 2019. Potential nematicidial active ity of silver nanoparticles against the root-knot nematode (Meloidogyne incognita). Online Journal of Complementary & Alternative Medicine, 2, 1–5.

Brar NS, Kumar T, Kaushik P (2020) Integration of Technologies Under Climate Change for Profitability in Vegetable Cultivation: An Outlook. https://doi.org/10.20944/ preprints202009.0542.v1

Brar NS, Saini DK, Kaushik P, Chauhan J, Kamboj NK (2019) Directing for Higher Seed Production in Vegetables. Agronomy

Burrough PA, McDonnell R, McDonnell RA, Lloyd CD. 2015. Principles of geographical information systems. Oxford university press

Chhipa, H. 2017. Nanopesticide: Current status and future possibilities. Agricultural Research and Technology, 5, 1–4.

Costan, A., Stamatakis, A., Chrysargyris, A., Petropoulos, S. A., & Tzortzakis, N. 2020. Interactive effects of salinity and silicon application on Solanum lycopersicum growth, physiology and shelf-life of fruit produced hydroponically. Journal of the Science of Food and Agriculture, 100(2), 732-743.

Das, C.K., Srivastava, G., Dubey, A., Roy, M., Jain, S., Sethy, N.K. and Das, M. 2016. Nano-iron pyrite seed dressing: A sustainable intervention to reduce fertilizer consumption in vegetable (beetroot, carrot), spice (fenugreek), fodder (alfalfa), and oilseed (mustard, sesamum) crops. Nanotechnology for Environmental Engineering, 1, 2.

Dias, J.S. (2011).World Importance, Marketing and Trading of Vegetables. Acta Horticulturae. Pp : 153-170.

El-Hamd, A.S.A., & Abd-Elwahed, A.H.M. 2017. Improving the growth and yield of okra plants (Abelmoschus esculentus L.) using Lithovit fertilizer. Academia Journal of Scientific Research, 6, 65–71.

El-Sawy, M.M., Elsharkawy, M.M., Abass, J.M., & Kasem, M.H. 2017. Antiviral activity of 2-nitromethyl phenol, zinc nanoparticles and seaweed extract against cucumber mosaic virus (CMV) in eggplant. Journal of Virology and Antiviral Research, 6, 1–6.,

Elshamy, M.T., El-Khallal, S.M., Husseiny, S.M., & Farroh, K.Y. 2019. Application of nano-chitosan NPK fertilizer on growth and productivity of potato plant. Journal of Scientific Research, 36, 424–441.

El-Waseif, A.A., Attia, M.S., & El-Ghwas, D.E. 2019. Potential effects of silver nanoparticles, synthesized from Streptomyces clavuligerus, for controlling of wilt disease caused by. Fusarium Oxysporum. Egyptian Pharmaceutical Journal, 18, 228–235.

Emissions mitigation, farm productivity and economics. Sustainability 9:1339

Franci, G., Falanga, A., Galdiero, S., Palomba, L., Rai, M., Morelli, G., & Galdiero, M. 2015. Silver nanoparticles as potential antibacterial agents. Molecules, 20, 8856–8874.

Goddek S, Delaide B, Mankasingh U, Ragnarsdottir KV, Jijakli H and Thorarinsdottir R. 2015. Challenges of sustainable and commercial aquaponics. Sustainability (Switzerland), 4199-4224p.

Gupta, D., & Chauhan, P. 2016. Fungicidal activity of silver nanoparticles against Alternaria brassicicola. 2nd International conference on emerging technologies: micro to nano. pp 1724. pp 1724.

Guzel, S., Odun, U. C., Cakmakci, T., Cakmakci, O., & Sahin, U. 2018. The effect of cucumber (cultivation in aquaponic and hydroponic systems on plant nutrient elements and antioxidant enzyme activity. Fresenius Environmental Bulletin, 27(1), 553-558.

Hayles, J., Johnson, L., Worthley, C., & Losic, D. 2017. Chapter 5- nanopesticides: A review of current research and perspectives. Academic Press, 193–225.

Hu, J., Wu, X., Wu, F., Chen, W., Zhang, X., White, J & Wang, X. 2020. TiO2 nanoparticle exposure on lettuce (Lactuca sativa L.): Dose-dependent deterioration of nutritional quality. Environmental Science Nano, 7, 501–513.

Huang, S., Wang, L., Liu, L., Hou, Y., & Li, L. 2015. Nanotechnology in agriculture, livestock and aquaculture in China. A Revirew. Agronomy for Sustainable Development, 35.

Islam Fahmida, Sk., Uddin, M.S. and Bansal, J.C. 2022. Harvesting Robots for Smart Agriculture. Computer Vision and Machine Learning in Agriculture, 2:1–13.

Jin, E., Cao, L., Xiang, S., Zhou, W., Ruan, R., & Liu, Y. 2020. Feasibility of using pretreated swine wastewater for production of water spinach (Ipomoea aquatic Forsk.) in a hydroponic system. Agricultural Water Management, 228, 105856.

Jones JB, Wolf N & Milla HA. 1991.Plant Analysis Handbook. Micro-Macro Publishing, Inc. Athens, GA.

Kalteh, M., Zarrin, T.A., Shahram, A., Maryam, M.A., & Alireza, F.N. 2014. Effect of silica nanoparticles on basil under salinity stress. Journal of Chemical Health Risks, 4, 49–55.

Kaushik P, Kumar S, Dhaliwal MS .(2018) .Usefulness of Three under Exploited Vegetable Legumes—A Review. International Journal of Current Microbiology and Applied Sciences, 7:861-870

Keraita B, Cofie O. Irrigation and soil fertility management. 2014.In: Drechsel P, Keraita B, editors. Irrigated Urban Vegetable Production in Ghana: Characteristics, Benefits and Risk Mitigation. 2nd ed. Colombo, Sri Lanka: International Water Management Institute (IWMI). pp. 74-88.

Kim, T.Y., Lee, S.H., Ku, H., & Lee, S.Y. 2019. Enhancement of drought tolerance in cucumber plants by natural nano carbon materials. Plants, 8, 446.

Klinger D and Naylor RL. 2012.Searching for Solutions in aquaculture: charting a sustainable course. The Annual Review of Environment and Resources,247-276.

Kumar, G.D., Raja, K., Natarajan, N., Govindaraju, K., & Subramanian, K.S. 2019. Invigouration treatment of metal and metal oxide nanoparticles for improving the seed quality of aged chilli seeds (Capsicum annum L.). Materials, Chemistry and Physics, 242, 122492.

Lakhiar, I. A., Gao, J., Xu, X., Syed, T. N., Chandio, F. A., Jing, Z., & Buttar, N. A. 2019. Effects of Various Aeroponic Atomizers (Droplet sizes) on the Growth, Total Polyphenol Content and Antioxidant Activity of Leafy Lettuce (Lactuca sativa L.). Transactions of the ASABE, 0.

Lu, L., Huang, M., Huang, Y., Corvini, X., Ji, P.F., & Zhao, L. 2020. Mn3O4 nanozymes boost endogenous antioxidant metabolites in cucumber (Cucumis sativus) plant and enhance resistance to salinity stress. Environmental Science Nano, 4, 1–12.

Ma, L., Zhang, M., Bhandari, B., & Ga, Z. 2017. Recent developments in novel shelf life extension technologies of fresh-cut fruits and vegetables. Trends in Food Science and Technology, 64, 23–38.

Merghany, M.M., Shahein, M.M., Sliem, M.A., Abdelgawad, K.F., & Radwan, A.F. 2019. Effect of nano-fertilizers on cucumber plant growth, fruit yield and its quality. Plant Archives, 19, 165–172.

Mohan, C., Kumar Srivastva, D., Verma, R., Kumar, V., Singh, V., Chaudhary, S. and Saini, P.N. 2021. Current Status, Challenges and Future Prospects of Vegetable Seed

Morales-Espinoza, M.C., Cadenas-Pliego, G., Perez- Alvarez, M., Hernandez-Fuentes, A.D., Da La Fuente, M.C., Benavides-Mendoza, A & Juarez-Maldonado, A. 2019. Se nanoparticles induce changes in the growth, antioxidant responses and fruit quality of tomato developed under NaCl stress. Molecules, 24, 3030.

Nafady, N.A., Alamri, S.A.M., Hassan, E.A., Hashem, M., Mostafa, Y.S., & Abo-Elyousr, K.A.M. 2019. Application of zinc oxide nanoparticles to manage Rhizopus soft rot of sweet potato and prolong shelf-life. Folia Horticulturae, 31, 319–329.

Osei, M.K, Ofori, P.A., Adjebeng-Danquah, J., Nketia, S.O., Frimpong-Anin, K., Osei-Bonsu, I., Agyare, R.Y., Opoku-Adoma, J., Ofori, P., Adjei, E.A., Newton, I., Gyau,J. and Puoza,D. 2022. Harnessing Technologies for Vegetable Cultivation: A Panacea for Food and Nutrition Insecurity in Ghana. In book: Vegetable Crops - Health Benefits and Cultivation.

Palmqvist, N.G.M., Seisenbaeva, G.A., Svedlindh, P., & Kessler, V.G. (2017). Maghemite nanoparticles acts as nanozymes, improving growth and abiotic stress tolerance in Brassica napus. Nano Express, 12, 631.

Pawar, V.A. & Laware, S.L. 2018. Seed priming: A critical review. International Journal of Scientific Research in Biological Sciences, 5, 94–101.

Preetha, S., Kannan, M., Lokesh, S., Viji, N., Prithiva, J.N., & Gowtham, V. 2018. Titanium dioxide (TiO2) nanoparticles as a novel insecticide against diamondback moth, Plutella xylostella L. in cauliflower. Trends in Biosciences, 11, 2999–3003.

Qadeer, A., Butt, S. J., Asam, H. M., Mehmood, T., Nawaz, M. K., & Haidree, S. R. 2020. Hydroponics as an innovative technique for lettuce production in greenhouse environment. Pure and Applied Biology (PAB), 9(1), 20-26.

Ray PK (2015) Hi-Tech Horticulture and Climate Change. Climate Dinamics in Horticultural Science, Principles and Applications; Choudhary, ML, Patel, VB, Siddiqui, MW, Mahdi, SS, Eds 1-22

Savassa, S.M., Duran, N.M., Rodrigues, E.S., Almeida, E., Gestel, C.A.M., Bompadre, T.F.V., & De Carvalho, H.W. P. (2018. Effects of ZnO nanoparticles on Phaseolus vulgaris germination and seedling development determined by X-ray spectroscopy. ACS Applied Nano Materials, 1, 6414–6426.

Schreinemachers, P., Simmons, E.B. and Woperies, M.C.S. 2018. Tapping the economic and nutritional power of vegetables. Elsevier, 16: 36-45.

Sekhon, B.S., Pedada, S. and Kamatyanatti, M. 2018. Recent Advances In Soil-Less Vegetable Production: A Review.JETIR , 5(11): 438-447.

Shafie, R.M., Salama, A.M., & Farroh, K.Y. 2018. Silver nanoparticles activity against Tomato spotted wilt virus. Middle East Journal of Applied Sciences, 7, 1251–1267.

Shang, Y., Hasan, M.K., Ahammed, G.J., Li, M., Yin, H., & Zhou, J. 2019. Applications of nanotechnology in plant growth and crop protection: A review. Molecules, 24, 1–23.

Sharma N, Acharya S, Kumar K, Singh N, Chaurasia OP. 2018. Hydroponics as an advanced technique for vegetable production: An overview. Journal of Soil and Water Conservation, 17(4):364-371.

Sharma, H., Dhirta, B., & Shirkot, P. 2017. Evaluation of biogenic iron nano formulations to control Meloidgyne incognita in okra. International Journal of Chemical Studies, 5, 1278–1284.

Shi Y, Thomasson JA, Murray SC, Pugh NA, Rooney WL, Shafian S, Rajan N, Rouze G, Morgan CL, Neely HL (2016) Unmanned aerial vehicles for high-throughput phenotyping and agronomic research. PloS one 11:e0159781

Siddiqui, Z.A., Parveen, A., Ahmad, L., & Hashem, A. 2019. Effects of graphene oxide and zinc oxide nano particles on growth, chlorophyll, carotenoids, proline contents and diseases of carrot. Scientia Horticulturae, 249, 374–382.

system in India. Journal of AgriSearch, 8(3): 188-200.

Taimooz, S.H. 2018. Behavior of some nanomaterials in improving the growth of onion plant, Allium cepa and its effect on Pythium aphanidermatum. Plant Archives, 18, 857–862.

Tripodi P, Massa D, Venezia A, Cardi T (2018) Sensing technologies for precision phenotyping in vegetable crops: current status and future challenges. Agronomy 8:57

Urkude, R. 2019. Application of nanotechnology in insect-pests management. The International Research Journal of Science and Engineering, 7, 151–156.

Vargas, E.R.L., Ortiz, H.O., Pliego, G.C., Romenus, K.D.A., Fuente, C.D.L., Mendoza, A.B., & Maldonado, A.J. 2018. Foliar application of copper nanoparticles increases the fruit quality and the content of bioactive compounds in tomatoes. Applied Sciences, 8, 1020.

Waiba, K.M., Sharma, P., Sharma, A., Chadha,S. and Kaur, M. 2020. Soil-less vegetable cultivation: A review. Journal of Pharmacognosy and Phytochemistry, 9(1): 631-636

Weerakkody, P. and Mawalagedera, M. 2020. Recent Developments in Vegetable Production Technologies in Sri Lanka. In book: Agricultural Research for Sustainable Food Systems in Sri Lanka. doi:10.1007/978-981-15-2152-29.

Wolska, J.C., Mazur, K., Niedzinska, M., Kowalczyk, K., & Zolnierczyk, P. 2018. The influence of foliar fertilizers on the quality and yield of sweet pepper (Capsicum annuum L.). Folia Horticulturae, 30, 183–190.

Yuan W (2019) A Multi-Sensor Phenotyping System: Applications on Wheat Height Estimation and Soybean Trait Early Prediction.

Zahedi, S.M., Karimi, M., & Da Silva, J.A.T. 2019. The use of nanotechnology to increase quality and yield of fruit crops. Society of Chemical Industry, 100, 25–31.

Zulfiqar, F., Navarro, M., Ashraf, M., Akram, N.A., & Bosch, S.M. 2019. Nanofertilizer use for sustainable agriculture: Advantages and limitations. Plant Sciences, 289, 1–11.

Chapter 24

Subtropical Fruit Crops Diseases and their Management Strategies

Shahid Ahamad, Vijay Kumar, Rakesh Kumar, Suza Nabi Qureshi, Anjum Ayoub, Vishal Gupta, Mahital Jamwal and Mohd. Naser

Sher-e- Kashmir University of Agricultural Sciences and Technology, Jammu, (J&K), India

Fruit crops are produced on nearly 50 million hectares worldwide. Fruit plants are susceptible to diseases at every step in their production and distribution, with economic losses occurring at the nursery, in producers' fields, during storage after harvest, and in the marketplace. As with other crops, the majority of diseases of fruit crops are caused by fungi. Fungi primarily cause fruit rots, leaf spots and blights, and stem cankers. However, several fungal genera affect roots of fruit crops, either individually (e.g., Armillaria) or as members of a disease complex (e.g., Fusarium and Rhizoctonia). Oomycetes (water molds) primarily cause root and crown rots, but can also cause fruit rots (e.g., strawberry leather rot caused by Phytophthora cactorum) and shoot blights (e.g., grape downy mildew caused by Plasmopora viticola). Consumers have high standards when selecting produce and usually pass over fruit with blemishes or other imperfections. Thus, there is nearly zero tolerance for diseases that directly affect fruit. Aesthetics are less important for fruit that will be processed rather than grown for the fresh market, but growers almost always receive a lower price for fruit destined for processing rather than the fresh market. Since the middle 1990s, the demand for high-quality fruit grown in accordance with organic standards has increased, and organically grown fruit typically is sold at a premium. While growing organic fruit might appear lucrative, managing diseases without synthetic fungicides can be a major challenge depending on the crop and climate where it is produced. As with other crops, managing diseases of fruit depends on manipulating the host, pathogen(s), and environment to favor plant health and inhibit pathogen growth and spread. Since most fruit plants are long-lived perennials, disease management presents unique challenges. The roles of host resistance, pathogen exclusion, cultural practices, and chemical and biological control in fruit disease management should be integrated to manage diseases.

Soil is the natural habitat of a number of microflora such as fungi, bacteria and nematodes many of which cause diseases of crop and fruit plants of economic importance. As such it becomes necessary to manage these diseases from the soil borne diseases, especially when the plants at the nursery stage. By growing fruits plant seedlings in sterilized soil filled in polythene bags simply leads to avoidance of many soil borne diseases. Soil is sterilized should be collected in a heap, drenched with formalin 40% (1litre formalin in 5 liter water), covered with polythene sheet for 4-5 days, thereafter raked for 3-4 days to let the fumes escape into air. This soil should be filled in polythene bags for raising seedlings for further use for budding/grafting. The plants so prepared are easy to transport without disturbing the earth-ball. The polythene covering is torn out gently with a sharp-edged blade before planting the seeding in the field. Such plant has a higher success rate in establishment in the field, besides being free from soil-borne diseases, which goes a long way in maintaining good plant health in the orchards thus established. This also helps in spreading soil-borne diseases from the sick nursery to the new orchard.

Mango

Anthracnose (Colletotrichum gloeosporioides)

Symptoms

The disease appears on young leaves, stem, inflorescence and fruits. Leaves show oval or irregular, greyish-brown spots which may coalesce to cover larger area of the leaf. The affected leaf tissues usually dry and shred. Leaves on infected petioles droop and fall. On young stem, grey-brown spots develop. These enlarge and cause girdling and drying of the affected area. The disease appears on young leaves, stem, inflorescence and fruits. Often, black necrotic areas develop on the twigs from the tip downwards causing a dieback. In humid weather, minute, black dots develop on the floral organs. The infected flower-parts ultimately shed resulting in partial or complete de-blossoming. Latent infections of fruit are established before harvest. The ripening fruits show typical anthracnose. Black spots appearing on skin of the affected fruits gradually become sunken and coalesce.

Management

The diseased twigs should be pruned and burnt along with fallen leaves. Spraying twice with Carbendazirn (Bavistin 0.1%) at 15 days interval during flowering controls blossom infection. Spraying of copper fungicides (0.3%) is recommended for the control of foliar infection. Postharvest disease of mango caused by anthracnose could be controlled by dip treatment of fruits in Carbendazim (0.1%) in hot water at 520C for 15 minutes.

Powdery Mildew (Oidium mangiferae)

Symptoms:

The characteristic symptom of the disease is the white superficial powdery fungal growth on leaves, stalk of panicles, flowers and young fruits. The affected flowers and fruits crop pre-maturely reducing the crop load considerably or might even prevent the fruit set. Fog or mist accompanied by cooler nights during flowering is congenial for the disease spread.

Management

Alternate spraying of Wettable sulphur 0.2% (2 g Sulfex/litre), Tridemorph O.1% (1 ml Calixin / litre) and carbendazim (Bavistin) @ 0.1 % (1 g/litre) at 15 days interval is recommended for effective control of the disease. The first spray is to be given at panicle emergence stage.

Malformation (Fusarium moniliforme)

Symptoms

Mango Malformation Disease is a fungal disease of mangoes caused by Fusarium moniliforme. Mango is the only known host of the disease. The disease spreads on a tree very slowly, but if left unchecked, can severely reduce yields. The main method of spreading MMD to new areas is through infected vegetative planting material. There is no evidence that the disease can spread on fruit or the seeds, or that it affects human health. It is usually associated with the bud mite, Aceria mangiferae but the mites have been shown to spread the disease. Abnormal development of inflorescence is known as floral malformation. Flowers appear in clusters. Witches broom like growth of shoot is known as bunchy top or vegetative malformation. These infected branches do not bear fruits.

Management

(1) Remove diseased parts. (2) Spray insecticides. (3) Spray Captan (0.2%) or Benlate (0.1%). (4) Spray naphthalene acetic acid (NAA) 200 ppm (1 ml /5 litre water) during first week of October followed by de-blossoming January-February.

Grape

Powdery Mildew (Erysiphe necator formerly Uncinula necator)

Symptoms

The disease is characterized by production of white powdery or dusty growth on the upper surfaces of the leaves and other green parts of the vine. Severely affected leaves turn brown and fall. Later in the season it attacks berry clusters and young berries just after flowering. As the berries enlarge fungus growth disappears but infected grapes become rough skinned, turn hard, remain undersized and may drop.

Management

Spray wettable sulphur 0.2% or 0.05% dinocap (Karathane) with first appearance of symptoms. Avoid sulphur sprays or dusts when day temperature is >35oC. Triazole fungicides such as penconazole 0.05% or hexaconazole 0.1% can also be used but avoid their repeated spray. Use of 0.1% carbendazim (Bavistin) may also be used.

Anthracnose (Elsinoe ampelina)

It causes heavy reduction in yield and quality. Spots on leaves are gray with darker margins. Severely infected leaves become distorted and curl down from the margins. Later, the centre of the lesion may drop. Fruit infections appear as circular spot with light-gray centres and reddish-brown borders, quite similar to bird's eye, hence often referred to as 'Bird's eye rot'. Stem lesions are similar in colour, elongated, sunken, with slightly raised borders.

Management

During annual pruning, remove the infected vines drastically; apply Bordeaux paste at cut ends and spray 0.5% Bordeaux mixture or 0.3% mancozeb or 0.3% copper oxychloride or 0.1% carbendazim. Burn the infected /pruned vines, left over berries, if any, and leaves. During the season, spray the vines with any of the fungicides twice or thrice. Carbendazim application should be used once only and not to be repeated. Avoid spraying during flowering.

Downey Mildew (Plasmopara viticola)

It is a major disease of grape and could be more damaging under prolonged humid spell. Small, pale yellow circular lesions appear first on the upper surface of the leaf. On the underside of the lesion downy growth of the fungus appear as whitish-greyish growth. The fungus produces sporangia, which liberate zoospores into a film of water and spread the infection during prolonged rains or dew. Later, the leaf tissue dies in the area of the lesion. Often numerous lesions coalesce, causing large areas of the leaf to turn dark brown or black. Badly damaged leaves fall prematurely.

Management

Apply fungicides Bordeaux Mixture/copper oxychloride/Ridomil MZ 0.25% with first appearance of symptoms or in the event of humid and wet weather as preventive spray or once before bloom and once after bloom. Avoid repetition of spray of Ridomil MZ.

Citrus

Citrus canker (Xanthomonas axonopodis pv. citri)

Symptoms

The disease can appear anywhere on the exposed surface of the plant. Lesions first appear as moist spots that enlarge and grow into raised white scabs that are a result of the bacterium stimulating cells to divide. The scabs darken and become cratered and surrounded by yellowed tissue or they may merge into large scabs. Lesions on fruit do not actually enter the flesh of the fruit, but the cratered appearance makes the fruits unmarketable.

Management

Resistant varieties of many citrus crops are available and canker on them can be controlled by early spraying with copper compounds. Windbreaks are also used to prevent wind-driven spread. Immediate and rapid destruction of infected trees and strict quarantine are essential to controlling the spread. Citrus canker is the most feared disease of citrus crops and when established it can take years to eradicate.

Melanose (Diaporthe citri)

Damage is superficial and does not affect internal fruit quality. On the fruit, leaves and small twigs, small, dark brown to black spots are produced which are raised and rough to touch. The spots are superficial and can be removed with your fingernail. The incidence of melanose usually increases as trees age and the amount of dead wood in the canopy increases. The melanose fungus also causes a wood rot which occurs when trees are stressed, such as by drought. It causes a cinnamon brown discolouration of bark with a well defined margin between healthy and diseased tissue, often with streaks of yellow gum. The melanose fungus also causes one form of stem end rot of fruit (Phomopsis stem-end rot). Leaf symptoms begin as tiny water-soaked specks that become depressed in the center and surrounded by a translucent yellow area that is not depressed. Later, the leaf cuticle ruptures and a gummy substance is exuded which turns brown and hardens. The yellowish margin disappears and the hardened gummed areas will have a sandpaper-like texture. Infected areas on the leaf may be scattered, aggregated, or in streaks, depending on where water transported the spores prior to infection (Figure 1). Heavily infected leaves become pale green to yellow, can be distorted and may fall from the tree.

Melanose is seldom severe on the spring growth flush. When it does occur on this flush, the pustules are usually few and little or no leaf drop occurs. On the summer growth flushes, melanose can be severe enough to cause defoliation, particularly in years following freeze-induced twig dieback.

Management

Melanose is usually not severe on the spring growth flush. Copper fungicides provide little protection when applied at this time. In practice, only melanose on fruit can be economically controlled. For copper fungicides to control melanose on fruit they have to be applied post-bloom, preferably late April, prior to anticipated increased frequency of attack. Several applications may be needed to protect the fruit from petal fall until it becomes resistant in late June or early July. Melanose can be reduced by pruning and burning dead wood which will reduce inoculum (spore supply). Regardless of how conscientiously the pruning and burning are carried out, fungicide spraying will still be necessary to produce blemish-free fruit. Overhead irrigation for up to 12 hours has shown to have little or no effect on melanose severity. Melanose is usually most serious in older groves. Spray with carbendazim (0.1%) or copper oxychloride (o.3%).

Gummosis /Foot Rot (Phytophthora spp.)

Symptoms

An early symptom of Phytophthora gummosis is sap oozing from small cracks in the infected bark, giving the tree a bleeding appearance. The gumming may be washed off during heavy rain. The bark stays firm, dries, and eventually cracks and sloughs off. Lesions spread around the circumference of the trunk, slowly girdling the tree. Decline may occur rapidly within a year, especially under conditions favorable for disease development, or may occur over several years. Phytophthora fungi are present in almost all citrus orchards. Under moist conditions, the fungi produce large numbers of motile zoospores, which are splashed onto the tree trunks. The Phytophthora species causing gummosis develop rapidly under moist, cool conditions. Hot summer weather slows disease spread and helps drying and healing of the lesions. Secondary infections often occur through lesions created by Phytophthora. These infections kill and discolor the wood, in contrast to Phytophthora infections, which do not discolor wood.

Management

Management of Phytophthora gummosis focuses on preventing conditions favorable for infection and disease development. All scion cultivars are susceptible to infection under the right environmental conditions. Plant trees high enough so that the first lateral roots are just covered with soil. Correcting any soil or water problems is essential for a recovery. In addition to improving the growing conditions, you can halt disease spread by removing the dark, diseased bark and a buffer strip of healthy, light brown to greenish bark around the margins of the infection. Allow the exposed area to dry out. You can also scrape the diseased bark lightly to find the perimeter of the lesion and then use a propane torch to burn the lesion and a margin of 1 inch (2.5 cm) around it. Recheck frequently for a few months and repeat if necessary. Cover the exposed area with Bordeaux paint. Drench root zone of the diseased plant with 0.1% metalaxyl (Ridomil-MZ) 3-4 times until the tree shows the signs of recovery.

Chapter 25

Malnutrition and Food-Based Strategies to Combat Malnutrition in India

Anjum Ayoub[*1], *Ankita Choudhary*[2], *Anjali Langer*[3], *Shahid Ahamad*[4], *Sukomaljot Kour*[5]

Sher-e- Kashmir University of Agricultural Sciences and Technology, Jammu-180 009, Jammu & Kashmir, India

**Corresponding author:anjumparay77@gmail.com*

Abstract

While often overlooked, malnutrition is believed to contribute to over one-third of child fatalities globally. Addressing the multifaceted challenges such as escalating food prices, economic downturns, heightened competition for natural resources, and climate change requires a concerted global effort in scientific and technological innovation. Additionally, conditions like cancer, oral health issues, and chronic renal failure further compromise nutritional status. However, advancements in food processing techniques, probiotic foods, and biofortification hold promise in overcoming malnutrition. A recently introduced home-based treatment for severe acute malnutrition has already begun saving the lives of hundreds of thousands of children annually. This article underscores critical food-based strategies to combat malnutrition and delves into its underlying causes.

Keywords: Malnutrition, Types, Combat, Strategies

Introduction

Sufficient nutrition is fundamental to human development. Malnutrition encompasses both undernutrition and overnutrition, indicating deficiencies, excesses, or imbalances in the consumption of energy, protein, and/or other essential nutrients. Insufficient intake of calories (required for energy), protein (essential for muscle development and overall health), iron (vital for proper blood cell function), and various nutrients can lead to malnutrition (Torpy *et al.*, 2004). The advantages of good health are not only evident at the individual level but also extend to societal and national levels. An individual's health is influenced by a complex interplay of social, economic, dietary, lifestyle, environmental, governmental policies, and political factors.

Nutrition stands as a cornerstone in fostering robust growth and cognitive development, while also significantly impacting a child's socio-economic standing. It serves as the bedrock of good health, pivotal in nurturing physical and cognitive faculties. The specter of malnutrition looms large, jeopardizing the trajectory of long-term human development. Its insidious effects can impede cognitive maturation and diminish labor productivity, particularly among families grappling with financial constraints that relegate them beneath the poverty threshold (Martin *et al.*, 2011). Consequently, inadequate nutrition, particularly among the economically disadvantaged, emerges as a formidable adversary to a nation's socio-economic fabric.

Classification of Malnutrition

As per the World Health Organization (WHO), 'malnutrition' denotes deficiencies, excesses, or imbalances in an individual's consumption of energy and/or nutrients (WHO, 2016). It can manifest as either 'undernutrition' or 'overnutrition'. Indications of malnutrition comprise physical and mental fatigue, low weight concerning height (wasting), stunted growth for age, reduced skin folds, accentuated skeletal features, and diminished skin elasticity.

Undernutrition manifests in four broad forms: wasting, stunting, underweight, and micronutrient deficiencies.

Wasting: Wasting, characterized by low weight for height, typically signifies recent and significant weight loss, though it can endure over extended periods. It commonly arises due to insufficient intake of nutritious food and/or prolonged illnesses. In children, wasting correlates with an increased risk of mortality if left untreated.

Stunting: Stunting is characterized by a diminished stature compared to age norms. This condition arises due to persistent or cyclical malnourishment, often linked with economic deprivation, inadequate maternal healthcare and nutrition, frequent illnesses, and/or suboptimal feeding and nurturing during infancy. Stunting hinders children from attaining their full physical and cognitive capacities.

Underweight: is defined as low weight for age. An underweight child may be stunted, wasted, or both.

Micronutrient: Deficiency refers to the insufficiency of vital vitamins and minerals crucial for bodily functions, including the synthesis of enzymes, hormones, and other substances necessary for optimal growth and development.

Malnutrition is a common ailment that patients and doctors must deal with. It is often underdiagnosed and undertreated and occurs both as a disease's source and effect (Chatterjee *et al.*, 2022). Malnutrition can be divided into three categories: under-nutrition, micronutrient deficiencies (often known as "hidden hunger"), and over-nutrition/obesity/overweight (Rosenbloom *et al.*, 2008).

Tripartite Classification of Malnutrition

This approach to triage malnutrition reflects the evolving nature and new realities of human diet and health over the past few decades. To distinguish those who had "enough" food in terms of dietary energy but were deficient in certain micronutrients due to poor dietary quality, a distinct group of micronutrient deficiencies was identified. To account for the alarming increase in obesity, the group "obesity" was added to broaden the concept of "malnutrition," which had previously only applied to a lack of food intake (WHO, 2000). The terms "double-burden of malnutrition" and "triple-burden of malnutrition" were first used in the late 1990s to draw attention to the fact that both under-nourishment and obesity often coexist in the same geographic areas, social groups, and even individuals.

The tripartite classification is summarised as follows in a recent World Health Organization factsheet on malnutrition. When an individual does not get enough of something, or gets too much of something else, they are said to be malnourished. Malnutrition encompasses three major categories:

- Undernutrition, including wasting (low weight-for-height), stunting (low height-for-age), and underweight (low weight-for-age);
- Micronutrient-related malnutrition, including micronutrient deficiencies (a lack of important vitamins and minerals) or micro-nutrient excess; and
- Overweight, obesity and diet-related noncommunicable diseases (such as heart disease, stroke, diabetes and some cancers).

Undernutrition

Undernutrition is characterized by an insufficient consumption of nutrients, primarily calories, along with protein and micronutrients. Symptoms of undernutrition include stunting, wasting, underweight, kwashiorkor, marasmus, and disorders related to micronutrient deficiencies. Undernutrition can generally be categorized into two subtypes: acute undernutrition, which leads to wasting, and chronic undernutrition, which results in stunting. Protein energy malnutrition (PEM) is a form of acute malnutrition that gives rise to metabolic disorders like kwashiorkor and marasmus. Micronutrient deficiencies are considered part of undernutrition and can contribute to conditions such as stunting. Individuals who lack micronutrients despite having sufficient energy intake are classified as micronutrient malnourished.

Micronutrient deficiencies

The term "micronutrient deficiency" refers to a nutritional situation in which the body does not receive enough of one or more specific micronutrients for appropriate growth and functioning, which in turn can cause a variety of diseases and health problems. Undernourishment is understood to include micronutrient deficiencies, and it is acknowledged that those who are considered to be undernourished may also be micronutrient deficient (Gupta *et al.*, 2017). In contrast to those who are chronically undernourished, those who are micronutrient deficient and so suffer from "hidden hunger" are distinguished in the tripartite classification because they receive sufficient energy but are deficient in certain micronutrients. When discussing micronutrient malnutrition, the focus shifts from quantity to quality, with "quality" being defined only in terms of the micronutrient content of the diet. The primary micronutrient deficiencies identified globally are iron, vitamin A, zinc and iodine.

Overnutrition

Overnutrition arises when there's an excessive intake of food compared to the body's dietary requirements for nutrients. It covers various aspects including overconsumption of foods rich in fats and sugars, leading to obesity and the onset of noncommunicable diseases. However, obesity or being overweight serves as the primary indicator of this type of malnutrition, with the term "overnutrition" often left out from its description. The nutritional assessment of obesity relies on body size measurements, typically indicated by a Body Mass Index (BMI) of 25 or 30 or higher. As per the energy-balance model of weight gain, overnutrition primarily stems from consuming more energy than the body expends, consequently leading to overweight or obesity (Lobstein *et al.*, 2017). Obesity and NCDs like diabetes, cardiovascular disease, and cancer are the main effects of overnutrition. Obesity increases NCD risk. Overnutrition leads to obesity and NCDs in the tripartite paradigm. Some obesity professionals and organisations increasingly consider obesity an illness.

Table 1: Classification for assessing severity of malnutrition by prevalence ranges among children under 5 years of age

Indicator	Severity of malnutrition by prevalence ranges			
	Low (%)	Medium (%)	High (%)	Very high (%)
Stunting	< 20	20–29	30–39	≥ 40
Underweight	< 10	10-19	20-29	≥ 30
Wasting	< 5	5-9	10-14	≥ 15

Major causes of malnutrition in India

- High levels of adult undernutrition affect one-third of the adults in the country.
- High levels of child undernutrition, affecting almost every second child.
- High levels of maternal undernutrition, lead to low birth-weight.
- Inappropriate and sub-optimal feeding and care practices for infants and young children.
- Issues related to disease prevention and access to healthcare.
- Low awareness of nutrition and the use of local nutritious foods including nutrient sources.
- Insufficient attention to the health and nutrition of adolescent girls.
- Inadequate access to safe drinking water and proper sanitation facilities

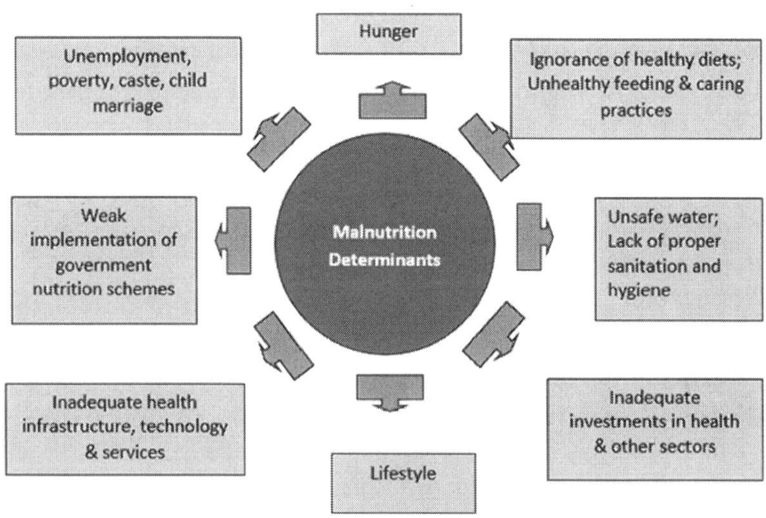

Figure 1: Common Reasons for Prevalence of Malnutrition

(*Source:* Rumi Aijaz, 2017)

Effects of Malnutrition

People who restrict their food intake, whether due to an eating disorder, illness, desire to lose weight, or lack of appetite, are more prone to experiencing a compromised immune system and a downward spiral of malnutrition. Those who are malnourished, impoverished, hospitalized, and either very young or elderly are particularly vulnerable. When medical examinations reveal a weakened immune system in a malnourished individual, the risk of mortality and morbidity significantly escalates. Malnutrition exacerbates existing illnesses, creating a vicious cycle wherein disease worsens due to malnutrition, which in turn exacerbates malnutrition. This destructive cycle often commences with compromised immunity, as illness impairs appetite, disrupts absorption and digestion, increases excretion, or alters metabolism, further compromising nutritional status (Finch 2007).

Remedies towards Eradication of Hunger and Malnutrition

Despite the nation's economic growth, the resurgence of malnutrition presents a grave dishonour. The policies intended to eradicate this crisis remain largely unexecuted, a matter of profound shame (Source: Tackling Child Malnutrition, 2012).

- Ensure the universalization and inclusive implementation of all nutrition-related programs and policies, committed to quality for all.
- Take immediate action to provide swift access to the most impoverished populations, expanding the antyodaya program to deliver standard, state-of-the-art nutrition and care to children suffering from severe malnutrition.
- Implement decentralization in service delivery, management, and response to ensure economically disadvantaged communities and farmers benefit from food availability through integrated child development services and the public distribution system.
- Revise the Integrated Child Development Services to encompass children under two years of age and intensify the focus on early childhood education.
- Foster community ownership in the management and monitoring of programs through involvement of Panchayati Raj institutions and village-level committees, promoting participatory planning.
- Strengthen efforts to enhance nutrition through a leadership and coordination mechanism, with appropriate authority and responsibility, operating from the local to the national level.
- Institute a comprehensive program integrating national nutrition education and behaviour change initiatives.
- Prioritize women's empowerment, providing opportunities for advancement and addressing socio-cultural patriarchal norms that hinder women's access to resources.
- Promote awareness of infant feeding practices and build the capacity of accredited volunteer social health activists as community ambassadors for women's rights.
- Expand the focus on the nutritional needs of women and children within existing national programs.

Food-based strategies to combat malnutrition

The nutrients and other bioactive components in the foods supplied by food systems are linked to the nutritional welfare and health of individuals and populations. Agriculture is the backbone of the food chain because most nutrients may be found in crops and animal products. Malnutrition occurs when food production fails to provide enough calories, protein, vitamins, and minerals to keep people healthy and able to work. For reducing malnutrition, focus should be on both increasing yields and the nutritional quality of crops produced (Miller and Welch, 2013).

In low-resource areas, supplements scheme distribution and sustainability are challenges. Hence, food-based techniques including fortification, dietary diversification/modification, nutrition education, and biofortification are becoming more popular. Some of these strategies are detailed below.

Agricultural Production Strategies

Growing evidence suggests that agricultural interventions play a critical role in programmes aiming at enhancing people's diets and health (Masset *et al.*, 2012). Changes in cropping patterns that incorporate more nutrient-dense crops will enhance the nutritional output of farming systems, according to a large body of evidence (Graham *et al.*, 2007). They proposed cropping system improvements and alternative crop combinations as strategies for meeting nutritional goals while maintaining or improving agronomic and societal acceptability within the farming systems. They came to the conclusion that, with the right information and financial incentives, farmers can design cropping systems that suit human needs despite limited access to resources.

Field fortification tactics, which involve adding enriched fertilisers to the soil, have been used by farmers to increase the micronutrient and trace element content of their crops. A study by Gibson et al in 2007 showed that by applying zinc fertiliser to zinc-deficient rice fields, the zinc intake of children in north-east Thailand was nearly doubled.

Agricultural biotechnology solutions offer a promising chance to sustainably increase crop yields and nutrient density across a wide range of nutrients. Plants with high nutrient concentrations in their seeds can be selected by simple plant breeding, and cross-breeding or genetic engineering can be used to alter a plant's characteristics.

Biofortification is a cutting-edge approach to permanently fixing the problem of micronutrient malnutrition. Plant breeding and agronomic techniques, such as the application of micronutrient fertiliser, are used to boost nutrient levels in staple food crops (Bouis *et al.*, 2011). The micronutrient status of the rural poor has improved as a result of biofortification activities in a number of nations. New orange flesh sweet potato (OFSP) types are being introduced in various African countries. These sweet potatoes have been biofortified with pro-vitamin A carotenoids (Low *et al.*, 2017).

Dietary Diversification

Micronutrient adequacy, or the intake of enough of these nutrients to promote growth and development, is positively correlated with dietary diversity in children (Li *et al.*, 2020). Food-based approaches, which emphasise dietary diversity, can fulfil these prerequisites of reducing malnutrition. Assessing dietary intake, increasing and diversifying food production, enhancing food processing, preservation, storage, and marketing, and bettering food preparation are all part of this strategy. A nutrition education initiative is required to complement this tactic (Tontisirin *et al.*, 2002).

The quality of a child's diet and how they are fed can be quantified by looking at their minimum dietary diversity (MDD) (WHO, 2008). If a baby or toddler has eaten something from five out of the eight food groups in the previous 24 hours, then they have met the minimum dietary requirements for development. These 8 groups are:

- Cereals, tubers, and roots.
- Beans and nuts.
- Dairy.
- Meats, poultry, and fish.
- Eggs.
- Fruits and vegetables high in vitamin A.
- Other fruits and vegetables and
- Breast milk.

High consumption of cereals, especially rice, and low consumption of vegetables, fruits, and pulses characterise diets with low diversity. Zero Hunger can be achieved in large part through agricultural diversification (Franzo *et al.*, 2013). However, current agriculture and food systems have restricted production options, leading to nutritional deficiencies due to poorly rounded meals. People in the Asia-Pacific region eat mostly rice, one of a small number of staple crops, which does not provide adequate nutrition when compared to the dietary guidelines (Fairulnizal *et al.*, 2015).In order to reduce hunger and malnutrition, it is important to have a varied diet, and a diverse food supply, which in turn helps mitigate the effects of climate change (Xuan and Siddique, 2018).

Food Processing Strategies

Both dietary diversity and the quantities of micronutrients in regularly consumed foods have considerable potential to be increased by food processing. Processing food helps maintain its freshness, safety, flavour, convenience, nutritional value, and efficiency (Floros *et al.*, 2010).

In addition to preventing spoiling and destroying infections, processing fresh fruits and vegetables may also aid in retaining micronutrients. Spinach, for instance, loses all of its vitamin C after just seven days when kept at room temperature, but just thirty percent when frozen immediately after harvest and kept at -20 °C for ten months (Rickman *et al.*, 2007a). Heat-sensitive vitamins, such as ascorbic acid and thiamine, can be severely degraded during the canning process, which entails boiling items sealed in cans to 121 °C for up to 30 minutes. Yet, once finished, the items can be stored at room temperature for long periods of time without deteriorating. Canned fruits and vegetables have a far lower rate of nutrient loss compared to fresh and frozen options (Rickman *et al.*, 2007b). Nutrient intakes are greatly bolstered by processed fruit and vegetable products.

Reducing Food Waste

Food losses and food waste have been acknowledged for a long time as significant causes to hunger and malnutrition. In fact, one of the initial missions of the United Nations Food and Agricultural Organization (FAO) in 1945 was to reduce food losses (Parfitt *et al.*, 2010). Reducing food losses and waste will not only increase the quantity and quality of food available for consumption, but it will also lower the amount of energy and water needed to produce more food to replace the lost food.

Food Fortification

Dietary supplementation is a subset of food processing. The Codex Alimentarius (1991) defines food fortification or enrichment as "the addition of one or more essential nutrients to a food whether or not it is normally contained in the food for the purpose of preventing or correcting a demonstrated deficiency of one or more nutrients in the population or specific population groups."

Iodization of salt, fortification of milk with vitamin D, fortification of vitamin B and iron to flour and other cereal product are some of the food fortification strategies to reduce malnutrition, micronutrient malnutrition in general.

Eichler *et al.*, (2012) conducted a systematic literature review on the effect of feeding fortified versus unfortified milk and cereal to children aged six months to five years. They reviewed 18 clinical trials involving more than 5,400 youngsters. They found that micronutrient-enriched diets were efficient in boosting serum concentrations of haemoglobin and retinol, but not zinc. The risk of anaemia was 57% lower in children who consumed fortified meals than in those who did not.

Economic and Consumer Education Strategies

In order to effectively combat malnutrition and promote optimal health, it is imperative that agricultural and food systems prioritize the production, preservation, and distribution of a diverse range of nutrient-rich foods. These foods should be readily available at retail outlets that are conveniently located for consumers. Furthermore, they must be both appealing and economically accessible to the public. Equally important is the need for consumers to possess the necessary resources, understanding, and enthusiasm to select and incorporate a variety of nourishing foods into their diets (Tontisirin *et al.*, 2002).

In developing countries, income is correlated with food quality. Poor individuals often rely on cereals and other cheap staples that are often deficient in important nutrients. The cost of food is another element influencing people's eating habits in third world countries. Since staple foods are the cheapest way to meet calorie needs, when their prices rise, as they have recently, more of people's incomes go towards purchasing them. As a result, people have less money to spend on nutrient-dense meals like meat and vegetables that aren't considered staples (Bouis *et al.*, 2011).

Governments have issued food-based dietary guidelines (FBDGs) to encourage healthy eating habits among their citizens for the past few decades. The consumption of foods rich in micronutrients can be increased through the effective implementation of food-based dietary guidelines at different levels. Pregnant and breastfeeding mothers, infants, and young children have special dietary needs that can be met by following specific food-based dietary guidelines (Tontisirin *et al.*, 2002).

Conclusion

To effectively tackle childhood undernutrition in India, a comprehensive program that combines essential nutritional interventions with initiatives addressing root causes is imperative for long-term sustainability.

References

Aijaz, R. Preventing hunger and malnutrition in India. 2017. Observer Research Foundation Issue Brief.

Alderman H, Hoogeveen H, Rossi M. 2009. Preschool nutrition and subsequent schooling attainment: longitudinal evidence from Tanzania., 57(2):239–60.

Bouis, H.E., Hotz, C., McClafferty, B., Meenakshi, J.V., Pfeiffer, W.H.2011. Biofortification: a new tool to reduce micronutrient malnutrition. Food and Nutrition Bulletin, 32(1): S31-40.

Chatterjee, K., Rajesh, S., Praveen, K. 2022.Finding sustainable solutions for childhood under nutrition in india: An Assessment Of Association Of Childhood Under Nutrition With Multiple Factors. International Journal of Community medicine and public health,9(5):2214-2224.

Eichler, K., Simon, W., Isabelle, R., and Urs, B.2012.Effects of micronutrient fortified milk and cereal food for infants and children: a systematic review. BMC Public Health, 12:506.

Fairulnizal, M,M., Norhayati, M.K., Zaiton, A., Norliza, A.H., Rusidah, S., Aswir, A.R., 2015. Nutrient content in selected commercial rice in Malaysia: an update of Malaysian food composition database. International Food Research Journal,22: 768–76.

Fanzo, J., Hunter, D., Borelli, T., Mattei, F. editors.2013. Diversifying Food and Diets: Using Agricultural Biodiversity to Improve Nutrition and Health. New York.

Finch C.E.2007.The biology of Human Longevity: Inflammation, Nutrition, and Aging in the Evolution of Lifespans. London: Academic press.

Floros, J. D., Rosetta, N., William, F. G, V., Barbosa, Canovas, H, C. C., Patrick, D. J., Bruce, G., Richard, L. H., Dennis, R.H., ´ Mukund, V. K., Stephen, J. K., Theodore, P. L., Daryl, B. L., Martina, N.M., James, L., Robinson, J. G. Sebranek, R. L., Shewfelt, W. F. T., Connie, M. W., and Gregory, R. Z.2010. Feeding the World Today and Tomorrow: The Importance of Food Science and Technology,9 572-599.

Graham, R.D., Welch, R.M., Saunders, D.A., Ortiz-Monasterio, I., Bouis, H.E., Bonierbale, M.2007.Nutritious subsistence food systems. Adv Agron 92:1–74.

Gupta A, Kapil U.2017. Undernutrition Amongst Under-five Children Belonging to High Income Group Communities in India. Indian Pediatrics, 54(8):686-687.

Lobstein, T.2017. Commentary: which child obesity definitions predict health risk? Italian Journal of Pediatrics, 43:20.

Low, J.W., Mwanga, R.O.M., Andrade,M., Carey, E., Ball, A.M.2017.Tackling vitamin A deficiency with biofortified sweetpotato in sub-Saharan Africa. Global Food Security. 14:23-30.

Martins VJB, Florencio TMMT, Grillo LP, do CarmoFranco MP, Martins PA, Clemente APG, San-

Martins VJB, Florencio TMMT, Grillo LP, do CarmoFranco MP, Martins PA, Clemente APG, Santos CDL, de Fatima Vieira M, Sawaya AL.2011.Long-lasting e⃞ects of Econ Dev Cult Change undernutrition. International journal of environmental health research, 8(6):1817–46.

Masset, E., Haddad, L., Cornelius, A. & Isaza-Castro, J. 2012. Effectiveness of agricultural interventions that aim to improve nutritional status of children: Systematic review. British Medical Journal., 344.

Miller, D. D., Welch, R. M.2013. Food system strategies for preventing micronutrient malnutrition. Food Policy 42:115–128.

Narayan, J., John, D. and Ramadas, N. 2018. Malnutrition in India: status and government initiatives. Journal of Public Health Policy, 40 (9859).

Public Health. 2011;8(6):1817–46.

Rickman, J.C, Barrett, D.M., Bruhn, C.M.2007a.Nutritional comparison of fresh, frozen and canned fruits and vegetables. Part 1. Vitamins B and C and phenolic compounds. Journal of the Science of Food and Agriculture, 87:930–944.

Rickman, J.C, Bruhn, C.M., Barrett, D.M. 2007b.Nutritional comparison of fresh. frozen and canned fruits and vegetables. Part 2. Vitamin A and carotenoids, vitamin E, minerals and fibre. Journal of the Science of Food and Agriculture, 87:1185–1196.

Rosenbloom, J. I. Dorit N. K, and Elliot M. B.2008. A Global Nutritional Index. Food and Nutrition Bulletin, 266-277.

Tackling Child Malnutrition. 2012. Retrieved November 16, 2014 from http://everyone.savethechildren.net/sites/everyone.savethechildren.net/files/library/CB_DA_INDIA_Lores11.pdf

Tontisirin K, Nantel G, Bhattacharjee L. 2002.Food-based strategies to meet the challenges of micronutrient malnutrition in the developing world. Proceedings of the Nutrition society. 61(2):243-50.

Torpy, J.M. , Lynm, C. and Glass, R.M. 2004. Malnutrition in children, 292 (5):648-648

tos CDL, de Fatima Vieira M, Sawaya AL. Long-lasting e□ects of undernutrition. Int J Environ Res

WHO. 2016. What is malnutrition? http://www.who.int/featu res/qa/malnu triti on/en/. Accessed 12 June 2018.

Xuan, L., Yadav, R., Siddique, K.H.M.2018. Neglected and Underutilized Crop Species: The Key to Improving Dietary Diversity and Fighting Hunger and Malnutrition in Asia and the Pacific. Frontiers in Nutrition, (19) 7:593711.

Chapter 26

Microbes in Management of Plant Disease for Sustainable Food Production

L. V. Ravishankar, Ranbir Singh, Bushra Rasool, Nisar Ahmad Dar and Sakshi Sharma

Sher-e-Kashmir University of Agricultural Science and Technology Faculty of Agriculture, Chatha,Jammu-18009

Abstract

Plant disease management is crucial in modern agriculture, and leveraging phyllosphere microorganisms had emerged as a promising strategy. The Phyllosphere, encompassing the above-ground plant parts, especially leaves, hosts a diverse microbial community influencing plant health. Interactions within this community provide sustainable and eco-friendly solutions for disease control. Phyllosphere microorganisms act as biological control agents, inhibiting plant pathogens. Bacteria like *Bacillus subtilis* produce antimicrobial compounds suppressing harmful fungi, while Trichoderma species combat pathogens on leaf surfaces. These microorganisms contribute to disease management through induced systemic resistance (ISR). Beneficial microbes, such as *Pseudomonas fluorescens,* stimulate plant defense mechanisms, enhancing resistance against pathogens. Understanding the phyllosphere microbiome dynamics is crucial for effective disease management. The leaf surface microbial community acts as a biological barrier, preventing pathogenic establishment. Studying these interactions allows identifying key microbial players for targeted disease control. Endophytic microorganisms within plant tissues produce antimicrobial secondary metabolites, protecting against pathogens. Exploring and utilizing endophytes offer potential for biocontrol agents targeting above-ground plant diseases. The phyllosphere hosts microorganisms capable of biological nitrogen fixation, contributes in plant nutrition and overall health. Nitrogen-fixing bacteria associated with leaves convert atmospheric nitrogen, enhancing plant resilience against diseases. Phyllosphere microorganisms indirectly contribute to disease management by promoting plant growth. They improve nutrient uptake, produce growth-promoting hormones, and enhance stress tolerance, creating an unfavourable environment for pathogens.

Keywords: Microbes, plant disease, Biocontrol.

Introduction

Soil, a critical component of the environment, not only serves as a valuable natural resource in agriculture but also sustains numerous life processes. Agricultural productivity is intricately linked, either directly or indirectly, to the diverse microbial communities inhabiting the soil. In recent years, plant growth-promoting bacteria (PGPB), especially endophytic bacteria, have emerged as key contributors to sustainable agriculture (Basu *et al.*, 2017; Riederer and Muller, 2008). Various forms of beneficial relationships exist between plants and microorganisms, including mutualism, commensalism, symbiosis, cohabitation, biofilms, and more. Among these, endophytes constitute a group of microorganisms residing within the host microenvironment, receiving protection from environmental stresses, gaining access to nutrients, and consequently facing reduced competition from other microbes. Microorganisms play a crucial role in plant disease management through various mechanisms that can either directly inhibit pathogens or enhance the plant's defense mechanisms (Dogra *et al.*, 2019).

These microorganisms play crucial roles in plant health, development, and ecology. Phyllosphere microorganisms are crucial for plant health and ecosystem functioning (Singha *et al.*, 2019). Understanding their classification and role in plant-microbe interactions is essential for developing sustainable agricultural practices and disease management strategies. The Phyllosphere refers to the above-ground parts of plants, particularly the leaves, where a diverse community of microorganisms, including fungi and bacteria, can be found. Classifying Phyllosphere fungi and bacteria involves categorizing them based on various characteristics such as morphology, physiology, and genetic traits (Gupta *et al.*, 2010). Recent cultivation-independent studies have provided valuable insights into the composition of microbial Phyllosphere communities. It is evident that these communities do not form randomly but undergo a selection process, resulting in relatively predictable microbial communities characterized by a limited number of dominant (sub-) phyla. This observation is based on assessments of community composition concerning plant host species, as well as temporal and spatial factors. While most research on the Phyllosphere has concentrated on bacteria and, to a lesser extent, fungi, it appears that archaea are not notably abundant in this environment. Endophytes play a more crucial role in the plant defense mechanism as they engage directly with the plant system compared to rhizosphere and Phyllosphere bacteria (Sturz *et al.*, 2000). Within plant systems, the intercellular spaces serve as optimal sites for the growth and multiplication of endophytic bacteria due to the abundance of essential nutrients such as potassium, calcium, sulfur, carbohydrates, as well as various amino acids and organic acids.

Various common plant diseases have a significant impact on global crop productivity, including blight, canker, leaf curl, leaf spot, powdery mildew, root rot, and wilt disease. Endophytes, comprising bacteria, fungi, and actinomycetes, play a crucial role in combatting phytopathogens. Endophytic bacteria, for example, produce antimicrobial compounds, siderophores, and induce systemic resistance through the synthesis of pathogenesis-related proteins (PRPs) and defense enzymes, inhibiting disease development by phytopathogens bacterial endophytes also exhibit promise as biocontrol agents. Plant microorganisms causing diseases adversely affect crop yields by significantly reducing plant performance and crop quality (Gupta *et al.*, 2014).

Endophytic bacterial strains such as Lysinibacillus sp., *Bacillus subtilis,* nitrogen-fixing bacteria (*Azotobacter chrococcum*), and *phosphate-solubilizing* bacteria (PSB) like *Pseudomonas cepacea,* as well as their combination as bio-fertilizers, have been demonstrated to inhibit bacterial wilt disease incidence in chili plants by up to 80% (Istifadah *et al.,* 2017). The endophytic bacterial strain *B. subtilis* exhibited the highest (80%) disease suppression and significantly promoted chili growth. Traditional methods for controlling phytopathogens involve chemical pesticides, but concerns about environmental contamination and the development of resistance over time have led to a search for alternative solutions. Endophytic bacteria, acting as biocontrol agents or bioinsecticides(Beattie, 2006), present an effective alternative to chemical pesticides. Numerous mechanisms contribute to the biocontrol of plant diseases, including direct antagonism through the production of antibiotics, siderophores, hydrogen cyanide, hydrolytic enzymes (chitinases, proteases, and lipases), and more(Vincent *et al.*, 2022). In a study conducted in Iran, endophytic bacteria isolated from healthy wild pistachio trees demonstrated biocontrol potential against bacterial plant pathogens. For instance, the *Pseudomonas protegens* Pb78 strain exhibited the highest inhibitory effects on bacterial plant pathogens, with *Pseudomonas syringae* pv. *Syringae* Pss 20 and *Pseudomonas tolaasii* Pt18 showing a lesser effect under in vitro conditions(Fu *et al.*, 2016). Plant diseases caused by microorganisms can significantly reduce crop yields and impact plant performance and quality.

Bio-fertilizers comprising endophytic bacterial strains such as Lysinibacillus sp., *Bacillus subtilis*, nitrogen-fixing bacteria (*Azotobacter chrococcum*), and phosphate-solubilizing bacteria (PSB) like *Pseudomonas cepacea* have demonstrated efficacy in inhibiting bacterial wilt disease in chili plants by up to 80%. The most effective strain, *B. subtilis*, not only exhibited an 80% disease suppression but also significantly enhanced chili plant growth. traditional control methods involving chemical pesticides had raised environmental contamination concerns and led to the development of resistance. Consequently, there is a growing interest in using endophytic bacteria as biocontrol agents or bio-insecticides, offering an environmentally friendly alternative to chemical pesticides.

The terms "Phyllosphere endophytes" and "epiphytes" refer to microorganisms that inhabit the surfaces of plants, particularly on leaves. However, there are distinctions between these two concepts:

Phyllosphere Endophytes

Definition: Phyllosphere endophytes are microorganisms that live within the internal tissues of plant leaves without causing any apparent harm to the host.

Location: These endophytes reside inside the leaf tissues, making them distinct from epiphytes, which are found on the external surfaces of leaves.

Relationship: Phyllosphere endophytes can form symbiotic relationships with plants, providing benefits such as improved resistance to pathogens or enhanced tolerance to environmental stress.

Epiphytes

Definition: Epiphytes are microorganisms, including fungi and bacteria, that live on the external surfaces of plant tissues, such as leaves.

Location: Unlike Phyllosphere endophytes, epiphytes do not invade the internal tissues of the plant; instead, they colonize the outer surfaces.

Relationship: Epiphytes can have various relationships with their host plants. Some may be commensal, where they neither benefit nor harm the plant, while others may be pathogenic, causing diseases. Some epiphytes may also have mutualistic relationships with plants.

Both Phyllosphere endophytes and epiphytes play important roles in plant health and ecology:

Phyllosphere Endophytes: They can contribute to plant defense mechanisms, produce bioactive compounds that deter herbivores or pathogens, and enhance the overall fitness of the host plant.

Epiphytes: While some epiphytes can be harmful and cause diseases, others may contribute to nutrient cycling, influence plant physiology, or have mutualistic relationships that benefit both the plant and the microorganism.

Role of the Endophytes in Plant Disease Management

Endophytes play a crucial role in enhancing plant defense mechanisms and managing stress in host plants, with various direct and indirect beneficial mechanisms. These microorganisms are associated with phytoremediation, bio-prospecting, phytochrome synthesis, and siderophore production. Additionally, endophytic bacteria exhibit a range of biocontrol mechanisms to directly counter plant pathogens and insects. They are known to promote plant growth and manage diseases through the production of phytohormones like abscisic acid (ABA), auxin, cytokinin, and the enzyme ACC deaminase. Moreover, they prevent infections through antifungal or antibacterial pathogens, siderophore production, nanoparticles synthesis, and by establishing the plant's systemic resistance. Endophytic bacteria, particularly as plant growth-promoting bacteria (PGPB), offer advantages over conventional chemical control methods due to their eco-friendly and non-toxic nature. Their application is sustainable for both the environment and human health. Furthermore, PGPB endophytes act against pathogens in ways that enhance crop growth and yields. For instance, four endophytic bacterial strains—Ochrobactrum sp. CB36/1, Pantoea agglomerans CC37/2, *Bacillus thuringiensis* CA41/1, and *Pseudomonas fluorescens* CC44—successfully suppressed bacterial spot disease effects on tomato and pepper plants caused by Xanthomonas euvesicatoria (Xe). Bacillus genus members are recognized as producers of biomolecules inhibiting the growth of multiple phytopathogens.

Lipopeptides, particularly, have shown efficacy as versatile weapons against various phytopathogens. Bacillus spp. isolated from different maize varieties produce antifungal lipopeptides, with extracts showing inhibitory action against *Fusarium moniliforme*. The endophytic bacteria in maize varieties may contribute to host protection by secreting antifungal lipopeptides that inhibit pathogens and induce the upregulation of pathogenesis-related genes in host plants. Additionally, endophytic bacteria from spinach roots and *Agave tequilana* plants demonstrated effective control against Fusarium wilt and antagonism against *Fusarium oxysporum* AC132, respectively.

Endophytic bacterial strain *Enterobacter cloacae* SM lO, isolated from spinach roots, significantly suppressed disease incidence by inhabiting xylem vessels. Similarly, isolates from healthy *Agave tequilana* plants, identified as Acinectobacter sp., *A. baumanii*, *A. bereziniae*, and others, exhibited plant growth-promoting activities and antagonism. Moreover, bacterial strains like *Bacillus cereus* and *B. subtilis*, isolated from different plant sources, demonstrated inhibition of various pathogens' growth and chitinase activity. These findings highlight the potential of endophytic bacteria as biocontrol agents for diverse plant diseases, contributing to sustainable agriculture and environmental health. These symbiotic relationships can be beneficial for both the plant and the endophyte. The classification of endophytes is primarily based on the type of organism and the specific plant host they inhabit. Here's a broad classification:

Endophytes are microorganisms, primarily fungi and bacteria, that live within the tissues of plants without causing any apparent harm to the host. These symbiotic relationships can be beneficial for both the plant and the endophyte. The classification of endophytes is primarily based on the type of organism and the specific plant host they inhabit.

Fungal Endophytes

Ascomycetes: These are a major group of fungi that form symbiotic relationships with plants. Examples include species of Clavicipitaceae, which are known for producing alkaloids that can protect the host plant from herbivores.

Basidiomycetes: Another group of fungi that includes endophytes like *Cryptococcus* and *Rhizoctonia.*

Bacterial Endophytes

Actinobacteria: Some endophytic actinobacteria are known for their production of bioactive compounds with potential agricultural and pharmaceutical applications.

Proteobacteria: Certain members of the Proteobacteria, such as nitrogen-fixing bacteria, can form endophytic associations with plants.

Based on the plant host:

Grass Endophytes: Many endophytic fungi are associated with grass species. For example, *Epichloë* species form endophytic relationships with grasses like ryegrass and fescue.

Tree Endophytes: Endophytes can also be found in trees, forming associations with various types of woody plants. Examples include *Colletotrichum* species in eucalyptus trees.

Crop Endophytes: Some endophytes are associated with important crop plants. For instance, species can form endophytic associations with crops like wheat and rice.

Fungal Endophytes in Plant Disease Management

The devastating impact of fungal pathogens on crops, leading to plant death, reduced yield, and diminished quality, as well as the production of harmful mycotoxins, poses significant challenges to global agriculture. In response, synthetic chemical fungicides have become a staple in agricultural practices, aiming to control these fungal threats. However, the use of such fungicides raises environmental concerns due to their non-target impacts, affecting beneficial fungi crucial for crop health, including mutualistic fungi like arbuscular mycorrhizae.

Arbuscular mycorrhizae play a pivotal role in enhancing plant fitness, and their loss due to extensive fungicide application can result in dramatic declines in plant health. Moreover, chemical fungicides often exhibit selectivity, harming non-target beneficial microorganisms more than the targeted pests. To address these ecological concerns, biocontrol endophytes, exemplified by Ampelomyces, offer environmentally friendly alternatives. Biocontrol endophytes effectively decrease the prevalence of pathogenic fungi while preserving mutualistic fungal relationships, presenting a sustainable approach to agricultural pest management.(Misci *et al.*, 2022)

The integration of biocontrol agents into pest management strategies aligns with the principles of integrated pest management (IPM), contributing to improved sustainability in agriculture and the maintenance or enhancement of soil health. This holistic approach also has the potential to mitigate or manage chemical pesticide resistance, a growing concern in modern agriculture. Secondary metabolites produced by endophytes have garnered considerable attention for their potential as agrochemicals. Studies have employed top-down approaches to extract and isolate diverse compounds from specific taxonomic orders of fungi, such as Xylariales, revealing a wealth of bioactive metabolites.

These compounds include glucosides, cytochalasans, azaphilones, terpenoids, non-ribosomal peptides, macrolide polyketides, benzenoids, and lactones. The identification and characterization of these compounds contribute to the exploration of natural products with agricultural applications. Researchers adopt various methods to identify specific antifungal compounds produced by endophytes. This includes characterizing fungal endophytic diversity within host plant species, selecting endophyte cultures for dual-culture assays to assess antagonism against known plant pathogens, and subsequent extraction and analysis of compounds. Liquid chromatography and gas spectrometry coupled with mass spectrometry provide valuable insights into the chemical composition of these compounds. The results aid in identifying candidate endophyte species with biocontrol potential, underscoring the intrinsic role of antagonists within the plant microbiome. Moreover, endophytes contribute to host plant resistance against fungal pathogens by inducing a systemic response upon colonization. This response involves the initiation of a defensive strategy by the plant, reinforcing cell walls through the deposition of defensive compounds. Although endophytes possess mechanisms, such as exoenzymes, to access these strengthened cells, the deposits act as a barrier, preventing pathogens from penetrating. Understanding these mechanisms sheds light on how endophytes enhance plant resistance and underscores their multifaceted role in promoting overall plant health. In summary, the intricate interactions between endophytes, plants, and fungal pathogens unveil a promising avenue for sustainable agricultural practices. The exploration of biocontrol endophytes, the identification of bioactive compounds, and the understanding of their mechanisms contribute to the development of environmentally friendly alternatives to synthetic chemical fungicides, fostering agricultural sustainability and ecosystem health.(Abo-Elyousr *et al.*, 2014).

Conclusion

The intricate relationships between plants and microbes play a pivotal role in the dynamic field of plant disease management. The devastating impact of fungal pathogens on crops and the associated risks of chemical fungicides have spurred a growing interest in harnessing the potential of plant-associated microbes for sustainable and eco-friendly solutions (Saxena *et al.*, 2016). Biocontrol endophytes, exemplified by fungi like Ampelomyces, emerge as promising alternatives to synthetic chemical fungicides. These microbes demonstrate the ability to reduce pathogen prevalence while preserving essential mutualistic relationships within the plant microbiome. The integration of biocontrol agents into integrated pest management approaches not only mitigates the environmental impact of chemical pesticides but also contributes to the overall sustainability of agricultural practices. Exploration into the secondary metabolites produced by endophytes adds another layer to the potential applications of plant-associated microbes. The identification of diverse and bioactive compounds holds promise for developing natural products that can serve as effective agrochemicals. This avenue of research not only expands our understanding of microbial diversity but also offers practical solutions for enhancing crop health and resilience against fungal pathogens. Moreover, the mechanisms through which endophytes enhance host plant resistance add depth to our comprehension of plant-microbe interactions. The induction of systemic responses, such as the reinforcement of cell walls, provides insights into the sophisticated strategies employed by plants and their associated microbes in the ongoing battle against pathogens. Understanding these mechanisms is crucial for optimizing the use of plant-associated microbes in disease management strategies. As we navigate the challenges of global agriculture, the holistic approach of integrated pest management, coupled with the judicious use of plant-associated microbes, stands out as a beacon for sustainable practices. The promotion of soil health, the reduction of chemical pesticide resistance and the preservation of essential microbial communities within the plant ecosystem all contribute to the overarching goal of achieving agricultural sustainability.

References

Abo-Elyousr, K. A., Abdel-Hafez, S. I., and Abdel-Rahim, I. R. (2014). Isolation of Trichoderma and evaluation of their antagonistic potential against Alternaria porri. *Journal of Phytopathology* 162, 567-574.

Basu, S., Rabara, R., and Negi, S. (2017). Towards a better greener future-an alternative strategy using biofertilizers. I: Plant growth promoting bacteria. *Plant Gene* 12, 43-49.

Beattie, G. (2006). Plant-associated bacteria: survey, molecular phylogeny, genomics and recent advances. *Plant-associated bacteria*, 1-56.

Dogra, N., Yadav, R., Kaur, M., Adhikary, A., Kumar, S., and Ramakrishna,

W. (2019). Nutrient enhancement of chickpea grown with plant growth promoting bacteria in local soil of Bathinda, Northwestern India. *Physiology and Molecular Biology of Plants* 25, 1251-1259.

Fu, S.-F., Sun, P.-F., Lu, H.-Y., Wei, J.-Y., Xiao, H.-S., Fang, W.-T., Cheng, B.-Y., and Chou, J.-Y. (2016). Plant growth-promoting traits of yeasts isolated from the phyllosphere and rhizosphere of Drosera spatulata Lab. *Fungal biology* 120, 433-448.

Gupta, N., Sahoo, D., and Basak, U. C. (2010). Evaluation of in vitro solubilization potential of phosphate solubilising Streptomyces isolated from phyllosphere of Heritiera fomes (mangrove). *Afr J Microbiol Res* 4, 136-142.

Misci, C., Taskin, E., Vaccari, F., Dall'Asta, M., Imathiu, S., Cocconcelli, P. S., and Puglisi, E. (2022). Valorization of African indigenous leafy vegetables: The role of phyllosphere microbiota. *Food Research International,* 111944.

Riederer, M., and Muller, C. (2008). "Annual plant reviews, biology of the plant cuticle," John Wiley & Sons.

Saxena, A., Raghuwanshi, R., Gupta, V. K., and Singh, H. B. (2016). Chilli anthracnose: the epidemiology and management. *Frontiers in microbiology* 7, 1527.

Singha, M., Srivastavaa, M., Kumarb, A., Singhc, A., and Pandeya, K. (2019). aCentre of Advance study in Botany, Institute of Science, Banaras Hindu University, Varanasi, India, bInstitute of Plant Sciences, Agricultural Research Organization, Volcani Center, Rishon LeZion, Israel, cKrishi Vigyan Kendra, Khalilabad, India. *Microbial Endophytes: Prospects for Sustainable Agriculture,* 61.

Sturz, A., Christie, B., and Nowak, J. (2000). Bacterial endophytes: potential role in developing sustainable systems of crop production. *Critical reviews in plant sciences* 19, 1-30.

Vincent, S. A., Ebertz, A., Spanu, P. D., and Devlin, P. F. (2022). Salicylic acid-mediated disturbance increases bacterial diversity in the phyllosphere but is overcome by a dominant core community. *Frontiers in Microbiology* 13, 809940.

Chapter 27

Genetic Resource Management of Nut Crops for Food and Nutritional Security

Muzafar Mir, Shahid Ahamad, Parshant Bakshi, Suja Qurashi and Mahital Jamwal*

Sher-e-Kashmir University of Agricultural Sciences & Technology of Jammu (J&K)-India.
**Corresponding author: drmuzaffarcqar@gmail.com*

Introduction

India faces many challenges with food and nutritional security including undernourishment, malnutrition, food price inflation and climate change. Nut crops play a significant role in the well-being of people as they provide supplemental food, a balanced diet, and enhance household income, in addition to their substantial contribution to the sustainability of ecosystems and environmental health. Temperate nut species come from locations with cold winters and well-defined climatic seasons and temperatures suitable for spring and summer growth. Edible nut species that are native to this region include almond (Prunus amygdalus), walnut (Juglans regia), pecan nut (Carrya illinoinensis), Hazel nut (Corylus avellana), Chest nut (Castanea spp) and Pine nut (Pinus pinea). Nut crop diversity is crucial for nutrition, livelihoods and ecosystem resilience around the world. The importance of these nut species are increasing day by day due to its contribution to food and nutrition, ecosystem and environment sustainability and empowerment of livelihoods. Since most of these nut crops are perennial trees species, the conservation of these tree species in the centre of origin and or centre of genetic diversity is essential for its continued evolution. Their conservation requires understanding which leads to their effective utilization in plant improvement, both for productivity and quality traits. However, the regions that host genetic diversity are threatened by loss of habitat, vandalism, urbanization, industrialization and modernization of agricultural practices. Several nut species are semi-wild and domestication of these tree species helps to inculcate a secondary or tertiary gene pool and evolved for a new set of genome adapted for the agro ecological condition and biotic/abiotic stresses. A better understanding of domestication across many crop species may provide crucial information that may be needed to breed better nut crops for the future.

Agriculture's Dependence on Genetic Resources

Agriculture and genetic resources are critically interdependent. All agricultural commodities, even modern varieties, descend from an array of wild and improved genetic resources from around the world. Furthermore, agricultural production depends on continuing infusions of genetic resources for yield stability and growth. In the India agricultural productivity has increased continuously over the last century (Ahearn *et al.*, 1998). Increases in crop yields have contributed to farmers' ability to produce more food and fiber with fewer inputs. Half the India's yield increases of the last 60 years have been attributed to such genetic improvements (Fuglie *et al.*, 1996). Genetic improvements have arisen in several ways. Before the development of modern varieties, farmers cultivated landraces. Landraces are varieties of crops that evolved and were improved by farmers over many generations, without the use of modern breeding techniques. These varieties are generally very diverse within species, because each was adapted to a specific environment. The pace of improvement accelerated as modern breeding techniques were developed that facilitated selection of specific desirable traits.

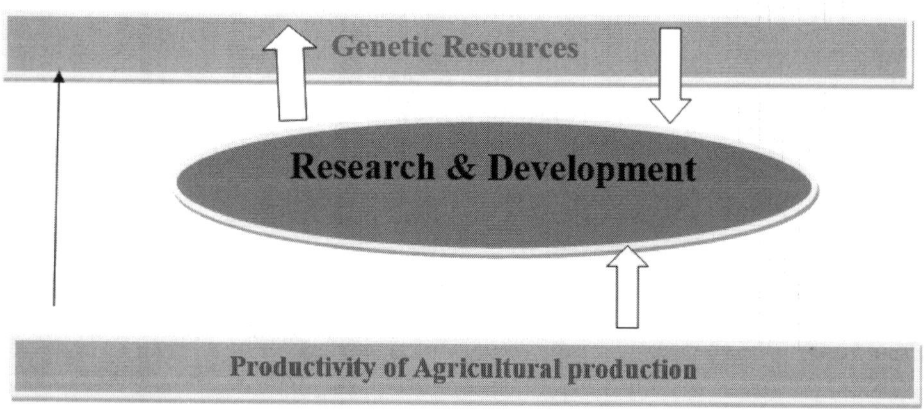

Figure-1. The relationship between genetic resources and agriculture

Conservation of Nut Genetic Resources

Germplasm once lost cannot be replaced so its conservation is very important for sustainable life. Conservation refers to the preservation of germplasm. The conservation of germplasm is achieved by two methods such as in-situ conservation and ex-situ conservation.

In Situ Conservation

Species preserved in situ remain in their natural habitat. Most of the world's genetic diversity is found in situ. For agriculturally important species, the greatest diversity in landraces and in wild relatives may be found near their centers of origin, or the places in which they were first domesticated. One of the first scientists to define centers of origin was the Russian botanist N. I. Vavilov in the early twentieth century. Conservation of temperate nut genetic resources in their place of occurrence through the maintenance of their habitat either through protection or promotion is among the most appropriate approaches for perennial nut species, especially in the case of wild and endemic species. In addition, most of the field gene banks (FGBs) end up maintaining only the elite material for use in the breeding programs, rather than conserving diversity. Hence, the in situ method aids in preserving genetic and species diversity and allows evolutionary forces to act on the material, either through natural selection or selection by farmers, based on their changing preferences. The in situ conservation approach is well suited for the long-term conservation of temperate nut species. Globally, crop wild relatives, including temperate nut crops, are poorly represented in ex situ collections, and they thus need special attention to be protected in identified "protected" natural habitats. Efforts must be increased to identify and earmark-specific sites of specific species and studies on population dynamics are needed as well. In India, there are 18 biosphere reserves, 104 national parks, and 551 wildlife sanctuaries (Ghosh-Harihar *et al.* 2019). The Western Himalayas region is top in temperate nut crop diversity, followed by the North-East region (Sharma and Kumar, 199)

Ex Situ Conservation

In this method conservation of germplasm is carried outside their native habitat in the form of seed, embryo, tissue, etc. Objective of ex-situ conservation is to maintain the accessions without any change in genetic constitution. The conservation of perennial nut crop genetic resources, especially elite material, is largely dependent on ex situ methods, such as botanical gardens, arboreta, and gene-banks. Gene banks are the dedicated organizational units, working to manage plant genetic resource collections for their conservation and ease of access for utilization. Typically, gene-banks conserve those resources in the form of seed, plant, tissue, pollen, or DNA in the seed gene-bank or, in vitro gene-bank, cryogene-banks, or DNA banks, or as live collections in the FGBs (Agrawal *et al.*2021). Even though botanical gardens and private orchards play a role in ex situ conservation, the genetic diversity conserved in these forms is not adequate to represent the existing variability, owing to the limited number of species that are planted in botanical gardens. Thus, gene-banks play a significant role in the management of plant genetic resources.

Field Gene banks (FGBs)

The field gene bank is an area where the growing plants are conserved. Most of horticultural genetic resources are either difficult or impossible to conserve as seeds because of being recalcitrant or they reproduce vegetatively and hence conserved in field gene banks. Germplasm of major fruits is being maintained in FGBs by horticultural institutes, NRCs, SAUs and NAGS. Conservation of true-to-type genetic material in fruit crops is feasible only through vegetative propagules, since most of the nut species are highly cross-pollinated. Seed banking is not a reliable method for nut genetic resources conservation, owing to highly recalcitrant seeds, which lose viability upon desiccation and storage; hence, most of these resources are conserved in FGBs. For effective conservation through FGBs, a sizeable number of genotypes of the given species is required, which entails a larger area and inputs, making it expensive. However, the FGBs are easy and readily accessible to breeders or curators for evaluation, documentation, and use in improvement programs. The major FGBs in the world for F&N conservation include 11 USDA repositories; the International Musa Germplasm Transit Centre (ITC), Belgium; the Brazilian Agriculture Research Corporation- Cassava and Fruits (EMBRAPA-CNPMF), Brazil; Seed Savers Exchange, USA; the National Agriculture and Food Research Organization (NARO), Japan; the Centre for Pacific Crops and Trees (CePaCT), Fiji; the German Fruit Genebank (GFG), Germany; etc. (Ranjan *et al.* 2022).

In India, the ICAR-NBPGR, New Delhi, manages the F&N genetic resources as the nodal agency with its seven regional stations, in collaboration with the various ICAR institutes, such as the Central Institute of Temperate Horticulture (temperate nuts, 398 acc.). For the material conserved in FGBs, appropriate protocols need to be established for producing plant material for exchange. Since most of the time only elite material of temperate nut crop is exchanged, it is usually in the nature of propagule for clonal propagation. In rare cases where representative materials for genetic diversity are exchanged, appropriate protocols for the regeneration of seed, based on their breeding system, need to be followed (FAO 2013).

Conservation of Tissues

In Vitro

Since seed conservation is not an option for most nut crop genetic resources as they produce recalcitrant seeds, researchers have developing and evaluation the storage of tissue or other propagules using different methods of storage. These methods are further elaborated here. In future the storage of DNA could be exploited as well.

In Vitro Gene-banks (IVG)

Utilizing tissue culture technology offers short- to medium-term conservation (2–3 years) of genetic material in many species; this is achieved through the maintenance of cultures at normal or reduced growth condition by manipulating the culture conditions or media supplements, thereby prolonging the subculture intervals and minimizing the chances of variation. In the case of vegetatively propagated crops like fruit crops, the IVAGs serve as a safety backup of accessions maintained in FGBs, using minimal space for larger numbers and also for the safe exchange of germplasm between countries. However, the development of sound in vitro regeneration protocol is a prerequisite, which is a an in vitro repository of different nut fruit crops. Although in vitro genebanks serve as a safety backup, the selection pressure or sub culturing involved in the technique can lead to somaclonal variants. Hence, under in vitro conditions, genetic material should be periodically monitored for genetic stability using morphological, biochemical, or molecular methods (Agrawal *et al.* 2021).

Cryogene Bank

Cryopreservation is a relatively new conservation method and protocols have been developed for conservation of various crops for long term storage at -196°C under liquid nitrogen or vapour phase (Gupta, 2010). Cryopreservation comprises of following steps:

a. Choice of explant: Explant includes shoot apex, somatic embyros, seed, excised zygotic embryo, dormant vegetative bud and pollen.

b. Pre-culture: Culture of cells, tissues, organs in the presence of amino acids like proline, sugars/ sugar alcohols like sucrose and mannitol or at low temperature prior to freezing initiates in them important physiological changes which increase their freezing tolerance. Plantlets may be hardened by growing them at -1ºC for 16 hours each day. Rapid dehydration is done by application of sucrose at concentration of 30-240g/l.

c. Cryoprotectant: Here Cryoprotectant like dimethyl sulphoxide (DMSO), glycerol or sorbitol are given to tolerate cryo-injury.

d. Cooling: the material is cooled 0.5-4°C/min up to -40 to -100°C for 20-45 minutes.

e. Storage in liquid nitrogen: Finally the explant is stored in liquid nitrogen at -196°C

In most temperate nut crops like chestnut (Castanea sp.), pecan nut (Carya sp.), walnut (Juglans sp.), Pecan nut (Carrya illinoinensis) etc., the seeds are highly recalcitrant and desiccation sensitive, and hence, embryos or embryonic axes can be cryopreserved with varying success. Long-term conservation of nut crop germplasm via rough cryopreservation is a promising strategy; however, the optimization of regeneration protocols in woody perennials is challenging as in the case of in vitro gene-banking. It must be noted that with increased pressure on land, cryopreservation holds the key for the conservation of most perennial species, including nut species, and hence serious attention needs to be given to research on cryopreservation of a much larger list of species, including nut crops.

Germplasm Characterization and Evaluation

The main aim of germplasm characterization and evaluation is to identify gene sources for resistance to biotic and abiotic stresses, earliness, dwarfness, and quality characters and to know the significance of individual germplasm line. Characterization and evaluation are the keys to assessing the potential of any germplasm. Characterization allows the morphological or biophysical identification of the qualitative traits, which are mono-oligogenic and highly heritable, while evaluation is related to the quantitative, oligo-polygenic traits that are susceptible to environment differences but are generally useful in crop improvement (Fideghelli *et al.* 2001). The germplasm collection must be characterized, using a list of descriptors (the identifiable and measurable trait of a plant accession) developed by the international or national organizations of PGR management. The most used descriptors are those of the International Union for the Protection of New Varieties of Plants (UPOV), the US Department of Agriculture (USDA), Bioversity International, etc. ICAR-NBPGR has also developed minimal descriptors for many fruit crops, which can be used for preliminary characterization. Wherever descriptors are not available, the curators must develop the descriptors, in consultation with the crop experts of the respective species. Adopting internationally recognized descriptor lists allows homogeneity in data collection, which can facilitate and rationalize germplasm and information exchange. The germplasm description includes three levels of data collection (Agrawal *et al.* 2021) Passport data: all basic information recorded at the time of collecting the sample, type of sample, site of collection, etc. and accession information which includes information recorded at the time of registering a sample as an accession in a genebank (Archak *et al.* 2022) characterization data (morphological traits, etc.) and (Arora and Nayar 1984) evaluation data (resistance to various stresses, biochemical and molecular traits, etc.). The morphological characterization, along with biochemical and molecular markers describes accessions almost in entirety. In the case of nut crops, in situ preliminary characterization at the collection site is carried out to identify and collect the elite genotypes, in addition to general genetic diversity, to overcome the limitations in establishing and maintaining large FGBs. Nut crops should be characterized using an appropriate statistical design that always includes a standard cultivar as a control. If the number of accessions in the collection is more, using accessions from such a large collection for research and crop improvement programs may be challenging. Evaluation helps to identify germplasm with a trait of interest for their future utilization in the breeding programs. Evaluation is carried out in the already characterized germplasm, with a multidisciplinary approach, at multiple locations, using suitable experimental designs.

Information and Documentation of PGR

Much of the world's PGR information is poorly documented. The passport data and characterization and evaluation data of germplasm accessions conserved in various genebanks across the globe are scattered at different places, in the form of crop catalogues, published articles or reports, etc., thus hindering the effective use of these data. They must be collected, collated, and exchanged for better use of conserved genetic resources. The national and international initiatives on such PGR informatics include the database and PGR portals of the Germplasm Resources Information Network (GRIN) of the USA; the National Agriculture and Food Research Organization (NARO) of the Japanese genebank; European Search Catalogue for Plant Genetic Resources (EURISCO); and Bioversity International's New World Fruit Database. In India, ICAR-NBPGR has also developed a PGR portal, which is a gateway to information on material conserved in the Indian National Gene Bank at ICAR-NBPGR, New Delhi. The mobile apps "Genebank" and "PGR map" have also been developed in PGR informatics, for easier access to the PGR information on germplasm collections in the country (Archak *et al.* 2022). The India Biodiversity Portal (https://indiabiodiversity.org/) has some information on cultivated and wild relatives of temperate nut crop species. Such database management systems make the PGR information easily available and readily accessible, facilitating the exchange of information and the effective utilization of the germplasm.

Utilization

The information on the characterization and evaluation of germplasm conserved in ex situ genebanks and/or in situ conservation sites will lead to the identification of elite genotypes having desirable traits, which can then be used in crop improvement programs through the incorporation of desirable trait(s) into cultivars selected. To facilitate the effective use of any germplasm, registration of elite germplasm with an appropriate national body or a publication is a prerequisite. ICAR-NBPGR facilitates the registration of unique germplasm and genetic stocks, which makes the germplasm information available in the public domain, for their use in the improvement programs. The developer of the germplasm must maintain and supply the registered germplasm to the bonafide users (Singh *et al.* 2018). In the case of temperate nut fruit crops, the landraces and wild relatives are the treasure house of many useful genes, and efforts to introgression these genes into the cultivated types are important in the present conditions of climate change. Breeding programs, involving intra or inter-specific and distant hybridization and other breeding techniques, coupled with biotechnological interventions (molecular markers, embryo rescue, somatic hybridization, etc.), can assist in better utilization of the identified elite germplasm in temperate nut crop improvement programs.

Conclusion

In the situation of climate change and depletion of natural resources, the challenges are more to feed growing population, so many efforts are required for exploring the unexplored areas for collection of horticultural biodiversity wealth for conservation and further utilization for the benefit of mankind. In the quest to meet the emerging challenges, the gains with respect to genetic resources have to be sustained and further collection of new genes are required to be looked and utilized for gains where the strong base of horticultural plant biodiversity have to be in driving seat for bringing gene revolution. As far as the temperate nut crop genetic resources management is concerned, the important actions to be intensified for tackling the various challenges posed by the changing climatic conditions and emerging pests and diseases, trait-specific germplasm introductions and gap analysis and targeted collecting must be carried out with inter-institutional and international collaborations.

References

Agrawal A, Srivastava V, Malhotra EV, Patil P, Singh K (eds) (2021) Training manual for virtual training course on management of fruit genetic resources, CAR-National Bureau of Plant Genetic Resources, New Delhi, ICAR-AII India Coordinated Research Project on Fruits, ICAR-Indian Institute of Horticultural Research, Bengaluru, p 140

Ahearn, M., J. Yee, E. Ball, and R. Nehring (1998). Agricultural Productivity in the United States. AIB No. 740. Economic Research Service, United States Department of Agriculture.

Archak S, Singh K, Agrawal RC (2022) PGR informatics: sustaining PGR management and supporting PGR policy. Indian J Plant Genet Res 35(3):218–220.

FAO (Food and Agriculture Organization) (2013). Gene-bank standards for plant genetic resources for food and agriculture. Food and Agriculture Organization of the United Nations, Rome.

Fuglie, K., N. Ballenger, K. Day, C. Klotz, M. Ollinger, J. Reilly, U. Vasavada, and J. Yee (1996). Agricultural Research and Development: Public and Private Investments Under Alternative Markets and Institutions, Agricultural Economic Report No. 735. Washington, D.C.: Economic Research Service, U.S. Department of Agriculture.

Ghosh-Harihar M, An R, Athreya R, Borthakur U, Chanchani P, Chetry D, Datta A, Harihar A, Karanth KK, Mariyam D, Mohan D. (2019). Protected areas and biodiversity conservation in India. Biol Conserv 237:114–124.

Gupta, S. 2010. Management of Temperate fruit genetic resources in India. Acta Horticulturae: 71-80.

Ranjan P, Brahmi P, Tyagi V, Ranjan JK, Srivastava V, Yadav SK, Singh SP, Singh S, Binda PC, Singh SK, Singh K (2022). Global interdependence for fruit genetic resources: status and challenges in India. Food Secur 14(3):591–619.

Sharma R L and Kumar K. 1994. Temperate fruit crop improvement in India. Progress in temperate fruit breeding. Vol. 1, ISBN: 978-94-010-4209-3.

Singh K, Kumar S, Gupta K, Pradheep K, Rajasekharan PE, Pandey C, Ranjan P, Gupta S, Gangopadhyay KK (2018). Genetic resource management and utilisation of horticultural crops in India–a perspective. Int J Innov Hortic 7(2):71–103

Chapter 28

Millets: The Smart Grain for Sustainable Agriculture

***Daman Preet Kour*[*], *Neeraj Gupta, Seerat Gupta and Vikas Gupta*[1]**

Division of Post Harvest Management, SKUAST-Jammu, Chatha 180009, India.
[1]Basmati Rice Research Centre, SKUAST-Jammu Chatha 180009, India.
****Corresponding author: damanreen323@gmail.com***

Introduction

Millets are important crops for dry land farmers; they are highly nutritious and climate-compliant. Overall millets consumption in India has declined over the years. In order to revive the demand for millets in India, government and non-government led a consortium funding to undertake interventions to bridge the gaps identified in the millet value chain: at on-farm production, processing diversification, nutritional certification, promotion and marketing. This is due to productivity gains in some varieties, with the production of millet showing some increase despite the shrinkage of the area (Saraswathi and Hameed, 2022). A significant food crop with a significant economic impact on developing nations is millet. Millets are advantageous because they are grains that are resistant to pests and drought. It is a small- to medium-sized crop that is grown all over the tropics and subtropics (Wang *et al.*, 2010).

Millets are among the earliest food sources known to humans, and they may have been the first grains that were used for domestic purposes. It belongs to the cereal crop plant in the Graminae grass family. Millet encompasses several types of small seeded annual grasses belonging to five genera in the tribe *Paniceae*, namely *Panicum, Setaria, Echinocloa, Pennisetum,* and *Paspalum,* and one genus *Eleusine* in the tribe *Chlorideae* (Chandrasekara and Shahidi, 2010). Pearl millet (*Pennisetum glaucum,* with synonyms of *P. americanum, P. typhoides,* and *P. typhoideum*), Foxtail millet (*Setaria italica*), Proso millet or white millet (*Panicum miliaceum*), and Finger Millet (*Eleusine coracana)* are among the major millet crops and the minor millets include Barnyard millet (*Echinochloa* spp.), Kodo millet (*Paspalum scrobiculatum*), Little millet (*Panicum sumatrense*), Guinea millet (*Brachiaria deflexa/Urochloa deflexa*), and Browntop millet (*Urochloa ramose/ Brachiaria ramose/Panicum ramosum*) (Chinchole *et al.,* 2017; ICRISAT, 2017). Millet-based foods are considered to be potential prebiotic and probiotic sources with potential health benefits (Palaniswamy and Govindaswamy, 2016). Millet grains are widely consumed as a source of traditional medicines and as a healthy food. Millet crops' antioxidant, anticancer and anti-inflammatory, antifungal, and blood clot inhibition properties may be attributed to phytonutrients and vitamins (Dykes and Rooney, 2007; Yadav *et al.,* 2013).

Technology for Millet Value Addition

Composite Flour

One approach to increasing the broader use of millet is by blending its grains with commonly consumed cereals such as wheat and maize, as well as nutrient-rich pulses. This process can help expand millet's application in various food products. Popular wheat-based items like cakes, pasta, macaroni, vermicelli, noodles, spaghetti, and flakes are widely eaten in both developed and developing countries. Typically, these products are made using refined wheat flour or semolina. By incorporating millet flour into these foods at specific ratios, the nutritional and functional qualities can be improved, while also altering the physicochemical characteristics of the products (Singh *et al.,* 2005). Studies suggest that blending millet flour in the range of 10–30% with wheat flour (70–90%) enhances the overall quality of the resulting product.

Puffed/Popped and Flaked Millets

The consumption of cereals in puffed and popped forms is widely practiced in both developed and developing regions. Since starch is the primary component of millet and other cereal grains, various processing methods, particularly thermal treatments, cause significant physicochemical and structural changes in the starch-protein matrix. These changes result in the expansion of the grains, producing puffed products. High-temperature short-time (HTST) treatment is commonly used to create expanded flakes and other popped products (Jaybhaye *et al.,* 2014). In regions like the North West Himalayas of India, coarse cereals such as finger millet are a key part of the local diet, often consumed as *roti* (unleavened flatbread). Unlike wheat-based *rotis,* finger millet *rotis* do not puff during cooking due to the absence of gluten in the flour, resulting in a harder texture and less consumer appeal. To address this, traditional methods in the Uttarakhand hills use the dried bark of a tree (*Boehmeria rugulosa*), locally known as "gethi," to improve the softness and puffability of finger millet *rotis*.

Extruded Products

Extrusion cooking is a technique used in post-harvest processing of starchy and protein-rich materials, where the food is subjected to high temperatures for a short period. This method improves protein digestibility, enhances the quality, and increases the versatility of processed food products (Dahlin and Lorenz, 1992). Heat is applied either directly via steam injection or indirectly through a jacket, with mechanical energy from the shearing process within the blend contributing to the cooking process. Studies have shown that ready-to-cook extruded products made by mixing pearl millet and finger millet flour has acceptable levels of nutrient content, color, texture, cooking quality, and sensory characteristics (Devi and Narayanasamy, 2013).

Baked Products

Baked goods, known for their extended shelf life, easy marketability, and attractive packaging, are widely popular across all age groups worldwide However, the baking industry is predominantly dominated by value-added wheat products. Due to the high gluten content, these products are unsuitable for individuals with celiac disease, especially in developed countries. Millet-based baked goods not only offer superior nutritional benefits but can also command a higher price in the market (Verma and Patel, 2013). Flour from foxtail millet and finger millet is ideal for making biscuits and muffins, while barnyard millet is flavoured for cakes. In recent years, the demand for finger millet-based baked products has surged in urban areas.

Fermented Products

While millets are generally a good source of protein, their protein quality is considered low due to an insufficient essential amino acid profile (Jaybhaye *et al.*, 2014). However, processes like probiotic fermentation and germination have been found to improve protein digestibility and increase the levels of lysine, thiamine, niacin, sugars, protein fractions, soluble fibers, and the bioavailability of micronutrients (Arora *et al.*, 2011). Fermented food items such as *idli* and *dosa* are popular and widely consumed breakfast options in many parts of India (Mbithi-Mwikya *et al.*, 2000).

Benefits of Millets for Farmers's Sustainable Approach

Millets offer several benefits for farmers and contribute to a sustainable agricultural approach. Here are the key advantages:

1. Climate Resilience

- **Drought Tolerance**: Millets are highly resilient to drought and can grow in arid and semi-arid regions where other crops may fail.
- **Heat and Pest Resistance**: These crops are resistant to high temperatures and common pests, making them ideal for areas affected by climate change.
- **Low Water Requirements**: Millets need significantly less water compared to crops like rice or wheat, making them a sustainable choice in water-scarce regions.

2. Low Input Costs

- **Minimal Fertilizer and Pesticide Use**: Millets require fewer chemical inputs such as fertilizers and pesticides, reducing costs for farmers and minimizing environmental damage.
- **Adaptability to Poor Soils**: They can thrive in marginal and nutrient-poor soils, where other staple crops would struggle to grow, making them suitable for smallholder farmers in resource-constrained areas.

3. Short Growth Cycle

- **Fast Maturity**: Millets have a shorter growing cycle compared to many other cereals, allowing farmers to harvest more frequently and reduce the risk of crop failure due to adverse weather conditions.
- **Multiple Harvests**: Some millet varieties allow for multiple harvests in a single growing season, increasing productivity.

4. Biodiversity and Soil Health

- **Promotes Crop Rotation**: Growing millets enhances crop diversity and encourages rotation, which improves soil fertility and reduces the risk of pest and disease buildup.
- **Natural Weed Suppression**: Millets are effective in controlling weeds, which reduces the need for herbicides and protects soil quality.

5. Economic Viability and Market Potential

- **High Market Demand**: With increasing awareness of their health benefits, millets are gaining popularity, especially in urban markets, providing farmers with higher income potential.
- **Value-Added Products**: Farmers can diversify income streams by processing millets into value-added products like flour, snacks, and beverages, which command premium prices in health-conscious markets.

6. Nutritional Benefits

- **Nutrient-Rich Crops**: Millets are rich in fiber, protein, vitamins, and minerals like calcium and iron, which contributes to food security and improved nutrition in farming communities.
- **Combatting Malnutrition**: By growing millets, farmers not only sustain their livelihoods but also help combat malnutrition, particularly in rural areas.

7. Support for Marginalized Farmers

- **Suits Smallholders and Women Farmers**: Millets are easy to grow and manage, making them an accessible crop for small-scale and women farmers who often have limited access to resources.
- **Traditional Knowledge Preservation**: Millets support the continuation of traditional agricultural practices, helping preserve indigenous knowledge systems and cultural heritage.

By incorporating millets into farming practices, farmers can achieve both economic stability and environmental sustainability, while contributing to global food security (Karuppasamy, P. 2015; Saleh *et al.*, 2013).

Conclusion

Millet grains offer a sustainable and holistic solution to the challenges of global nutritional food security. Their ability to thrive in harsh climates, low water requirements, and minimal reliance on chemical inputs make them a climate-resilient crop for the future. Rich in essential nutrients such as fibre, proteins, vitamins, and minerals, millets provide a healthier alternative to staple crops like wheat and rice, addressing malnutrition and dietary deficiencies in many regions. By integrating millets into agricultural practices, we can support smallholder farmers, promote biodiversity, and ensure a sustainable food system that not only feeds growing populations but also safeguards environmental health. Embracing millets as part of a broader strategy for smart agriculture will enhance food security, improve rural livelihoods, and contribute to global sustainability efforts.

References

Arora, S., Jood, S. and Khetarpaul, N. 2011. Effect of germination and probiotic fermentation on nutrient profile of pearl millet-based food blends. *Br Food J*, 113 (4): 470–481.

Chandrasekara, A. and Shahidi, F. 2011. Determination of antioxidant activity in free and hydrolyzed fractions of millet grains and characterization of their phenolic profiles by HPLC□DAD□ESI□*MSn.Journal of Functional Foods*, 3: 144–158.

Chinchole, M., Pathak, R. K., Singh, U. M. and Kumar, A. 2017. Molecular characterization of EcCIPK24 gene of finger millet (*Eleusine coracana*) for investigating its regulatory role in calcium transport. *Biotechnology*, 7: 267. coracana). *Emir J Food Agric*, 25:169–176.

Dahlin, K. and Lorenz, K. 1992. Protein digestibility of extruded cereal grains. *Food Chem*, 48: 13–18.

Devi, M.P. and Narayanasamy, S. 2013. Extraction and dehydration of millet milk powder for formulation of extruded product. *J Environ Sci Toxicol Food Technol*, 7: 63–70.

Dykes, L. and Rooney, L. W. 2007. Phenolic compounds in cereal grains and their health benefits. *Cereal Foods World*, 52 (3): 105–111.

ICRISAT. 2017. Small millets. Retrieved from http://www.icrisat.org/

Jaybhaye, R.V., Pardeshi, I.L., Vengaiah, P.C. and Srivastav, P.P. 2014. Processing and technology for millet-based food products: a review. *J Ready Eat Food*, 2:32–48.

Karuppasamy, P. 2015. Overview on Millets. *Trends in Biosciences*, 8 (13):3269-3273.

Mbithi-Mwikya, S., Ooghe, W., Van-Camp, J., Nagundi, D. and Huyghebaert, A. 2000.Amino acid profile after sprouting, autoclaving and lactic acid fermentation of finger millet (Eleusine coracana) and kidney beans (Phaseolus vulgaris L.). *J Agric Food Chem*, 48: 3081–3085.

Palaniswamy, S. K. and Govindaswamy, V. 2016. In□vitro probiotic characteristics assessment of feruloyl esterase and glutamate decarboxylase producing *Lactobacillus spp*. isolated from traditional fermented millet porridge (*kambu-koozh*). *LWT□Food Science and Technology*, 68: 208–216.

Saleh A., Zhang Q., Chen J. and Shen Q. 2013. Millet grains: nutritional quality, processing, and potential health benefits. *Compr Rev Food Science Food Safety*, 12: 281–295.

Saraswathi, R and Hameed R.S. 2022. Value Addition Influenced in Millet Products. *Journal of Nutrition and Food Sciences*, 10 (1): 1-6.

Singh, P., Singh, G., Srivastava, S. and Agarwal, P. 2005. Physico-chemical characteristics of wheat flour and millet flour blends. *J Food Sci Technol*, 42: 340–343.

Verma, V. and Patel, S. 2013. Value added products from nutri-cereals: finger millet (Eleusine

Wang, B., Jiang, K., Jing, M., Wang, M. and Zhang, F. 2010. Selecting high-quality millet by quantification of indicators. *Chinese Agricultural Science Bulletin*, 26 (1): 62–66.

Yadav, N., Chaudhary, K., Singh, A. and Gupta, A. 2013. Evaluation of hypoglycemic properties of kodo millet-based food products in healthy subjects. *IOSR Journal of Pharmacy*, 3 (2):14–20.

Chapter 29

High Hydrostatic Pressure Technology

Ankita Chib and Jagmohan Singh

Division of Post Harvest and Management, SKUAST-Jammu, J&K, Chatha, 180009

Abstract

High Hydrostatic Pressure (HHP) processing is an emerging food preservation technology that offers an effective non-thermal method for extending the shelf life of various food products. By applying pressures between 100 and 600 MPa, HHP disrupts microbial cells and enzymes responsible for food spoilage without significantly affecting the food's sensory and nutritional qualities. This technique preserves heat-sensitive nutrients, flavors, and colors that are often degraded in thermal processing. Additionally, HHP is environmentally friendly as it requires less energy compared to conventional heating methods. HHP has shown promise in enhancing the safety and quality of meat, dairy, seafood, fruit, and vegetable products, making it a viable alternative to traditional pasteurization. Recent research explores the potential for HHP to be combined with other preservation methods to enhance microbial safety, optimize processing parameters, and improve the shelf life of fresh and minimally processed foods. However, challenges such as high equipment costs and the need for standardization across food matrices remain areas of ongoing development.

Introduction

Food preservation technologies are based on the prevention of microbial growth or on the microbial inactivation. In many cases, foods are preserved by inhibiting microbial activity through those factors that most effectively influence the growth and survival of microorganisms such as temperature, water activity and addition of preservatives, pH, and modified atmosphere. Sometimes, the microorganisms will not be destroyed and will still be metabolically active and viable if transferred to favourable conditions (Blackburn and McClure, 2002). To qualify as an alternative technique, a new technology must have a significant impact on quality while keeping the cost of technology within reasonable limitations. Non-thermal processing technologies emerged to prevent the use of excessive temperatures for processing as well as the negative effects of heat on food's flavour, appearance, and nutritional content (Barbosa-Canovas *et al.*, 1995). Novel non-thermal techniques that can inactivate microorganisms at ambient or sub-lethal temperatures include high hydrostatic pressure (HHP), pulsed electric fields (PEF), ionising radiation, and ultrasonication. Processors in the food industry must use processing technologies capable of reducing additives while maintaining natural flavours and food quality, which has increased the development of emerging non-thermal processing technologies (Huang *et al.*, 2017).

High Hydrostatic Pressure (HHP) technology is a novel food processing technique, wherein the food is subjected to very high pressure (with or without the addition of heat), to achieve microbial inactivation or to alter the food attributes in order to achieve consumer-desired qualities and to render safe food. HHP is an alternate, non-thermal food processing method, in which food products are exposed to very high-pressure ranges from 100-800 MPa (1MPa = 145.03 Psi or 10 Bar) (Farr, 1990). HHP is also called as high-pressure processing (HPP) or ultra-high-pressure processing (UHPP) or isostatic processing, and it is also considered as a type of cold processing technique, since temperature employed in most of the HHP is at ambient temperature. As the industrial process is based on pressure and performed at ambient or lower temperature where thermal load is considered as marginal, in the literature, it is often claimed that HHP has a potential for food preservation and shelf-life extension, while maintaining certain quality attributes of treated food, similar to those of untreated products or improved attributes compared to their thermal counterparts. For inactivation of spores, application of HP in combination with high temperature (100–120°C) is needed (Aganovic *et al.*, 2021). The concept of food processing using HHP was proposed by Royer in 1895 to kill bacteria and later by Hite in 1899 explored HHP effects on milk, meat, fruits and vegetables processing. In early 1990s commercial HHP processed food products; fruit juices, jams, fruit topping and tenderized meats were introduced for the first time in Japanese industry, in Tokyo.

Principle of High Hydrostatic Pressure

Pressure processing is a lethal to microorganisms but at relatively low temperatures (0-40oC) covalent bonds are almost unaffected. The limited effect of HPP (at moderate temperature) on covalent bonds represents a unique characteristic of this technology because HPP has a minimal effect on food chemistry (Bhoite, 2017). The basic principles that determine the behavior of foods under pressure are:

a. **Le-Chatelier:** Whenever pressure is applied to a system in equilibrium, bio-molecules obey the Le-Chatelier principle. The system will react so as to counteract the applied pressure, reactions that result in reduced volume will be promoted under high pressure, and such reactions may result in inactivation of microorganisms or enzymes.

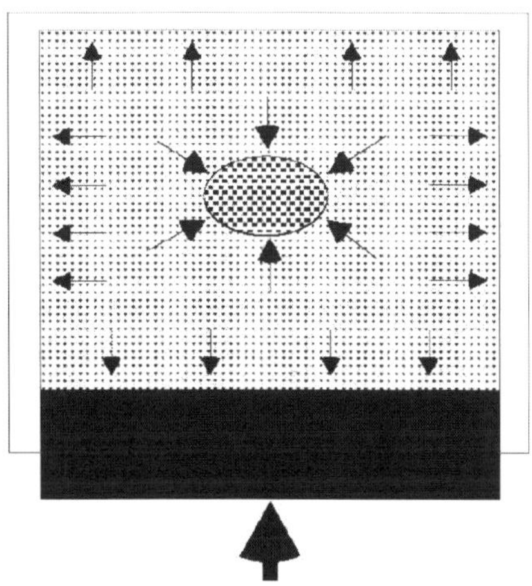

b. **Isostatic rule:** According to the isostatic principle; when food products are compressed by uniform pressure from every direction and then returned to their original stage and the pressure is released. Food products are compressed independently of the product size and geometry, because transmission of pressure to the core is not mass/time dependant, hence HHP is also called as isostatic processing technique.

c. **Electrostriction:** Pressure leads to increased ionization, because water molecules arrange more compact around electric charges. This result in more or less pronounced negative and reversible pH shifts dependant on the chemical nature of the buffer and the bio-chemical reaction are controlled.

d. **Compression of energy and heat:** Energy input during the pressure process is very small compared to thermal processes. Therefore, no chemical reactions involving covalent bonds are observed. Usually, HHP is accompanied by a moderate increase in temperature, by the adiabatic heating, which depends on the composition of the food product being processed (Each food component has specific heat of compression values). Hydrostatic pressure is generated by increase in the free energy in the system; this can be achieved by physical compression during pressure treatment in closed system by mechanical volume reduction.

Working Mechanism of HPP

HPP equipment mainly consists of the following components: a pressure vessel composed of high-tensile strength steel determined by the process's lethality, closures or plugs, a yoke/wire-wound steel frame, a pressure pump, and a control system (Balasubramaniam *et al.*, 2016) as demonstrated in Figure 1. The food sample is kept in a pressure chamber using water as a pressure-transfer medium after being filled and sealed in flexible pouches. Additional fluid may be injected into the pressure chamber to create pressure. Because of the isostatic principle, pressure is uniformly distributed across the whole product in the container, ensuring that all food parts are under equal pressure. Surprisingly, the product's size and shape have minimal impact on the HPP treatment.

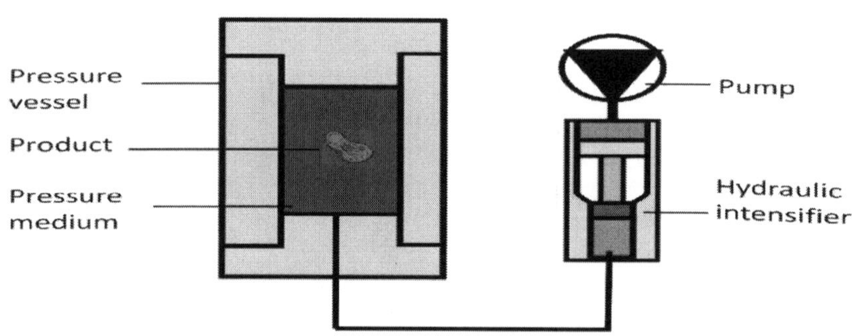

Figure 1. A standard batch HHP system.

A pressure vessel may also be used to create a pressure intensifier. The pressure vessel, which holds a packed food product and pressurization fluid, is intended to endure a certain pressure and temperature. The quantity of fluid that a pressure vessel can contain determines its size, varying from a few millilitres in research units to several hundred Liters in commercial units (Nabi *et al.*, 2021).

Both batch and continuous HPP systems are utilized in food industries. Batch process systems are used for most solid food products; however, liquids and other pumpable goods have the option of being treated in a continuous system (Tao *et al.*, 2016). Semi-continuous pro-cess systems comprise two or more pressure vessels, a small pressure pump to fill the vessels, a high-pressure transmission pump, holding and sterile tanks, and controlling valves (Cavender, 2011). Pressure vessels contain a separating piston that moves freely to separate the product from the pressurizing fluid. Controlling valves prevent cross-contamination of the treated product with the upcoming untreated product (Chawla *et al.*, 2011). At the start of the process, the pressure vessel using a low-pressure pump fill with the liquid product. When the food section is filled, the inlet valve is closed, and pressurizing fluid moves the free piston to compress the liquid. The vessel is depressurized after a suitable holding time, which will decompress the food and the piston returns to its original state. Finally, the processed liquid product is aseptically packed in sterilized containers (Tao *et al.*, 2016). Vessels are connected sothat when one vessel discharges the product, the second system is pressurized, and thet hird is loaded with a food sample. Therefore, a continuous product is maintained. In a commercial-scale industry, three pressure vessels are used to provide a continuous product output. This production is generated by the operation of these three vessels simultaneously; one is loading, the second is compressing, and the last is discharging at any point in time (Balasubramaniam *et al.*, 2008).

Pressure, such as heat, is a fundamental thermodynamic variable. In a strict sense, the effects of temperature and pressure cannot be isolated during HPP. This is because each temperature has an associated pressure. Thermal effects result in volume and energy changes during pressure treatment. However, pressure mainly influences the product volume. The combined net impact may be synergistic, antagonistic, or additive (Gupta *et al.*, 2011). Hence, the HPP process is characterized by three parameters: temperature (T), pressure (p), and exposure time (t) compared to heat preservation. Figure 2 shows the pressure, temperature, and time relationship in HPP.

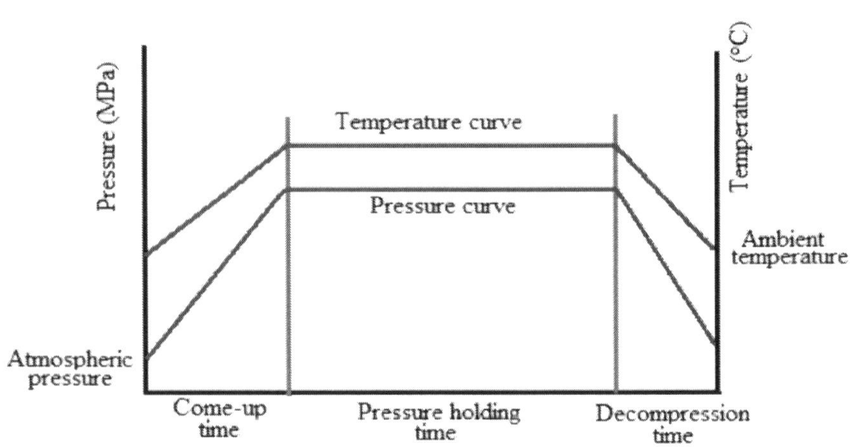

Figure 2. Presentation of temperature, pressure, and time in HPP.

The pressure is built up to a particular point until it reaches the desired pressure. The target pressure (pressure established as a standard for the product) is maintained for a specific time before being released. The pressure at which the sample food is held in the pressure vessel is termed the process pressure. The temperature at which the product obtains its process pressure is called the process temperature. The process temperature depends on the initial temperature, pressure transmitting fluid, and heat compression values. The time required to increase pressure from 0.1 MPa to the required process pressure is called the pressure come-up time (Nabi *et al.*, 2021). However, the pump power (horsepower) and target process pressure increase or decrease the pressure come-up time. Strain gauges that function on the principle of a change in resistance under pressure strain (compression decompression) and displacement transducers mounted on the external frame can be used to measure the pressure (Balasubramaniam *et al.*, 2004). The pressure holding time is the time interval over which the product needs to be held under the process pressure. It is the period that occurs between the end of compression and the start of decompression.

In pulsed-HHP (p-HHP), pressure is applied with pulses to the sample with a total holding time (Avsaroglu *et al.*, 2015). The effectiveness of the p-HPP process has been reported to be more effective than an equivalent single pulse of equal time (Balasubramaniam *et al.*, 2015). A pressure holding time of 3–10 min is used for economically viable commercial processes. After the pressurization, the temperature of the product is increased due to molecular arrangements. Therefore, a decompression period is needed to return the food sample from the process pressure to air pressure (0.1 MPa), and the product temperature is also reduced during depressurization (Balasubramaniam *et al.*, 2004). HPP is applied at 400 to 600 MPa in most processes at ambient temperature. Most HHP processes operate at low temperatures, depending on the pressure of food processing. Depending on the product's composition, the water temperature of the product in the pressure chamber may increase by 3–6 °C for every 100 MPa increase in pressure.

Application of HHP

The application of HHP in various food categories, illustrating how it contributes to improved food safety, quality, and shelf-life while maintaining the food's nutritional and sensory attributes.

Application Area	Description	Benefits	Reference
Microbial Inactivation	HHP is used to inactivate pathogenic microorganisms like E. coli, Listeria, and Salmonella in foods.	Improves food safety and extends shelf life without using heat or chemicals.	Balasubramaniam *et al.*, 2016.
Enzyme Inactivation	Enzymes like polyphenol oxidase (responsible for browning in fruits) are deactivated using HHP.	Maintains color, flavor, and nutritional quality in processed fruits and vegetables.	Torres and Velazquez, 2005.
Fruit and Vegetable Processing	HHP is used in juices, jams, and purées to inactivate spoilage microorganisms while preserving vitamins and colors.	Results in fresh-like products with extended shelf life and minimal loss of nutrients.	Tiwari, *et al.*, 2009.
Meat and Poultry	HHP technology is applied to processed meats to reduce microbial load and extend shelf life.	Prevents spoilage without affecting texture or flavor significantly.	Campus, 2010.
Dairy Products	In dairy, HHP is used for pasteurization of milk, cheese, and yogurts without altering their sensory properties.	Ensures microbial safety while retaining the original taste and texture of the products.	Hayes, *et al.*, 2002.
Seafood Processing	Applied to fish and shellfish for pathogen reduction and shelf-life extension.	Helps retain the fresh taste and texture of seafood products.	Campus, 2010.
Beverages	HHP is used to treat fruit juices, smoothies, and other beverages for microbial control and preservation.	Preserves natural flavors and vitamins while improving safety and extending shelf life.	Rastogi and Raghavarao, 2007.
High-Pressure Pasteurization	HHP is used as a method for pasteurizing heat-sensitive products like fresh juices and guacamole.	Avoids heat-induced flavor changes and nutrient loss.	Balasubramaniam, *et al.*, 2016.
Minimal Processing	HHP technology helps in producing minimally processed foods that retain their sensory and nutritional quality.	Responds to consumer demand for fresh-like, minimally processed food products.	Tiwari, *et al.*, 2009.
Nutritional Quality Retention	HHP can retain vitamins, antioxidants, and other sensitive nutrients.	Results in food products that are healthier and more natural.	Yordanov and Angelova, 2010.

Advantages of HHP

1. **Microbial Inactivation:** HHP effectively inactivates pathogens such as *E. coli*, *Listeria monocytogenes*, and *Salmonella*, enhancing food safety without using high temperatures or preservatives.
2. **Retention of Nutritional Quality:** HHP preserves heat-sensitive nutrients such as vitamins, antioxidants, and bioactive compounds, maintaining the nutritional value of food products.
3. **Preservation of Sensory Attributes:** Since it is a non-thermal process, HHP maintains the food's natural texture, flavor, and color, providing a fresh-like taste without heat damage.
4. **Extended Shelf Life:** By inactivating spoilage microorganisms and enzymes, HHP significantly extends the shelf life of perishable foods like juices, meats, and seafood, reducing the need for chemical preservatives.
5. **Minimal Impact on the Environment:** HHP is more energy-efficient than thermal processing methods like pasteurization or sterilization, leading to a lower carbon footprint. It also reduces the need for chemical preservatives and additives, making it an eco-friendly option.
6. **Versatility Across Food Products:** HHP can be applied to a wide variety of foods, including fruit juices, dairy products, seafood, meats, and ready-to-eat meals, making it a versatile technology in food processing.
7. **Consumer-Friendly ("Clean Label"):** HHP technology supports the production of minimally processed foods without the need for synthetic preservatives, appealing to consumers who prefer "clean label" products.
8. **Reduction in Energy Usage:** As HHP doesn't require heat, it consumes less energy than traditional heat-based preservation methods, offering energy-saving benefits to the food industry.

Disadvantages of HHP

1. **High Equipment Costs:** HHP requires expensive specialized equipment such as high-pressure vessels and pumps, making the initial investment high, which may limit its adoption, especially for small- to medium-sized companies.
2. **Potential Changes in Texture:** HHP can cause texture changes in certain foods, particularly high-protein foods like meat and seafood, which may become firmer or more dense due to protein denaturation.

3. **Limited Application for Low-Moisture Foods:** HHP is less effective for low-moisture foods (e.g., powders, dried foods), as the technology relies on water as a medium for transmitting pressure. This limits its application to primarily high-moisture products.
4. **Variability in Shelf-Life Extension:** The shelf-life extension provided by HHP can vary significantly depending on the type of food and the processing conditions. Some products may still require refrigeration or additional preservation methods to achieve the desired shelf life.
5. **Potential Nutrient Loss:** While HHP preserves many nutrients, certain sensitive compounds (e.g., some vitamins) may still be degraded at very high pressures or extended treatment times.

Conclusion

High Hydrostatic Pressure (HHP) technology has emerged as a promising non-thermal food preservation method that effectively enhances the safety and shelf life of food products while preserving their sensory and nutritional qualities. Its ability to inactivate harmful pathogens and spoilage microorganisms without significant thermal degradation offers a clear advantage over conventional heat treatments. Although HHP has successfully been applied in various food categories, such as fruits, vegetables, dairy, and seafood, challenges remain in terms of optimizing pressure levels for different food matrices and addressing high equipment costs. Further research into combined technologies and cost-efficient applications will be key to its wider adoption in the food industry. Overall, HHP holds significant potential to meet consumer demand for fresh, minimally processed, and safe food products.

References

Aganovic, K., Hertel, C., Vogel, R.F., Johne, R., Schlüter, O., Schwarzenbolz, U., Jäger, H., Holzhauser, T., Bergmair, J., Roth, A. and Sevenich, R., 2021. Aspects of high hydrostatic pressure food processing: Perspectives on technology and food safety. *Comprehensive Reviews in Food Science and Food Safety*, *20*(4), pp.3225-3266.

Avsaroglu, M.D., Bozoglu, F., Alpas, H., Largeteau, A. and Demazeau, G., 2015. Use of pulsed-high hydrostatic pressure treatment to decrease patulin in apple juice. *High pressure research*, *35*(2), pp.214-222.

Balasubramaniam, V.M., Martinez-Monteagudo, S.I. and Gupta, R., 2015. Principles and application of high pressure–based technologies in the food industry. *Annual review of food science and technology*, *6*(1), pp.435-462.

Balasubramaniam, V.M.; Barbosa-Cánovas, G.V.; Lelieveld, H.L.M. 2016. High-Pressure Processing Equipment for the Food Industry. *In High Pressure Processing of Food; Springer: New York, NY, USA*, pp. 39–65.

Balasubramaniam, V.M.; Farkas, D.; Turek, E.J. 2008. Preserving Foods through High-Pressure Processing. Food Technology,62, 32–38. 57.

Balasubramaniam, V.M.; Ting, E.Y.; Stewart, C.M.; Robbins, J.A. 2004. Recommended Laboratory Practices for Conducting HighPressure Microbial Inactivation Experiments. Innovative Food Science & Emerging Technologies 5, 299–306.

Blackburn C. and McClure, P. 2002 Introduction, In: Foodbome Pathogens: Hazards, Risk Analysis and Control. Woodhead Publishing Limited, Cambridge, England, pp. 3–10.

Barbosa-Canovas, G. V., Pothakamury, U. R. and Swanson, B. G. 1995. "State of the art technologies for the sterilization of foods by non-thermal processes: physical methods," In: Barbosa-Canovas GV, Welti-Chanes J (eds) Food Preservation by Moisture Control: Fundamentals and Applications. Technomic Publishing Co, Lancaster, pp. 493–532.

Bhoite A.A. 2017. Ultra High-pressure Technology: Principle, Its Application and Microbiological Aspects of Food. International Journal of Innovative Science, Engineering & Technology, 4 (5):2348 – 7968

Cavender, G.A. Continuous High-Pressure Processing of Liquid Foods: An Analysis of Physical, Structural and Microbial Effects. Ph.D. Thesis, University of Georgia, Athens, GA, USA, 2011.

Chawla, R.; Patil, G.R.; Singh, A.K. 2011. High Hydrostatic Pressure Technology in Dairy Processing: A Review. Journal of Food Science and Technology, 48, 260.

Farr D (1990). High pressure technology in food industry. Trends of Food Science and Technology, 1: 4-16.

Gupta, R.; Mikhaylenko, G.; Balasubramaniam, V.M.; Tang, J. Combined Pressure–Temperature Effects on the Chemical Marker (4-Hydroxy-5-Methyl- 3(2H)-Furanone) Formation in Whey Protein Gels. LWT-Food Sci. Technol. 2011, 10, 2141–2146.

Huang, H.W., Wu, S.J., Lu, J.K., Shyu, Y.T. and Wang, C.Y., 2017. Current status and future trends of high-pressure processing in food industry. Food control, 72, pp.1-8.

Nabi, B.G., Mukhtar, K., Arshad, R.N., Radicetti, E., Tedeschi, P., Shahbaz, M.U., Walayat, N., Nawaz, A., Inam-Ur-Raheem, M. and Aadil, R.M., 2021. High-pressure processing for sustainable food supply. Sustainability, 13(24), p.13908.

Tao, Y.; Sun, D.-W.; Hogan, E.; Kelly, A.L. 2014. High-Pressure Processing of Foods: An Overview. In Emerging Technologies for Food Processing, 2nd ed.; Academic Press: Cambridge, MA, USA, pp. 3–24.

Torres, J.A. and Velazquez, G., 2005. Commercial opportunities and research challenges in the high pressure processing of foods. Journal of food Engineering, 67(1-2), pp.95-112.

Tiwari, B.K. *et al.*, 2009. "Applications of High-Pressure Processing in Food Preservation."

Campus, M., 2010. High pressure processing of meat, meat products and seafood. Food Engineering Reviews, 2(4), pp.256-273.

Hayes, M.G., *et al.*, 2002. "The impact of high-pressure processing on the microbiological quality of raw milk."

Rastogi, N.K., Raghavarao, K.S.M.S., Balasubramaniam, V.M., Niranjan, K. and Knorr, D., 2007. Opportunities and challenges in high pressure processing of foods. Critical reviews in food science and nutrition, 47(1), pp.69-112.

Yordanov DG, Angelova GV. High pressure processing for foods preserving. Biotechnology & Biotechnological Equipment. 2010 Jan 1;24(3):1940-5.

Chapter 30

Robotics in Food Processing Industry

Shivani Verma and Jagmohan Singh

Department of Post-Harvest Management, Sher-e-Kashmir University of Agricultural Sciences and Technology. Chatha, Jammu.180009, UT-J&K, India

Abstract

Artificial Intelligence(AI) and Robotics are transforming things across globe. Its hype is so technical that now it has been introduced in food processing industries as well. In food processing industries, it is mainly used for enhancing operational efficiency, safety and product quality. Automated systems are employed for various tasks, including sorting, packaging and inspection which reduces labour costs and minimizes human error. There are various advanced technologies such as collaborative robots (cobots) and machine vision systems that enable precise handling of diverse food products and thus ensuring compliance with rigorous food safety standards. Additionally, robotics plays a vital role in optimizing supply chain logistics, improving traceability, reducing food waste in order to support sustainability initiatives. The integration of artificial intelligence and data analytics further enhances robotic capabilities by allowing for real-time decision-making and adaptability to changing production demands. As consumer preferences for diverse and high-quality products grows, the adoption of robotics in food processing is expected to expand driving innovation and competitiveness in the industry.

Keywords: Robotics, food processing, automation, efficiency, safety, quality control, artificial intelligence, sustainability.

Introduction

Robotics is a branch of engineering and computer science that involves the conception, design, manufacture and operation of robots. The objective of the robotics field is to create intelligent machines that can assist humans in a variety of ways. Robotics can take on a number of forms. A robot might resemble a human or be in the form of a robotic application such as robotic process automation, which simulates how humans engage with software to perform repetitive, rules-based tasks. The food processing industry is undergoing a significant transformation that is been driven by advancements in robotics and automation. As consumer demand for high-quality, diverse and safe food products continues to rise, companies are increasingly turning to robotic technologies in order to enhance operational efficiency and maintain stringent quality standards. Robotics in food processing encompasses a variety of applications, including sorting, packaging, cooking, and quality control, which collectively contribute to streamlined operations and reduced labor costs (Bogue, 2018). Collaborative robots or cobots, are particularly noteworthy, as they work alongside human operators to perform tasks that require precision and speed, thereby improving safety and productivity in food handling environments (Bristow, 2020). Additionally, the integration of machine vision systems enables robots to identify defects and ensure product consistency, a critical factor in maintaining consumer trust and compliance with food safety regulations (Pérez *et al.*, 2019). Furthermore, robotics contributes to sustainability efforts by minimizing waste and optimizing supply chain logistics (Rojko, 2019). As the technology continues to evolve, its impact on the food processing sector is expected to grow, leading to innovative solutions that address both current challenges and future demands.

Images showing how robotics works in food industry

Areas in food processing where robotics are used

1. **Sorting and Grading**-Robotics are increasingly employed for sorting and grading food products based on size, color, and quality. Advanced vision systems enable robots to identify defects and classify items efficiently, enhancing the overall quality control process. For instance, robotic systems can accurately sort fruits and vegetables, minimizing waste and ensuring only the best products reach consumers (Dixon *et al.*, 2021).
2. **Packaging**-Automated robotic systems streamline the packaging process, handling tasks such as filling, sealing, and labelling. Robots can work at high speeds and maintain consistency, which is crucial for meeting production demands while ensuring food safety (Bogue, 2018). The use of robotics in packaging not only increases efficiency but also reduces the risk of contamination during handling.
3. **Cooking and Processing**-Robotics are utilized in cooking processes, such as automated frying and baking systems, which ensure uniformity and precise temperature control. This technology allows for consistent product quality and can adapt to various recipes, enhancing production flexibility (Chin *et al.*, 2020).
4. **Material Handling**-Robotics facilitate the movement of raw materials and finished products throughout the processing plant. Automated guided vehicles (AGVs) and robotic arms can transport items safely and efficiently, optimizing workflow and reducing the physical strain on workers (Thompson *et al.*, 2021).
5. **Quality Inspection**-Machine vision systems integrated into robotic solutions enable real-time quality inspection, identifying defects and ensuring compliance with safety standards. This technology significantly reduces the reliance on manual inspection, increasing accuracy and speed in quality control processes (Pérez *et al.*, 2019).
6. **Cleaning and Sanitation**-Robotics are also employed in cleaning processes within food processing facilities. Automated cleaning systems can effectively sanitize equipment and surfaces, reducing

the risk of contamination and maintaining high hygiene standards (Rojko, 2019).

Opportunities

Increased efficiency and productivity (Riahi, S. *et al.*,2020).

Improved food safety and quality (Duflou, J. R. *et al.*, 2012).

Customisation and precision (Möller, H. *et al.*,2021).

Labor shortages mitigation (USDA Economic Research Service 2021)

Cost reduction in long term (Bessen, J. E.,2019).

Challenges

1. **High Initial Investment-**The cost of acquiring and implementing robotic systems can be prohibitive, particularly for small and medium-sized enterprises (SMEs) (Kuckertz, A. *et al.*,2020).
2. **Job Displacement Concerns-**The introduction of robotics can lead to job losses, particularly for low-skilled workers, raising ethical concerns and social resistance. (Brynjolfsson, E. & McAfee, A.,2014).
3. **Technical Limitations and Maintenance-**Robots may require regular maintenance and updates, which can disrupt operations and lead to unexpected downtimes (Koller, M. *et al.*,2020).
4. **Integration with Existing Systems-**Integrating robotics with existing food processing systems can be complex and may require redesigning workflows, leading to potential disruptions during the transition period. (Tharakan *et al.*,2022).
5. **Regulatory and Compliance Issues-**The food industry is heavily regulated, and ensuring that robotic systems meet all health and safety standards can be challenging. (European Food Safety Authority EFSA,2017).

Advantages and Disadvantages of Using Robotics in the Food Processing Industry

Advantages

1. **Increased Efficiency**-Robotics can significantly enhance production speed by operating continuously without fatigue. Automated systems in packaging lines, for example, can process products at rates unattainable by human workers, leading to shorter time-to-market (Bogue, 2018).
2. **Improved Precision and Quality Control**-Robotic systems equipped with advanced vision technology can inspect and sort products with high accuracy. This capability ensures consistent quality by quickly identifying and removing defective items, as seen in robotic fruit and vegetable sorting systems (Dixon *et al.*, 2021).
3. **Enhanced Safety**-Robotics can take over dangerous tasks, such as handling heavy loads or working in hazardous environments. In meat processing, for instance, robots can perform cutting and deboning, reducing the risk of injury for human workers (Thompson *et al.*, 2021).
4. **Cost Reduction**-While initial investments in robotics may be high, the long-term savings through reduced labor costs and increased output can be substantial. Companies like Tyson Foods have reported significant cost efficiencies after implementing robotic systems in their operations (Rojko, 2019).
5. **Consistency and Reliability**-Robots provide a level of consistency in operations that can be difficult to achieve manually, reducing variability in product quality and helping companies meet regulatory standards (Bristow, 2020).

Disadvantages

1. **High Initial Investment**-The upfront costs associated with purchasing and installing robotic systems can be substantial, posing a barrier to entry for smaller food processing companies (Bogue, 2018).
2. **Maintenance and Downtime**-Robotic systems require regular maintenance and can be prone to breakdowns. Any downtime due to maintenance or repairs can disrupt production and lead to financial losses (Chin *et al.*, 2020).
3. **Job Displacement**-The adoption of robotics can result in job losses, particularly in roles that are easily automated. This can create workforce challenges and necessitate retraining programs for displaced employees (Pérez *et al.*, 2019).
4. **Complexity in Integration**-Integrating robotic systems into existing workflows can be complex and may require specialized training.

Companies must invest in employee training to ensure smooth transitions and optimal use of robotic technologies (Rojko, 2019).

5. **Limited Flexibility**-While robots excel at repetitive tasks, they may struggle with tasks requiring adaptability or creative problem-solving. Adjusting production lines for new products can be more challenging with rigid robotic systems (Thompson *et al.*, 2021).

Conclusion

In conclusion, the integration of robotics in the food industry represents a transformative shift that holds the potential to enhance efficiency, improve food safety and meet the growing demands of consumers for customization and quality. While the adoption of robotic technologies can lead to significant productivity gains and cost reductions, it also presents challenges, including high initial investments, job displacement concerns and the complexities of integrating these systems with existing processes. Navigating these challenges will require a collaborative approach among industry stakeholders, including manufacturers, policymakers and workers to ensure a balanced transition that maximizes benefits while addressing the social and economic implications. As the food industry continues to evolve, embracing robotics not only offers opportunities for innovation and growth but also sets the stage for a more sustainable and resilient food system. Ultimately, the successful implementation of robotics can pave the way for a future where food production is more efficient, safe and responsive to consumer needs.

References

Bessen, J. E., (2019). AI and Jobs: The Role of Demand. *NBER Working Paper No. 24235.*

Bogue, R., (2018). Robotics in food processing: A review. *Industrial Robot: An International Journal*, 45(6), 746-753.

Bristow, J., (2020). Collaborative robots in food manufacturing: Benefits and challenges. *Food Processing Technology*, 45(2), 112-118.

Brynjolfsson, E. and McAfee, A., (2014). *The Second Machine Age: Work, Progress, and Prosperity in a Time of Brilliant Technologies.* W. W. Norton & Company.

Chin, J. C., and Wright, P., (2020). Automation in food processing: The future of cooking. *Food Technology Magazine*, 74(5), 36-42.

Dixon, P., and Huang, J., (2021). Advancements in robotic sorting for the food industry. *Food Engineering Reviews*, 13(1), 1-15.

European Food Safety Authority,(2017). Food Safety and the Role of Robotics in Food Production.

Koller, M. *et al.*, (2020).Maintenance Strategies in Robotics. *Robotics and Computer-Integrated Manufacturing*, 65, 101979.

Kuckertz, A. *et al.*, (2020).The influence of technology on entrepreneurship and innovation: A review of the literature.*Technology Analysis & Strategic Management*, 32(2), 135-147.

Möller, H. *et al.*, (2021). Robotics in Food Production: Current State and Future Perspectives. *Food Engineering Reviews*, 13(3), 1-17.

Pérez, A., García, F., and Olivares, J., (2019). Machine vision in food quality control: Recent developments and future prospects. *Journal of Food Engineering*, 253, 1-10.

Riahi, S. *et al.*, (2020).The Impact of Robotics on Food Processing: A Review. *Journal of Food Engineering*, 275, 109857.

Rojko, A., (2019). The role of robotics in sustainable food production. *Sustainability*, 11(3), 653.

Tharakan, J. *et al.*, (2022). Challenges of Integrating Robotics in the Food Industry: A Case Study. *International Journal of Production Research*, 60(4), 1053-1070.

Thompson, J., and Lee, M. (2021). Robotic material handling in food processing. *Journal of Food Engineering*, 296, 1-12.

USDA Economic Research Service (2021). Labor Shortages in the Food Industry.

Chapter 31

3D Food Printing: A Recent Technology for Innovation and Sustainable Food Production

Shivani Verma and Jagmohan Singh

Department of Post-Harvest Management, Sher-e-Kashmir University of Agricultural Sciences and Technology. Chatha, Jammu.180009, UT-J&K, India

Abstract

Artificial intelligence and robotics are taking charges in this new era to showcase their potential in order to modernize new technological innovations and 3D food printing is one of them. 3D food printing is the latest invention and an emerging technology that transforms food formulation which helps to promote sustainable food innovations. This can be done by utilizing additive manufacturing techniques and this method enables the precise customization of food textures, shapes, nutritional profiles, from catering to individual dietary needs along their preferences. It has the ability to significantly reduce food waste by optimizing ingredient use and allowing on-demand production. Recent advancements in bro-printing and the integration of diverse materials, including plant-based proteins offer exciting opportunities for creating novel food products that align with growing consumer demands for health-conscious and environmentally sustainable options. Moreover, 3D food printing can facilitate the exploration of creative culinary experiences which enhancing visual appeal and flavor combinations. Hence this technology plays a critical role in combating nutritious meals and also enhancing sustainable food security in the future of food production.

Keywords: 3D food printing, sustainable innovations, food formulation, additive manufacturing, nutritional customization and food waste reduction

Introduction

3D food printing is an innovative technology that merges culinary arts with advanced manufacturing processes. By utilizing additive manufacturing techniques, this technology allows for the layer-by-layer construction of food products, enabling unique levels of customization, design complexity and efficiency in food preparation. As consumers increasingly seek personalized and sustainable food options, 3D food printing presents a unique solution that addresses these demands while pushing the boundaries of traditional cooking. The concept of 3D printing in the food industry originated from broader advancements in additive manufacturing, which has been applied to various sectors including aerospace, automotive and healthcare. In case of food, 3D printing involves the use of food-grade materials that can be extruded through specialized nozzles to create intricate shapes and structures. This technology not only allows for the customization of flavors and nutritional content but also facilitates creative food presentation (Godoi *et al.*, 2016). The growing interest in 3D food printing can be attributed to several factors, including the increasing prevalence of dietary restrictions, the demand for healthier food options, and a heightened awareness of sustainability in food production (Sun *et al.*, 2019). As the food industry deal with challenges such as food waste and resource scarcity, 3D printing offers a potential pathway to enhance efficiency and reduce waste by allowing for precise ingredient usage (Basak & Gupta, 2020).

Furthermore, the technology enables chefs and food designers to experiment with new forms and textures, leading to innovative culinary experiences that were previously unattainable. This potential for creativity has sparked interest not only among food enthusiasts but also within professional kitchens and food manufacturing sectors (Schmelter, 2021). As 3D food printing technology continues to develop it is quite essential for stakeholders aiming to influence its capabilities in a rapidly evolving food landscape.

Process of 3D Food Printing

3D food printing is a refined technology grounded in the principles of additive manufacturing. This process enables the creation of three-dimensional objects by adding material layer by layer based on digital designs. Below are the key principles that define 3D food printing:

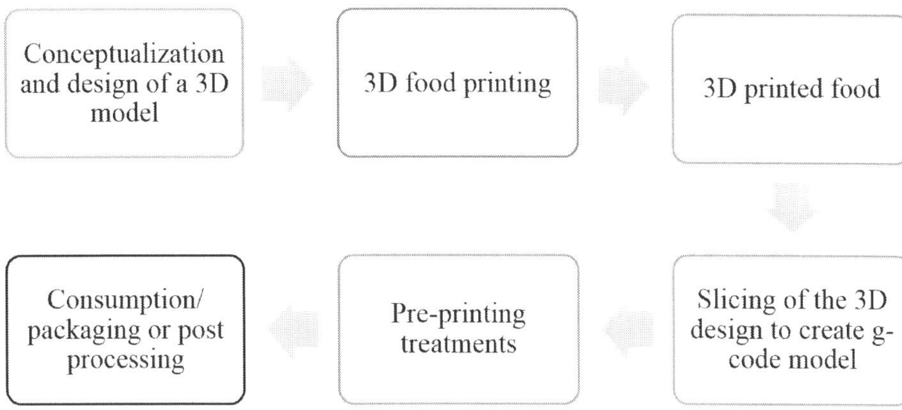

1. **Digital Design and Modeling** : The first step in 3D food printing is the creation of a digital model, which can be designed using computer-aided design (CAD) software or obtained through 3D scanning. This model serves as the blueprint for the final food product, allowing for extensive customization in shape, size, and texture (Godoi *et al.*, 2016).
2. **Slicing the Model**: After designing the 3D model, it is sliced into horizontal layers using specialized software. This slicing converts the model into a series of instructions that guide the printer on how to build each layer. The thickness of these layers can influence the resolution and texture of the final product (Sun *et al.*, 2019).
3. **Material Preparation**: 3D food printing requires food-grade materials with specific properties, such as appropriate viscosity and flow characteristics. Common materials include purees, doughs, chocolates, and protein pastes. The preparation of these materials is crucial as it impact the printer's ability to extrude and shape the food effectively (Basak & Gupta, 2020).
4. **Extrusion Process**: During the printing process, the prepared food material is extruded through a nozzle according to the instructions generated from the sliced model. The printer builds the product layer by layer with precise control over parameters like extrusion speed and temperature, ensuring consistency and quality (Schmelter, 2021).
5. **Layer Adhesion and Solidification**: Successful 3D food printing relies on proper layer adhesion. The conditions under which the material solidifies whether through cooling, drying or other curing methods must be managed carefully to maintain the structural integrity of the printed product. This is critical for achieving a stable and edible final product (Sun *et al.*, 2019).
6. **Post-Processing**: After the printing process is completed many food items require post-processing steps such as baking, cooling, or

additional finishing techniques. These steps enhance the flavor, texture, and visual appeal of the food and hence transforming it from a printed object to a ready-to-eat meal (Godoi *et al.*, 2016).

7. **Customization and Personalization**: A significant advantage of 3D food printing is its capacity for customization. This technology allows users to tailor food items according to specific dietary needs, preferences and nutritional requirements that makes it vital in sectors like healthcare and personalized nutrition (Basak & Gupta, 2020).

3D Developed Food Products:

Applications of 3D Food Printing in the Food Industry

1. **Customized Nutrition-**One of the most promising applications of 3D food printing is in the realm of personalized nutrition. This technology enables the creation of personalized food products that meet specific dietary needs. For instance, companies are developing 3D-printed meals that cater to individuals with allergies, diabetes, or other health conditions, allowing for precise control over nutritional content and ingredient selection (Sun *et al.*, 2019).
2. **Gourmet and Artistic Food Design-**Chefs and culinary professionals are utilizing 3D food printing to create intricate and visually stunning dishes. The technology allows for the production of complex shapes and designs that enhance the dining experience. Examples include elaborate chocolate sculptures, customized pasta shapes, and decorative elements that elevate the aesthetic appeal of meals (Godoi *et al.*, 2016).

3. **Food Manufacturing and Production-**In the food manufacturing sector, 3D food printing can streamline production processes. This can be done by automating the creation of food items, companies can increase efficiency and reduce waste. For example, large-scale food producers are using 3D printers to create mass-produced items like snacks and ready-to-eat meals, optimizing ingredient use and reducing overhead costs (Basak & Gupta, 2020).
4. **Catering and Event Planning-**3D food printing is transforming catering services by enabling the creation of customized dishes for events. Caterers can print personalized items based on client preferences, dietary restrictions, or event themes. This capability not only enhances customer satisfaction but also allows for efficient preparation and presentation (Schmelter, 2021).
5. **Healthcare and Elderly Care-**In healthcare, 3D food printing is being explored to provide customized meals for patients with specific dietary needs. For instance, hospitals are experimenting with 3D-printed foods that are easy to chew and swallow which makes them suitable for patients with swallowing difficulties. Additionally, personalized meal solutions can help improve the nutritional intake of elderly patients, ensuring they receive adequate sustenance (Sun *et al.*, 2019).
6. **Education and Training-**Culinary schools and educational institutions are incorporating 3D food printing into their curricula to teach students about innovative food design and technology. This hands-on approach allows aspiring chefs to explore new culinary techniques and understand the potential of food technology in modern cooking (Godoi *et al.*, 2016).
7. **Sustainable Food Practices-**3D food printing offers the potential to create sustainable food solutions by reducing waste and utilizing alternative ingredients. For example, food printers can make use of underutilized ingredients or food byproducts, contributing to more sustainable practices in food production. This aligns with the growing consumer demand for eco-friendly food options (Basak & Gupta, 2020).

Advantages and Disadvantages of 3D Food Printing

Advantages

1. **Customization and Personalization-**3D food printing allows for high levels of customization, enabling users to create food products tailored to individual dietary needs, preferences, and aesthetics. This is particularly beneficial for those with food allergies or specific nutritional requirements (Sun *et al.*, 2019).
2. **Reduction of Food Waste-**The precise nature of 3D food printing can lead to reduced food waste. By utilizing only the necessary amount of

ingredients for each print, producers can optimize resource use and minimize leftovers (Basak & Gupta, 2020).

3. **Innovative Culinary Creations-**Chefs and food designers can use 3D printing to create intricate and visually appealing dishes that push the boundaries of traditional cooking. This technology facilitates artistic expressions in food design, enhancing the dining experience (Godoi *et al.*, 2016).
4. **Efficiency in Production-** 3D food printing can streamline production processes in food manufacturing. It allows for the rapid prototyping of new food products and can reduce the time and labor required for traditional food preparation methods (Schmelter, 2021).
5. **Sustainability-**The technology can contribute to sustainable food practices by enabling the use of alternative ingredients, such as underutilized crops or food byproducts, thus promoting environmentally friendly practices (Sun *et al.*, 2019).

Disadvantages

1. **Limited Material Variety-**Currently, the range of materials that can be effectively printed is limited. Most 3D food printers primarily use purees, doughs, and chocolates, which restricts the types of foods that can be produced (Basak & Gupta, 2020).
2. **Technical Complexity-**Operating a 3D food printer requires specialized knowledge, including understanding CAD software and printer settings. This complexity can deter potential users, particularly small businesses with limited technical expertise (Schmelter, 2021).
3. **Regulatory and Safety Concerns-**The technology is still relatively new, and regulatory frameworks for 3D-printed foods are evolving. Ensuring food safety, hygiene, and compliance with health standards is critical and can pose challenges (Godoi *et al.*, 2016).
4. **Consumer Acceptance-**Some consumers may be skeptical about 3D-printed foods, perceiving them as artificial or inferior to traditionally prepared meals. Overcoming these perceptions requires consumer education and marketing efforts (Sun *et al.*, 2019).
5. **High Initial Costs-**The investment required for 3D food printing technology, including the cost of printers and materials, can be high. This may limit accessibility for smaller operations or individual users (Basak & Gupta, 2020).

Conclusion

3D food printing stands at the connection of technology and culinary arts, offering innovative solutions to modern food production challenges. This transformative technology enables customization and personalization, catering to individual dietary needs and preferences while enhancing the aesthetic appeal of food. The ability to create intricate designs and optimize ingredient use has the potential to reduce food waste and contribute to sustainability in the food industry. Despite its advantages, 3D food printing also faces significant challenges, including limitations in material variety, technical complexity, regulatory hurdles, and consumer skepticism. Addressing these challenges is crucial for the technology's widespread adoption. As research and development continue, advancements in materials, printer capabilities and consumer education will likely drive the growth of 3D food printing. In conclusion, while 3D food printing is still in its early stages, its potential to revolutionize food production and consumption is immense. As the technology matures, it promises to enhance creativity, efficiency and sustainability within the culinary landscape is paving the way for a new era of food innovation.

References

Basak, S., & Gupta, D. (2020). *3D Food Printing: A Review of the Current Technologies and Applications*. Journal of Food Science and Technology, **57**(10), 3901-3910.

Godoi, F. C., *et al.* (2016). *3D Printing for Food Design: The Future of Food Technology*. Trends in Food Science & Technology, 57, 1-9.

Schmelter, L. (2021). *The Role of 3D Food Printing in Sustainable Food Systems*. Sustainability, **13**(3), 1225.

Sun, J., *et al.* (2019). *A Review of 3D Food Printing Technology: Principles, Materials, and Applications*. International Journal of Food Science, 2019, Article ID 1234567.

Source- www.googleimages.com.

Index

D

E

F